Ultra-dense Networks for 5G and Beyond

Ultra-dense Networks for 5G and Beyond

Modelling, Analysis, and Applications

Edited by

Trung Q. Duong
School of Electronics, Electrical Engineering and Computer Science
Queen's University Belfast
UK

Xiaoli Chu
University of Sheffield
UK

Himal A. Suraweera
University of Peradeniya
Sri Lanka

This edition first published 2019

Registered Offices
John Wiley & Sons, Inc., 111 River Street, Hoboken, NJ 07030, USA
John Wiley & Sons Ltd, The Atrium, Southern Gate, Chichester, West Sussex, PO19 8SQ, UK

Editorial Office
The Atrium, Southern Gate, Chichester, West Sussex, PO19 8SQ, UK

For details of our global editorial offices, customer services, and more information about Wiley products visit us at www.wiley.com

Wiley also publishes its books in a variety of electronic formats and by print-on-demand. Some content that appears in standard print versions of this book may not be available in other formats.

Library of Congress Cataloging-in-Publication Data

Names: Duong, Trung Q., editor. | Chu, Xiaoli, editor. | Suraweera, Himal A., editor.
Title: Ultra-dense networks for 5G and beyond : modelling, analysis, and applications / edited by Trung Q. Duong, School of Electronics, Electrical Engineering and Computer Science, Queen's University Belfast, UK, Xiaoli Chu, University of Sheffield, UK, Himal A. Suraweera, University of Peradeniya, Sri Lanka.
Description: Hoboken, NJ : John Wiley & Sons, Inc., [2019] | Includes bibliographical references and index. |
Identifiers: LCCN 2018044566 (print) | LCCN 2018047948 (ebook) | ISBN 9781119473725 (Adobe PDF) | ISBN 9781119473718 (ePub) | ISBN 9781119473695 (hardcover)
Subjects: LCSH: Wireless communication systems–Technological innovations. | Network performance (Telecommunication)
Classification: LCC TK5103.2 (ebook) | LCC TK5103.2 .U48 2019 (print) | DDC 621.3845/6–dc23
LC record available at https://lccn.loc.gov/2018044566

Cover Design: Wiley
Cover Image: © Busakorn Pongparnit/Getty Images

Set in 10/12pt WarnockPro by SPi Global, Chennai, India

Printed in the UK

Contents

List of Contributors

Symeon Chatzinotas
Interdisciplinary Centre for Security
Reliability and Trust
University of Luxembourg
Luxembourg

Jie Deng
School of Electronic Engineering and
Computer Science
Queen Mary University of London
UK

Ming Ding
Data61, CSIRO
Australia

Trung Q. Duong
School of Electronics, Electrical
Engineering and Computer Science
Queen's University Belfast
UK

Lorenzo Galati-Giordano
Nokia Bell Labs
Dublin
Ireland

Adrian Garcia-Rodriguez
Nokia Bell Labs
Dublin
Ireland

Sumit Gautam
Interdisciplinary Centre for Security
Reliability and Trust
University of Luxembourg
Luxembourg

Giovanni Geraci
Department of Information and
Communication Technologies
Universitat Pompeu Fabra
Spain

Weisi Guo
School of Engineering
University of Warwick
UK

Zhu Han
Department of Electrical and Computer
Engineering
University of Houston
USA

Tiep M. Hoang
Queen's University Belfast
UK

Een-Kee Hong
College of Electronics and Information
Kyung Hee University
South Korea

Won-Joo Hwang
Department of Information and
Communication Engineering
Inje University
South Korea

Amir Hossein Jafari
Samsung Research America
USA

Markku Juntti
Centre for Wireless Communications
University of Oulu
Finland

Dani Korpi
Nokia Bell Labs
Espoo
Finland

Marios Kountouris
Mathematical and Algorithmic Sciences Lab
Paris Research Center
Huawei Technologies Co. Ltd.
France

Maria Liakata
Department of Computer Science
University of Warwick
UK

David López-Pérez
Nokia Bell Labs
Dublin
Ireland

Mbazingwa E. Mkiramweni
Xidian University
China

Guillem Mosquera
Mathematics Institute
University of Warwick
UK

Hien Quoc Ngo
School of Electronics, Electrical Engineering and Computer Science
Queen's University Belfast
UK

Huy T. Nguyen
Department of Information and Communication Engineering
Inje University
South Korea

Nam-Phong Nguyen
School of Electronics, Electrical Engineering and Computer Science
Queen's University Belfast
UK

Van Minh Nguyen
Mathematical and Algorithmic Sciences Lab
Paris Research Center
Huawei Technologies Co. Ltd.
France

Björn Ottersten
Interdisciplinary Centre for Security, Reliability and Trust
University of Luxembourg
Luxembourg

Weijie Qi
Department of Electronic and Electrical Engineering
University of Sheffield
UK

Tony Q.S. Quek
Information Systems Technology and Design Pillar
Singapore University of Technology and Design
Singapore

Taneli Riihonen
Laboratory of Electronics and Communications Engineering
Tampere University of Technology
Finland

Le-Nam Tran
School of Electrical and Electronic Engineering
University College Dublin
Ireland

Hoang D. Tuan
School of Electrical and Data Engineering
University of Technology Sydney
Australia

Mikko Valkama
Laboratory of Electronics and Communications Engineering
Tampere University of Technology
Finland

Nguyen-Son Vo
Institute of Fundamental and Applied Sciences
Duy Tan University
Vietnam

Quang-Doanh Vu
Centre for Wireless Communications
University of Oulu
Finland

Thang X. Vu
Interdisciplinary Centre for Security, Reliability and Trust
University of Luxembourg
Luxembourg

Chungang Yang
School of Telecommunications Engineering
Xidian University
China

Howard H. Yang
Information Systems Technology and Design Pillar
Singapore University of Technology and Design
Singapore

Jie Zhang
Department of Electronic and Electrical Engineering
University of Sheffield
UK

Qi Zhang
The Jiangsu Key Laboratory of Wireless Communications
Nanjing University of Posts and Telecommunications
China

Preface

We are observing an ever-increasing number of connected devices and the rapid growth of bandwidth-intensive wireless applications. The number of wirelessly connected devices is anticipated to exceed 11.5 billion by 2019, i.e. nearly 1.5 mobile devices per capita. In addition, it is expected that we will witness a 10 000-fold growth in wireless data traffic by the year 2030. Such unprecedented increases in mobile data traffic and network loads are pushing contemporary wireless network infrastructures to a breaking point. These predictions have raised alarm to the wireless industry and mobile network operators who are faced with the challenges of provisioning high-rate, low-delay, and highly reliable connectivity anytime and anywhere without significantly increasing energy consumption at the infrastructure, such as base stations, fronthaul and backhaul networks, and core networks.

The above challenges demand a paradigm shift to the wireless network infrastructure. In this context, ultra-dense networks, which are characterized by a very high density of low-power radio access nodes with different transmit power levels, radio frequency coverage areas, and signal/data processing capabilities, such as small cell (e.g. pico- or femto-cells) access points, remote radio heads, and relay nodes, have attracted a lot of interest from the wireless industry and research community. Compared to a high-power macrocell, each small cell has a much smaller coverage radius (from several meters to several hundred meters) and can assign a larger amount of radio resources to each user within its coverage area than in conventional cellular networks, thus improving users' quality-of-service (QoS) and quality-of-experience (QoE). Moreover, the low transmission power of small cells makes it possible for them to operate in an unlicensed spectrum; for example, the large available bandwidth in the millimeter-wave (mm-Wave) frequency range from 30 GHz to 300 GHz, to mitigate the scarcity of licensed spectrum in mobile networks.

While the commercial consensus is fuzzy regarding which technologies will be used for fifth generation (5G) wireless communications, ultra-dense networks have been widely considered as a key enabler for 5G and beyond in wireless communication network implementation. In addition to their potential in provisioning ubiquitous high-capacity wireless connectivity, ultra-dense networks also offer numerous opportunities and flexibility to be incorporated with other 5G candidate technologies, such as mm-Wave communications, massive multiple-input multiple-output (MIMO), non-orthogonal multiple access, in-band full-duplex operation, simultaneous wireless information and power transfer (SWIPT), device-to-device (D2D) communications, and distributed caching, to enable the realization of 5G technologies and systems' full potential.

Dense coexistence of various neighboring and/or overlapping radio access nodes, dynamic network topologies, complicated co-channel interference scenarios, backhaul provisioning, energy consumption, QoS provisioning, and security issues bring new challenges for ultra-dense network deployment. These technical challenges need to be addressed in order to exploit the true potential of ultra-dense networks. Although some research attempts have been made toward understanding the theoretical and practical performance limits, a comprehensive mathematical methodology to capture the dynamic topologies, large-scale heterogeneous interference, and the high level of randomness in ultra-dense networks essential to the system-level analysis and design of 5G networks is missing in the existing literature. In light of the ongoing densification of radio access nodes and wireless connected devices toward 5G wireless networks, there is an urgent need for the research community, industry, and even end users to better understand the fundamental technical details as well as the achievable performance gains of ultra-dense networks.

Ultra-dense Networks for 5G and Beyond: Modelling, Analysis, and Applications provides a comprehensive and systematic exposition on the state-of-the-art of ultra-dense networks and their applications in 5G cellular and other wireless networks, with a delicate balance between mathematical modeling, theoretical analysis, and practical design. It contains cutting-edge tutorials on the theoretical and technical foundations that underpin ultra-dense networks, as well as insightful surveys of ultra-dense network-related emerging technological trends that will be interesting and informative to readers of all backgrounds. The book is written by researchers currently leading the research and development of ultra-dense networks, covering a wide spectrum of topics, ranging from system modeling and performance analysis, network performance optimization, radio resource management, wireless self-backhauling, massive MIMO to unlicensed spectrum, energy efficiency, big data analytics, physical layer security, SWIPT, distributed caching, and cooperative video streaming, etc.

The book is organized into 12 chapters, which are grouped into three parts. In the following, we provide a brief tour through the book's parts and chapters to show how this book addresses the challenges faced by ultra-dense networks from various aspects.

Part I, including Chapters 1 to 3, presents the background information and fundamental knowledge necessary for understanding the theoretical and technical foundations of ultra-dense networks from three main disparate aspects.

Chapter 1 explores the fundamental performance limits of ultra-dense networks due to physical limits of radio wave propagation. Focusing on the impact of network node density on the network performance, the authors model the spatial distribution of network nodes using Poisson point processes (PPP). As a result of the close proximity of base stations and mobile devices in ultra-dense networks, the elevated base station height and dual-slope line-of-sight (LoS) and non-line-of-sight (NLoS) propagations are considered in the modeling of small-scale and large-scale fading. Under this system model, analytical expressions are derived for ultra-dense network downlink performance in terms of coverage probability, throughput, and average transmission rate. Then, under general multi-slope pathloss and channel power distribution models, an analytical framework is presented to enable the analysis of asymptotic performance limits (i.e. scaling laws) of network densification.

Chapter 2 presents a general framework for performance analysis of ultra-dense networks, particularly focusing on the impact of pathloss and multipath fading. Under a pathloss model that incorporates both LoS and NLoS propagations and a distance-dependent multi-path Rician fading model with a variant Rician K-factor, the analytical expressions for both coverage probability and area spectral efficiency are derived. Comparing the performance impact of LoS and NLoS transmissions in interference-limited ultra-dense networks under Rician fading with that under Rayleigh fading, the analytical and simulation results show that pathloss dominates the overall system performance of ultra-dense networks, while the impact of multi-path fading is quite limited and does not help to mitigate the performance loss caused by the many LoS interfering links in ultra-dense networks, especially in single-input single-output systems.

Chapter 3 considers the network densifications in wireless user devices, radio access nodes, and cloud edge nodes, and presents a comprehensive introduction to mean field game theory and tools, which is useful in the design and analysis of ultra-dense networks with spatial-temporal dynamics. Following that, the authors present a survey of the latest applications of mean field games in the 5G era in general (including device-to-device communications and cloud-edge networks) and in 5G ultra-dense networks in particular, with a focus on resource management problems such as interference mitigation, energy management, and caching.

Chapters 4 to 8 are grouped into ***Part II***, which exploits the synergy between ultra-dense networks and other key 5G candidate technologies.

Chapter 4 studies different schemes of wireless self-backhauling in ultra-dense networks, with a focus on the recently proposed in-band full-duplex self-backhauling. The deployment of wireless backhauling, in general, helps to resolve the scalability problems faced by wired backhaul connections, because wireless backhauling does not require physical cables. The use of inband wireless self-backhauling is even more intriguing than traditional wireless backhauling, because it enables the reuse of spectral resources between radio access links and backhaul links. In this respect, no additional or dedicated frequency resources are required for the backhaul, and thus the in-band wireless self-backhauling will be commercially beneficial. Three different in-band wireless self-backhauling schemes based on full-duplex or half-duplex operations are analyzed and compared.

Chapter 5 aims to optimally leverage the dense deployment of small cells and massive MIMO in future mobile networks. Using analytical tools from stochastic geometry, the authors derive a tight approximation of the achievable downlink rate for a two-tier heterogeneous cellular network and use it to compare the performance between densifying small cells and expanding base station antenna arrays. The results show that increasing the density of small cells improves the downlink rate much faster than expanding antenna arrays at base stations. However, when the small cell density exceeds a certain threshold, the network capacity may start to deteriorate. On the contrary, the network capacity keeps increasing with the expansion of base station antenna arrays until it reaches an upper bound, which is caused by pilot contamination. This upper bound surpasses the maximum network capacity achieved by the dense deployment of small cells. Moreover, the authors provide practical design insights into the tradeoff between the dense deployment of small cells and massive MIMO in future high-capacity wireless networks.

Chapter 6 introduces a promising 5G technology termed as cell-free massive MIMO, where radio access points distributed in an area coherently serve mobile users in the area using the same time/frequency resource. It is called cell-free because there is no boundary among cells, unlike as seen in traditional cellular networks. Cell-free massive MIMO networks complement ultra-dense networks by connecting closely located access points together. The connection of access points is feasible because only a limited amount of necessary information is exchanged and the central processing unit is not required to have a high computing capability. In particular, the authors study the physical layer security in cell-free massive MIMO networks under a pilot spoofing attack, where an active eavesdropper attacks the uplink training, and present a simple counterattack scheme based on the transmit power control at all the involved radio access points.

Chapter 7 proposes to solve the spectrum crunch in ultra-dense networks by employing massive MIMO in unlicensed spectrum and by leveraging the spatial interference suppression capabilities of multi-antenna systems. The authors discuss the motivation and justification of using unlicensed spectrum bands in ultra-dense networks, describe the technical fundamentals of massive MIMO in an unlicensed spectrum, and identify the main use cases and the associated challenges. The results show that massive MIMO in an unlicensed spectrum is capable of significantly boosting the performance of ultra-dense networks in both outdoor and indoor environments.

In Chapter 8, a set of advanced optimization tools is introduced to maximize the energy efficiency for ultra-dense networks, where the total power consumption may become a critical concern as the number of network nodes and connected devices increases. The chapter begins with the introduction of optimization techniques that are useful for maximizing energy efficiency, including concave–convex fractional programming, non-tractable fractional programming, and the alternating direction method of multipliers for distributed implementation. The chapter then moves on to demonstrate how these optimization techniques can be applied to maximizing the energy efficiency of spectrum-sharing dense small cell networks.

In ***Part III***, which includes Chapter 9 to Chapter 12, the promising applications of ultra-dense networks in the 5G era are presented and discussed.

In Chapter 9, the authors provide an insightful discussion on how big data methods can be used to improve the deployment of and the QoS in ultra-dense networks, covering both structured and unstructured data analytics. It is demonstrated that having access to consumer data and being able to analyse it in the wireless network context would allow effective user-centric deployment and operations of ultra-dense networks. In particular, the chapter outlines two big data approaches for improving ultra-dense network deployment: (1) identify the spatial-temporal traffic and service patterns to aid targeted deployment of dense networks; and (2) identify social community patterns to assist ultra-dense peer-to-peer and D2D networking.

In Chapter 10, the authors look into physical layer security opportunities and challenges for ultra-dense networks and show how physical layer security technologies can be applied to mitigate issues caused by unreliable wireless backhaul links in ultra-dense networks. Given that the wireless backhaul links are not reliable at all times, secrecy outage probability for a spectrum sharing ultra-dense network under the impact of unreliable backhaul links is evaluated.

Chapter 11 discusses the application of SWIPT in energy harvesting enabled ultra-dense networks with a distributed caching architecture. Without loss of generality, the system model consists of a source node and a destination node, which communicate with each other with the help of multiple relay nodes. Each relay is equipped with a cache memory and energy harvesting capability. Based on the time-splitting architecture, the authors have focused on the problem of relay selection to maximize the data throughput between the relay and the destination subject to the amount of harvested energy at the relays. A separate optimization problem is formulated to maximize the energy stored at the relays subject to given QoS constraints.

Chapter 12 investigates the application of cooperative video streaming with the support of D2D caching in ultra-dense networks. The ever-increasing demand for video streaming services at a high data rate and high QoE is increasingly likely to cause traffic congestion at the backhaul links in ultra-dense networks. The authors propose a joint rate allocation and description distribution optimization algorithm, where both the cache storage and downlink resources of mobile devices are exploited, to enable densely deployed base stations in order to provide high QoE video streaming services to mobile users while saving energy at the same time.

Part I

Fundamentals of Ultra-dense Networks

1

Fundamental Limits of Ultra-dense Networks

Marios Kountouris and Van Minh Nguyen

Mathematical and Algorithmic Sciences Lab, Paris Research Center, Huawei Technologies Co. Ltd., France

1.1 Introduction

Mobile traffic has significantly increased over the last decade, mainly due to the stunning expansion of smart wireless devices and bandwidth-demanding applications. This trend is forecast to be maintained, especially with the deployment of fifth generation (5G) and beyond networks and machine-type communications. A major part of the mobile throughput growth during the past few years has been enabled by the so-called *network densification*, i.e. adding more base stations (BSs) and access points and exploiting spatial reuse of the spectrum. Emerging 5G cellular network deployments are envisaged to be heterogeneous and dense, primarily through the provisioning of small cells such as picocells and femtocells. Ultra-dense networks (UDNs) will remain among the most promising solutions to boost capacity and to enhance coverage with low-cost and power-efficient infrastructure in 5G networks. The underlying foundation of this expectation is the presumed linear capacity scaling with the number of small cells deployed in the network. In other words, doubling the number of BSs doubles the capacity the network supports in a given area and this can be done indefinitely. Nevertheless, in this context, several important questions arise: how close are we to fundamental limits of network densification? Can UDNs indefinitely bring higher overall data throughput gains in the network by just adding more infrastructure? If the capacity growth arrives to a plateau, what will cause this saturation and how the network should be optimized to push this saturation point further? These are the questions explored in this chapter.

The performance of wireless networks relies critically on their spatial configuration upon which inter-node distances, fading characteristics, received signal power, and interference are dependent. Cellular networks have been traditionally modeled by placing the base stations on a regular grid (usually on a hexagonal lattice), with mobile users either randomly scattered or placed deterministically. Tractable analysis can sometimes be achieved for a fixed user location with a small number of interfering BSs and Monte Carlo simulations are usually performed for accurate performance evaluation. As cellular networks have become denser, they have also become increasingly irregular. This is particularly true for small cells, which are deployed opportunistically and in hotspots and dense heterogeneous networks (HetNets). As a result, the widely used deterministic grid model has started showing its limitations and cannot be used for general and

Ultra-dense Networks for 5G and Beyond: Modelling, Analysis, and Applications,
First Edition. Edited by Trung Q. Duong, Xiaoli Chu and Himal A. Suraweera.

tractable performance analysis results in UDNs. Although more real-world deployment data are needed to make conclusive statements on which is a better model for UDNs, stochastic spatial models are often a more appropriate model versus a deterministic one. In a random spatial model, the BS locations are modeled by a two-dimensional spatial point process, the simplest being the Poisson Point Process (PPP). This model has the advantages of being scalable to multiple classes of overlaid BSs and accurate to model location randomness, especially that of UDNs and HetNets. Additionally, powerful tools from stochastic geometry can be used to derive performance results, such as coverage, average rate, and throughput, for general dense multi-tier networks in closed form, which was not even possible for macrocellular (single-tier) networks using a deterministic grid model. Moreover, the suitability of this mathematical abstraction can be reinforced in particular for UDNs due to the fact that network planning of a large number of small base stations is complex, making the resulting network closer to a random distribution model. The employment of PPPs allows the capture of the spatial randomness of real-world UDN deployments (often not fully coordinated) and, at the same time, obtains precise and tractable expressions for system-level performance metrics [1, 2]. Furthermore, it has been shown that using random spatial models does not introduce important discrepancy to a regular/deterministic model. For instance, the PPP case has nearly the exact same signal-to-interference ratio (SIR) statistics as a very wide class of spatial BS distributions, including the hexagonal grid, with just a small fixed SIR shift (e.g. 1.53 dB) [3].

Recent work has also considered more general models with inhibition (e.g. cluster models) or repulsion (e.g. determinantal point process) [2, 4, 5]. Point process models like the Matern hardcore process, Ginibre process, Strauss process, Cox process, and the perturbed lattice are more realistic point process models than the PPP and the hexagonal grid models since they can capture the spatial characteristics of the actual network deployments better. However, more general point processes typically result in less tractable expressions that include integrals that must be numerically evaluated, which, however, is still much simpler than an exhaustive network simulation.

There has been noticeable divergence on the conclusions of various network studies using spatial models, according to which densification is not always beneficial to the network performance. Recent and often conflicting findings based on various modeling assumptions have identified that densification may eventually stop delivering significant throughput gains at a certain point. First, it has been shown that the coverage probability does not depend on the network density and thus the throughput can grow linearly with the BS density in the absence of background noise for both the closest [1] and the strongest BS association [6]. These results assume simple models—mostly for tractability reasons—in which (i) BSs are located according to a homogeneous PPP and are placed at the same height as the UEs, and (ii) the signal propagation is modeled using the standard single-slope unbounded pathloss and Rayleigh distribution for the small-scale fading. By contrast, using a dual-slope pathloss model, Rayleigh fading, and nearest BS association, [7] shows that both coverage and capacity strongly depend on the network density. More precisely, the coverage probability, expressed in terms of signal-to-interference-plus-noise ratio (SINR), is maximized at some finite BS density and there exists a phase transition on the asymptotic potential network throughput with ultra-densification (i.e. network density goes to infinity). If the near-field pathloss exponent is less than one, the potential throughput goes to zero with denser network deployment, whereas if it is greater than one, the potential throughput grows unboundedly as the network becomes increasingly denser.

In [8] and [9], the performance of millimeter wave systems is analyzed using stochastic geometry and considering scenarios with line-of-sight (LOS) and non-line-of-sight (NLOS) propagation for the pathloss attenuation. More comprehensive models can be found in [10] - [12], where the pathloss exponent changes with a probability that depends on the distance between BSs and UEs. In [13], the authors consider strongest cell association with bounded pathloss and lognormal shadowing and show that the coverage attains a maximum point before starting to decay when the network becomes denser. In [14], similar conclusions are obtained under Nakagami fading for the LOS propagation and both nearest and strongest BS association. Based on multi-slope pathloss and smallest pathloss association, [15] shows that the network coverage probability first increases with BS density, and then decreases. Moreover, the area spectral efficiency will grow almost linearly as the BS density goes asymptotically large. Optimal densification in terms of maximum SINR-coverage probability is investigated in [16]. In [17], interference scaling limits in a Poisson field with singular power law pathloss and Rayleigh fading are derived. Moreover the authors in [18] provide spectral efficiency scaling laws with spatial interference cancellation at the receiver. It is shown that linear scaling of the spectral efficiency with network density can be obtained if the number of receive antennas increases super-linearly with the network density (or linearly in the case of bounded pathloss).

In this chapter, we aim at providing an answer to whether there are any fundamental limits to 5G UDNs due to physical limits arising from electromagnetic propagation. To tackle this question, we follow two approaches. First, we derive analytical expressions for the downlink performance of UDNs under a system model that combines PPP-distributed elevated BSs and dual-slope LOS/NLOS propagation affecting both small-scale and large-scale fading. Second, we investigate the asymptotic performance limits (scaling laws) of network densification under general multi-slope pathloss and channel power distribution models. Using tools from extreme value theory [19], and in particular regular variation analysis [20], we present a general framework that allows us to derive the scaling regimes of the downlink SINR, coverage probability, potential throughput, and average per-user and system-wide rate.

Notation The distribution function of X is denoted by $F_X(x)$ and $\overline{F}_X(x) = 1 - F_X(x)$. In addition, for real functions f and g, we say $f \sim g$ if $\lim_{x\to\infty}(f(x)/g(x)) = 1$ and $f = o(g)$ if $\lim_{x\to\infty}(f(x)/g(x)) = 0$. We also use notation $\xrightarrow{d}, \xrightarrow{p}, \xrightarrow{a.s.}$ to denote the convergence in distribution, convergence in probability, and almost sure (a.s.) convergence, respectively. A positive, Lebesgue measurable function h on $(0, \infty)$ is called *regularly varying* with index $\alpha \in \mathbb{R}$ at ∞ if $\lim_{x\to\infty}(h(tx)/h(x)) = t^{\alpha}$ for $0 < t < \infty$. In particular, h is called *slowly varying* (respectively *rapidly varying*) (at ∞) if $\alpha = 0$ (respectively if $\alpha = -\infty$). We denote by $\mathcal{R}_{\alpha}$ the class of regularly varying functions with index α. $\mathbb{P}(\cdot)$ and $\mathbb{E}(\cdot)$ are respectively the probability and the expectation operators. $\mathbb{P}^0$ is the Palm probability with respect to the point process (in PPP $\mathbb{P}^o = \mathbb{P}$ – Slivnyak's Theorem) and the expectation $\mathbb{E}^0$ is taken with respect to the measure $\mathbb{P}^o$. The Gauss hypergeometric function is denoted by ${}_2F_1(a, b, c, z)$ and $\Gamma(z) = \int_0^{\infty} x^{z-1}e^{-x}\mathrm{d}x$ denotes the gamma function.

1.2 System Model

1.2.1 Network Topology

We consider a downlink dense wireless network, in which the locations of transmitters (BSs) are modeled as a homogeneous PPP $\Phi = \{x_i\} \subset \mathbb{R}^d$ of intensity λ, where d is the dimension of the network. Users are distributed according to some independent and stationary point process Φ_u (e.g. PPP), whose intensity λ_u is sufficiently larger than λ in order to ensure that all BSs are active, i.e. every BS has at least one active user associated within its coverage. The case of partially loaded networks can be straightforwardly applied by considering that the network intensity is $\vartheta\lambda$, where $0 \leq \vartheta \leq 1$ is the network load ratio. The analysis is performed for a typical user located at the origin $o = (0, 0)$; hence the link between the origin and its associated BS is a typical link. Since the network domain is limited (i.e. not the entire $\mathbb{R}^2$ is considered) and points far away from the origin generate weak interference to the typical user due to pathloss attenuation, the distance from the user to any node is upper bounded by an arbitrarily large constant $0 < R_\infty < \infty$. Each node transmits with a power level that is independent of the others but is not necessarily constant. Lastly, we assume that all BSs may be elevated at the same height $h \geq 0$, measured in meters, whereas the typical UE is at the ground level; alternatively, h can be interpreted as the elevation difference between BSs and UEs if the latter are all placed at the same height.

1.2.2 Wireless Propagation Model

Radio links in a wireless medium are susceptible to time-varying channel impediments, interference, and background noise. These include long-term attenuation due to pathloss, medium-term variation due to shadowing, and short-term fluctuations due to multi-path fading.

The small-scale fading between node $i \in \Phi$ and the typical user is denoted by g_i and is assumed independent across time and space. In the most general model, g_i includes all propagation phenomena and link gains except pathloss, such as transmit power, fast fading, shadowing, antenna gains, etc. We may refer to g_i as *channel power*. Fast fading is usually modeled by a Rayleigh distribution of the channel amplitude (i.e. g_i is exponentially distributed), while shadowing is modeled by a lognormal distribution of the channel power. Given node location $\{x_i\}$, the variables $\{g_i\}$ are assumed to be non-zero and independently distributed according to some distribution F_g. Incorporating the channel fading into the spatial model, $\tilde{\Phi} \triangleq \{(x_i, g_i)\} \subset \mathbb{R}^2 \times \mathbb{R}^+$, forms an independently marked PPP [5].

The large-scale pathloss function is denoted by $\ell(x) : \mathbb{R}^d \to [0, \infty]$ and is assumed to be a non-decreasing function of $\|x\|$. For a realistic and practically relevant pathloss model, we have that a pathloss function $\ell(\cdot)$ is *bounded* if and only if (iff) $1/\ell(r) < \infty, \forall r \in \mathbb{R}^+$, and *unbounded* otherwise. Furthermore, the pathloss function $\ell(\cdot)$ is said to be *physical* iff $1/\ell(r) \leq 1, \forall r \in \mathbb{R}^+$. Let $r_x \triangleq \|x\|$ denote the horizontal distance between x and the typical UE, measured in meters. We consider a distance-dependent LOS probability function $p_{\text{LOS}}(r_x)$, i.e. the probability that a BS located at x experiences LOS propagation depends on the distance r_x. We use $\Phi_{\text{LOS}} \triangleq \{x \in \Phi : x \text{in LOS}\}$ and $\Phi_{\text{NLOS}} \triangleq \Phi \setminus \Phi_{\text{LOS}}$ to denote the subsets of BSs in

LOS and in NLOS propagation conditions, respectively. We remark that each BS is characterized by either LOS or NLOS propagation independently from the others and regardless of its operating mode as serving or interfering BS.

We consider two different pathloss models: in the first part of this chapter where we aim at deriving exact analytical expressions for the network performance, we adopt the standard power-law (non-bounded) model; in the second part where we focus on asymptotic limits when $\lambda \to \infty$, we consider a more generic pathloss model.

For the first part, we consider a standard power law pathloss model for $d = 2$ and define the pathloss functions

$$\ell_{\mathrm{LOS}}(r_x, h) \triangleq (r_x^2 + h^2)^{\beta_{\mathrm{LOS}}/2}, \quad \text{if } x \in \Phi_{\mathrm{LOS}}, \tag{1.1}$$

$$\ell_{\mathrm{NLOS}}(r_x, h) \triangleq (r_x^2 + h^2)^{\beta_{\mathrm{NLOS}}/2} \quad \text{if } x \in \Phi_{\mathrm{NLOS}}, \tag{1.2}$$

with NLOS and LOS pathloss exponents satisfying respectively $\beta_{\mathrm{NLOS}} \geq \beta_{\mathrm{LOS}} > 2$.

For the second part, we consider a general pathloss model, where different distance ranges are subject to different pathloss exponents (see Figure 1.1). The pathloss function for $h = 0$ is then modeled as

$$\ell(r) = \sum_{k=0}^{K-1} A_k r^{\beta_k} \mathbb{1}\{R_k \leq r < R_{k+1}\}, \tag{1.3}$$

where $\mathbb{1}\{\cdot\}$ is the indicator function, which takes the value 1 when the statement $\{\cdot\}$ is true and 0 otherwise, $K \geq 1$ is a given constant characterizing the number of pathloss slopes, and R_k are constants satisfying

$$0 = R_0 < R_1 < \cdots < R_{K-1} < R_K = R_\infty, \tag{1.4}$$

β_k denote the pathloss exponents satisfying

$$\beta_0 \geq 0, \tag{1.5a}$$

$$\beta_k \geq d - 1, \quad \text{for } k = 1, \dots, K-1, \tag{1.5b}$$

$$\beta_k < \beta_{k+1} < \infty, \quad \text{for } k = 0, \dots, K-2, \tag{1.5c}$$

and A_k are constants to maintain continuity of $\ell(\cdot)$, i.e.

$$A_k > 0 \text{ and } A_k R_{k+1}^{\beta_k} = A_{k+1} R_{k+1}^{\beta_{k+1}}, \tag{1.6}$$

for $k = 0, \dots, K-2$. For notational simplicity, we also define $\delta_k = d/\beta_k$, for $k = 0, \dots, K-1$.

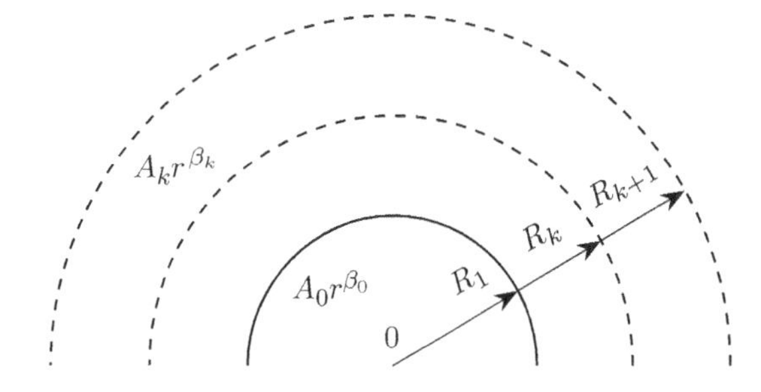

Figure 1.1 Multi-slope pathloss model. Reprinted with permission from [28].

The above general multi-slope model captures the fact that the pathloss exponent varies with distance, while remaining unchanged within a certain range. Condition (1.5a) is related to the near field (i.e. it is applied to the distance range $[0, R_1]$). In principle, free-space propagation in $\mathbb{R}^3$ has a pathloss exponent of 2 (i.e. $\beta = d - 1$), whereas in realistic scenarios, pathloss models often include antenna imperfections and empirical models usually result in the general condition (1.5b) for far-field propagation. Condition (1.5c) models the physical property that the pathloss increases faster as the distance increases. Notice, however, that this condition is not important in the subsequent analytical development.

The pathloss function as defined above has the following widely used special cases:

- $K = 1$, $\beta_0 \geq d$: $\ell(r) = A_0 r^{\beta_0}$, which is the standard singular (unbounded) pathloss model;
- $K = 2$, $\beta_0 = 0$: $\ell(r) = \max(A_0, A_1 r^{\beta_1})$, which is the bounded pathloss recommended by 3GPP [21], in which A_0 is referred to as the minimum coupling loss.

Notice that the pathloss function (1.3) is bounded iff $\beta_0 = 0$ and is physical iff $\beta_0 = 0$ and $A_0 \geq 1$.

1.2.3 User Association

The performance of wireless networks is defined with respect to the user's serving cell, which in turn depends on the underlying user association scheme. In this chapter, we consider two widely used association mechanisms: (i) nearest BS association, in which the user is connected to the geographically closest BS, and (i) strongest cell (highest SINR) association, in which the user is connected to the BS that provides the best signal quality (the strongest signal strength).

Notice that $\mathbb{P}(\text{SINR}_{\text{near}} \geq t) \leq \mathbb{P}(\text{SINR}_{\text{max}} \geq t), \forall t \in \mathbb{R}$, where SINR_{max} and $\text{SINR}_{\text{near}}$ denote the user's downlink SINR under strongest cell association and nearest BS association, respectively. Equality holds only if pathloss is not decreasing with the distance and the channel power is a deterministic constant (implying no fading or that fading is neglected).

The above association schemes can be presented in a unified framework as follows:

$$f_{r_x}(r) \triangleq \begin{cases} 2\pi\lambda e^{-\pi\lambda r^2} r, & \text{closest BS} \\ 2\pi\lambda r, & \text{strongest BS,} \end{cases} \tag{1.7}$$

$$v(r) \triangleq \begin{cases} r, & \text{closest BS} \\ 0, & \text{strongest BS,} \end{cases} \tag{1.8}$$

where $f_{r_x}(r)$ in Eq. (1.7) is the probability density function (PDF) of the distance r_x between the serving BS and the typical user.

1.2.4 Performance Metrics

The building block metric is the received signal-to-interference-plus-noise ratio (SINR) on one or more links, as each of the key performance metrics follows directly from it.

The SINR of the typical user associated to the ith BS is given by

$$\mathrm{SINR}_i = \frac{g_i/\ell(r_x, h)}{I_i + W} \tag{1.9}$$

where $I_i = \sum_{j\in\Phi\setminus\{i\}} P_j$ is the aggregate interference power received by the user served by the ith node and W denotes the power of additive background thermal noise, which is assumed to be Gaussian with average power W. The signal power that the user receives from the ith node is expressed as $P_i = g_i/\ell(||x_i||)$. In addition, we define

$$I_{\text{tot}} = \sum_{j\in\Phi} P_j$$

to be the total interference power, i.e. $I_{\text{tot}} = I_i + P_i, \forall i \in \Phi$.

For the analysis, we focus on the performance of the typical node, which allows us to derive spatial averages of the individual node characteristics. For notation convenience in the remainder of the chapter, the receive SINR and aggregate interference of the typical node are denoted by SINR and I, respectively.

The first key metric is the *coverage probability* of the typical user, defined as the probability that the receive SINR is larger than a target θ, i.e.

$$\mathrm{P_c}(\lambda, \theta) \triangleq \mathbb{P}^o(\mathrm{SINR} > \theta). \tag{1.10}$$

The user *average rate* is defined as the Shannon rate (in nats/s/Hz) assuming Gaussian codebooks and Gaussian interference (worst-case performance), i.e.

$$C(\lambda) = \mathbb{E}^o(\log(1 + \mathrm{SINR})). \tag{1.11}$$

Furthermore, we consider two system performance metrics, namely the *coverage density* $\mathcal{D}(\lambda, \theta)$ (in BSs/m^2) and the *area spectral efficiency* (ASE) $\mathcal{E}(\lambda)$ (in nats/s/Hz/m^2), which are, respectively, defined as

$$\mathcal{D}(\lambda, \theta) = \lambda\mathbb{P}^o(\mathrm{SINR} \geq \theta) \text{ and } \mathcal{E}(\lambda) = \lambda\mathbb{E}^o(\log(1 + \mathrm{SINR})). \tag{1.12}$$

The coverage density allows us to compute the *potential throughput*, defined as $\mathcal{T}(\lambda, \theta) \triangleq \lambda\mathrm{P_c}(\lambda, \theta)\log(1 + \theta) = \mathcal{D}(\lambda, \theta)\log(1 + \theta)$, measured in (bps/Hz/m^2). The coverage density gives an indication of the cell splitting gain [7], i.e. the achievable data rate growth from adding more BSs due to the fact that each user shares its BS with a smaller number of users; it provides the potential throughput if multiplied by the rate $\log_2(1 + \theta)$. When λ and λ_u scale at the same rate, the scaling of ASE $\mathcal{E}(\lambda)$ and $\mathcal{D}(\lambda, \theta)$ will be the same. Note also that $\mathcal{T}(\lambda, \theta) \leq \mathcal{E}(\lambda)$.

1.3 The Quest for Exact Analytical Expressions

In this section, we investigate the downlink performance of UDNs in terms of coverage probability and potential throughput. We present a general framework based on stochastic geometry that combines elevated BSs and dual-slope LOS/NLOS propagation affecting both small-scale and large-scale fading and that accommodates both closest and strongest BS association.

We assume that the channel amplitudes (small-scale fading) are Nakagami-m distributed for LOS propagation conditions and Rayleigh distributed for NLOS propagation conditions; the effect of shadowing is omitted for analytical tractability. The commonly used Rician distribution for LOS small-scale fading is well approximated by the more tractable Nakagami-m distribution with the shape parameter m computed as $m \triangleq (K+1)^2/(2K+1)$, where K is the Rician K-factor representing the ratio between the powers of the direct and scattered paths. Note that the value of m is rounded to the closest integer. Hence, the channel power fading gain g_i follows a gamma distribution $\Gamma(m, 1/m)$ if $x \in \Phi_{\text{LOS}}$, with the complementary cumulative distribution function (CCDF) given by

$$\overline{F}_{\text{LOS}}(z) \triangleq 1 - \frac{\gamma(m, mz)}{\Gamma(m)} = e^{-mz} \sum_{k=0}^{m-1} \frac{(mz)^k}{k!} \tag{1.13}$$

where the last equality holds when the shape parameter m is an integer; on the other hand, g_i is exponentially distributed with mean 1 if $x \in \Phi_{\text{NLOS}}$ and its CCDF $\overline{F}_{\text{NLOS}}(z)$ can be obtained from $\overline{F}_{\text{LOS}}(z)$ in Eq. (1.13) by simply setting $m = 1$. Although the shape parameter m should intuitively depend on the link distance, a fixed value of m is considered here for analytical tractability. Simulation results will show that the performance obtained with a distance-dependent m is very well captured by a fixed m in our UDN scenario.

The aggregate interference I is defined as

$$I \triangleq \sum_{y \in \Phi_{\text{LOS}} \setminus \{x\}} g_y / \ell_{\text{LOS}}(r_y, h) + \sum_{y \in \Phi_{\text{NLOS}} \setminus \{x\}} g_y / \ell_{\text{NLOS}}(r_y, h). \tag{1.14}$$

Note that downtilted antennas can be incorporated into the framework by multiplying each signal coming from BS $y \in \Phi$ by the term $|\cos^{\xi}(\arctan(h/r_y) - \vartheta)|$, where ξ depends on the vertical antenna pattern and ϑ is the downtilt angle [22]. For the sake of simplicity, we consider the interference-limited case, i.e. $I \gg W$, and we thus focus on the signal-to-interference ratio (SIR). Extending the results to include the background noise is easy, at the expense of more involved expressions.

1.3.1 Coverage Probability

Let us use $\mathcal{L}_I^{\text{LOS}}(s)$ and $\mathcal{L}_I^{\text{NLOS}}(s)$ to denote the Laplace transforms of the interference when the LOS probability $p_{\text{LOS}}(r) = 1$ and $p_{\text{LOS}}(r) = 0$, $\forall r \in [0, \infty)$, respectively, which correspond to the cases of LOS or NLOS interference, respectively:

$$\mathcal{L}_I^{\text{LOS}}(s) = \exp\left(-2\pi\lambda \int_{v(r)}^{\infty} \left(1 - \frac{1}{(1 + s/m\ell_{\text{LOS}}(t,h))^m}\right) t \, \text{d}t\right), \tag{1.15}$$

$$\mathcal{L}_I^{\text{NLOS}}(s) = \exp\left(-2\pi\lambda \int_{v(r)}^{\infty} \left(1 - \frac{1}{1 + s/\ell_{\text{NLOS}}(t,h)}\right) t \, \text{d}t\right). \tag{1.16}$$

The Laplace transform of the interference is calculated using the probability generating functional (PGFL) of the PPP [5].

1. *General LOS/NLOS Model*

Theorem 1.1 [23] The coverage probability is given by

$$\mathrm{P_c}(\lambda,\theta) = \int_0^\infty \left(p_{\mathrm{LOS}}(r) \sum_{k=0}^{m-1} \left[\frac{(-s)^k}{k!} \frac{\mathrm{d}^k}{\mathrm{d}s^k} \mathcal{L}_I(s) \right]_{s=m\theta\ell_{\mathrm{LOS}}(r,h)} + (1-p_{\mathrm{LOS}}(r))\mathcal{L}_I(\theta\ell_{\mathrm{NLOS}}(r,h)) \right) f_{r_x}(r)\mathrm{d}r \tag{1.17}$$

where

$$\mathcal{L}_I(s) \triangleq \mathcal{L}_I^{\mathrm{NLOS}}(s) \exp\left(-2\pi\lambda \int_{v(r)}^\infty p_{\mathrm{LOS}}(t) \left(\frac{1}{1+s/\ell_{\mathrm{NLOS}}(t,h)} - \frac{1}{(1+s/m\ell_{\mathrm{LOS}}(t,h))^m} \right) t\, \mathrm{d}t \right) \tag{1.18}$$

represents the Laplace transform of the interference I in Eq. (1.14), with $\mathcal{L}_I^{\mathrm{NLOS}}(s)$ defined in Eq. (1.16).

Due to the contribution from the interfering BSs under LOS propagation conditions, we have that $\mathcal{L}_I(s) \leq \mathcal{L}_I^{\mathrm{NLOS}}(s)$, with $\mathcal{L}_I^{\mathrm{NLOS}}(s)$ given in Eq. (1.16). This can be equivalently seen from the argument of the exponential function in Eq. (1.18), which is always negative since $\beta_{\mathrm{NLOS}} > \beta_{\mathrm{LOS}}$ and $m \geq 1$. On the other hand, the possibility of having LOS for the desired signal enhances the coverage probability in Eq. (1.17). Observe that $\mathcal{L}_I(s)$ in Eq. (1.18) reduces to $\mathcal{L}_I^{\mathrm{LOS}}(s)$ in Eq. (1.15) if $p_{\mathrm{LOS}}(r) = 1, \forall r \in [0,\infty)$ and to $\mathcal{L}_I^{\mathrm{NLOS}}(s)$ in Eq. (1.16) if $p_{\mathrm{LOS}}(r) = 0, \forall r \in [0,\infty)$.

Let $\mathrm{P_c^{NLOS}}(\lambda,\theta)$ (respectively $\mathrm{P_c^{LOS}}(\lambda,\theta)$) denote the coverage probability in the presence of NLOS (respectively LOS) propagation. Assuming that the LOS probability function $p_{\mathrm{LOS}}(r)$ is monotonically decreasing with r, then $\lim_{\lambda\to 0} \mathrm{P_{cov}}(\lambda,\theta) = \mathrm{P_{cov}^{NLOS}}(\theta)$ and $\lim_{\lambda\to\infty} \mathrm{P_{cov}}(\lambda,\theta) = \mathrm{P_{cov}^{LOS}}(\theta)$.

A useful upper bound for this type of expressions is provided in the following proposition.

Proposition 1.1 [23] For any $\mathcal{L}_X(z) \triangleq \mathbb{E}_X[e^{-zX}]$ and $N > 1$, the following inequality holds:

$$\sum_{n=0}^{N-1} \left[\frac{(-s)^n}{n!} \frac{\mathrm{d}^n}{\mathrm{d}s^n} \mathcal{L}_X(s) \right]_{s=z} < \sum_{n=1}^{N} (-1)^{n-1} \binom{N}{n} \mathcal{L}_X(n(\Gamma(N+1))^{-1/N} z).$$

2. *3GPP LOS/NLOS Model*

A widely used distance-dependent LOS/NLOS model is the ITU-R UMi model [21] (referred to as the 3GPP LOS/NLOS model in the following), which is characterized by the following LOS probability function:

$$p_{\mathrm{LOS}}(r) = \min\left(\frac{18}{r}, 1\right)(1-e^{-r/36}) + e^{-r/36}. \tag{1.19}$$

Using Eq. (1.19), the propagation is always in LOS conditions for $r \leq 18$ meters. In practice, this implies that for BS densities above $\lambda = 10^{-2}$ BSs/m^2 and the closest BS association, the probability of LOS coverage is very close to one and, as a consequence, some NLOS terms in Eqs. (1.17) and (1.18) can be neglected.

Following this line of thought, a simplified LOS/NLOS model that is suited for analytical calculations can be obtained by means of the LOS probability function

$$p_{\mathrm{LOS}}(r) = \begin{cases} 1, & r \in [0, D) \\ 0, & r \in [D, \infty) \end{cases} \tag{1.20}$$

with D being the critical distance below which all BSs are in LOS propagation conditions. The system performance resulting from Eq. (1.20) with $D = 18$ meters in terms of coverage probability very accurately approximates that obtained with the original 3GPP LOS probability function (1.19), as we will show below in numerical results. In this scenario, the coverage probabilities for closest and strongest BS association, given in the general form (1.17), simplify respectively to

$$\begin{aligned} \mathrm{P}_{\mathrm{c}}^{(\mathrm{C})}(\lambda, \theta) = 2\pi\lambda \Bigg(& \int_0^D \sum_{k=0}^{m-1} \left[\frac{(-s)^k}{k!} \frac{\mathrm{d}^k}{\mathrm{d}s^k} \tilde{\mathcal{L}}_I(s) \right]_{s=m\theta\ell_{\mathrm{LOS}}(r,h)} e^{-\pi\lambda r^2} r\mathrm{d}r \\ & + \int_D^\infty \mathcal{L}_I^{\mathrm{NLOS}}\left(\theta\ell_{\mathrm{NLOS}}(r,h)\right) e^{-\pi\lambda r^2} r\mathrm{d}r \Bigg), \end{aligned} \tag{1.21}$$

$$\begin{aligned} \mathrm{P}_{\mathrm{c}}^{(\mathrm{S})}(\lambda, \theta) = 2\pi\lambda \Bigg(& \int_0^D \sum_{k=0}^{m-1} \left[\frac{(-s)^k}{k!} \frac{\mathrm{d}^k}{\mathrm{d}s^k} \tilde{\mathcal{L}}_I(s) \right]_{s=m\theta\ell_{\mathrm{LOS}}(r,h)} r\mathrm{d}r \\ & + \int_D^\infty \tilde{\mathcal{L}}_I\left(\theta\ell_{\mathrm{NLOS}}(r,h)\right) r\mathrm{d}r \Bigg), \end{aligned} \tag{1.22}$$

with

$$\begin{aligned} \tilde{\mathcal{L}}_I(s) \triangleq \mathcal{L}_I^{\mathrm{NLOS}}(s) \exp\Bigg(-2\pi\lambda \int_{v(r)}^\infty \Bigg(& \frac{1}{1 + s/\ell_{\mathrm{NLOS}}(t,h)} \\ & - \frac{1}{(1 + s/m\ell_{\mathrm{LOS}}(t,h))^m} \Bigg) t\mathrm{d}t \Bigg). \end{aligned} \tag{1.23}$$

The coverage probabilities (1.21) and (1.22) can be evaluated via numerical integration and differentiation, although the latter can be cumbersome in practice, especially for large values of m. Thus, to make numerical evaluation more efficient, one can use Proposition 1.1 to obtain tractable upper bounds with no derivatives.

Considering a shape parameter $m = 10$ (which corresponds to a Rician K-factor $K \simeq 13$ dB), single pathloss exponent $\beta = 4$, non-elevated BSs ($h = 0$), and SIR threshold $\theta = 0$ dB, Figure 1.2 illustrates the coverage probability based on the LOS probability function (1.19) against the BS density λ. The coverage probability based on the simplified LOS probability function (1.20) with $D = 18$ meters and the corresponding upper bound, obtained by applying Proposition 1.1 followed by numerical integration, are also plotted. The coverage probability corresponds to the NLOS case for low BS densities (i.e. $\lambda \leq 10^{-4}$ BSs/m^2) and to the LOS case for high BS densities (i.e. $\lambda \geq 10^{-2}$ BSs/m^2): in particular, these two cases coincide for strongest BS association, as stated in Theorem 1.2(b) in a subsequent subsection. Furthermore, the upper bound is remarkably tight for low BS densities and, in general, tighter for closest BS association than for strongest BS association.

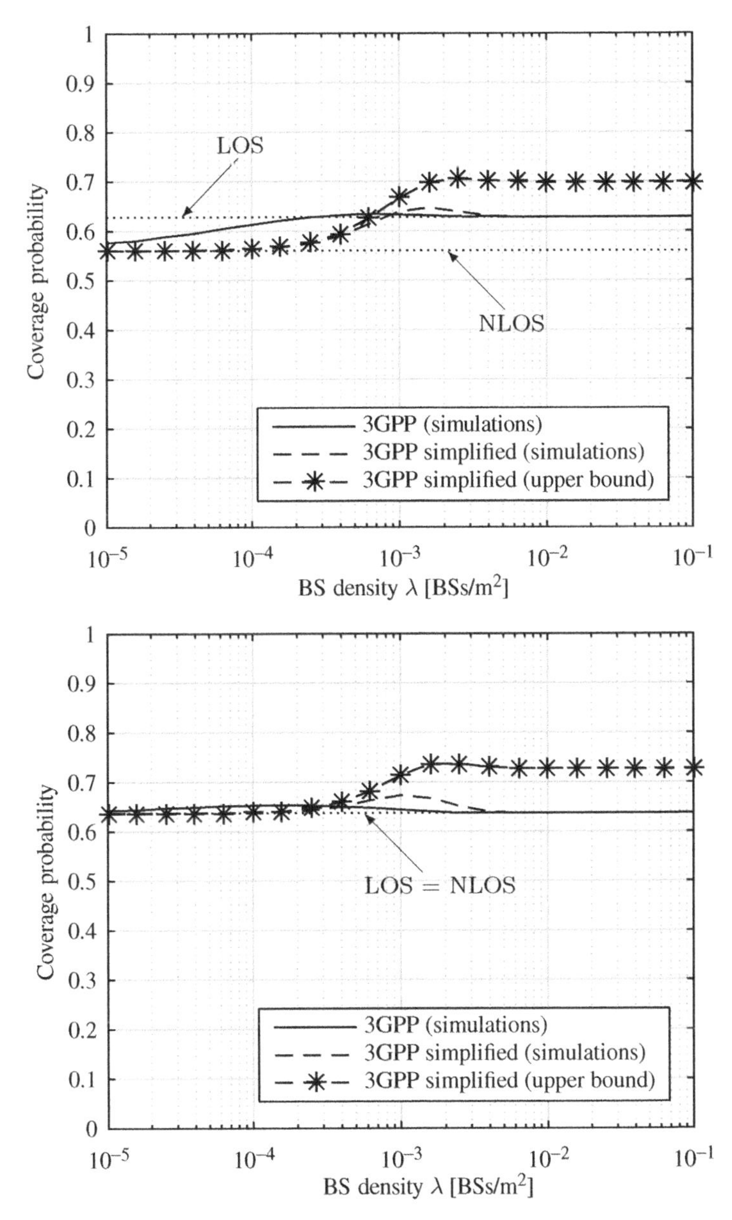

Figure 1.2 Coverage probability with the 3GPP LOS/NLOS model, $m = 10$, $\beta = 4$, and non-elevated BSs against BS density λ for closest (top) and strongest (bottom) BS association. Reprinted with permission from [23].

3. *LOS/NLOS Model with Randomly Placed Buildings*

In this subsection, we propose a practical model for $p_{\mathrm{LOS}}(r)$ that takes into account the combined influence of the link distance and the BS height through the probability of the link being blocked by a building. Other options exist in the literature: for instance,

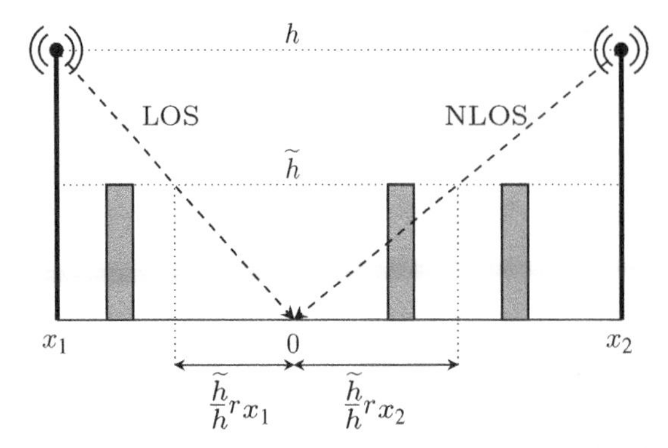

Figure 1.3 LOS/NLOS model with buildings randomly placed between the BSs and the typical UE. Reprinted with permission from [23]

in [24], the BS height, the link distance, and the pathloss exponent are related through the effect of the ground-reflected ray.

Given a BS located at x, we assume that buildings with fixed height $\tilde{h}$, measured in meters, are randomly placed between x and the typical UE. If the straight line between the elevated BS at x and the typical UE does not cross any buildings, then the transmission occurs in LOS propagation conditions; alternatively, if at least one building cuts this straight line, then the transmission occurs in NLOS propagation conditions. A simplified example is illustrated in Figure 1.3, where the typical UE is located at the origin. Note that, in this context, the probability of x being in LOS propagation conditions depends not only on the distance r_x but also on the parameter $\tau \triangleq \min(\tilde{h}/h, 1)$. More precisely, the LOS probability corresponds to the probability of having no buildings in the segment of length τr_x next to the typical UE.

If the location distribution of the buildings follows a one-dimensional PPP with density $\tilde{\lambda}$, measured in buildings/m, the LOS probability function is given by $p_{\mathrm{LOS}}(r_x, \tau) = e^{-\tilde{\lambda}\tau r_x}$. In this setting, $p_{\mathrm{LOS}}(r_x, \tau) = 1$ (all links are in LOS propagation conditions) when $\tilde{\lambda} = 0$ or $\tilde{h} = 0$, whereas $p_{\mathrm{LOS}}(r_x, \tau) = 0$ (all links are in NLOS propagation conditions) when $\tilde{\lambda} \to \infty$. The advantage of this model is that it has only one tuning parameter, i.e. the building density $\tilde{\lambda}$ in the line connecting transmitter and receiver.[1]

The parameters used for obtaining the numerical results are the following: pathloss exponents $\beta_{\mathrm{LOS}} = 3$ and $\beta_{\mathrm{NLOS}} = 4$, building height $\tilde{h} = 10$ m, building densities $\tilde{\lambda} = 10^{-4}$ buildings/m and $\tilde{\lambda} = 10^{-1}$ buildings/m, and SIR threshold $\theta = 0$ dB.

The coverage probability against the BS height h with $m = 1$ is illustrated in Figure 1.4. Evidently, the coverage probability is always monotonically decreasing with h for the closest BS association, since the pathloss increase due to the elevated BSs is more significant for the desired signal than for the interfering signals (which correspond to more distant BSs). On the other hand, for strongest BS association, the coverage probability may not be monotonically decreasing for $h > \tilde{h}$: in fact, a tradeoff between a higher probability of the serving BS being in LOS propagation conditions and a stronger pathloss for the desired signal arises when increasing the BS height (this can be observed in Figure 1.4 for $\tilde{\lambda} = 10^{-1}$ buildings/m and $\lambda = 10^{-3}$ BSs/m^2). Let us now consider the case of multi-antenna BSs, where maximum ratio transmission

1 Rectangular obstacles randomly distributed in a two-dimensional space are considered in [25], whereas the cumulative effect of multiple obstacles is studied in [26].

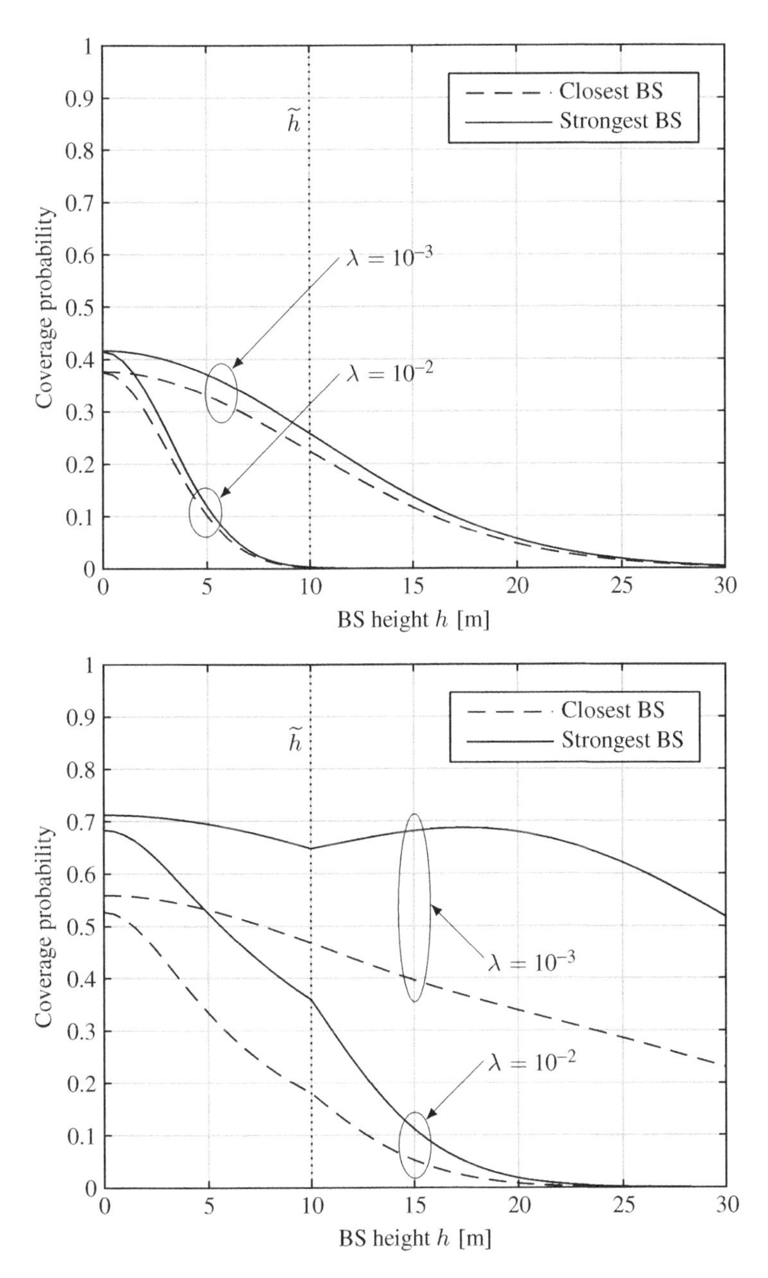

Figure 1.4 Coverage probability with the LOS/NLOS model with randomly placed buildings, $\beta_{\mathrm{LOS}} = 3$, $\beta_{\mathrm{NLOS}} = 4$, $m = 1$, and $\tilde{h} = 10$ m against BS height h for $\tilde{\lambda} = 10^{-4}$ (top) and $\tilde{\lambda} = 10^{-1}$ (bottom). Reprinted with permission from [23].

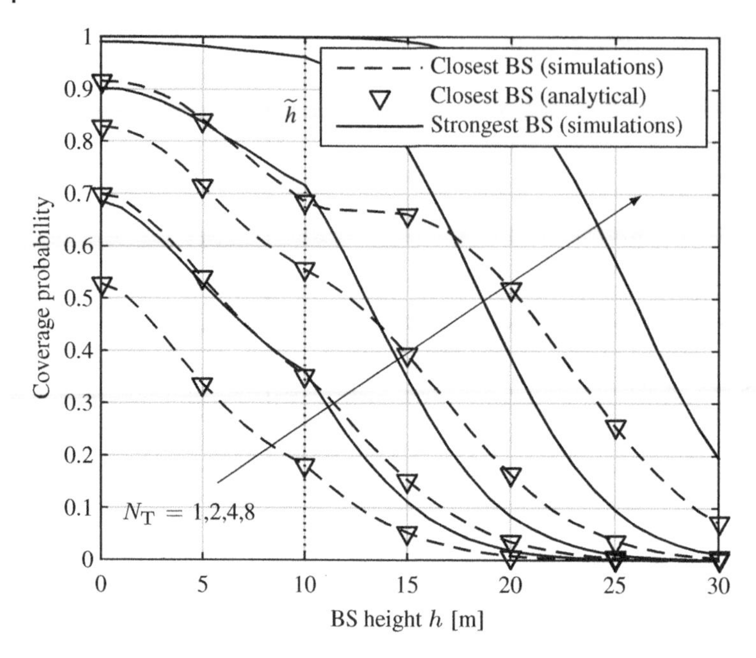

Figure 1.5 Coverage probability with the LOS/NLOS model with randomly placed buildings, $\beta_{\text{LOS}} = 3$, $\beta_{\text{NLOS}} = 4$, $m = 1$, and $\tilde{h} = 10$ m against BS height h for different numbers of transmit antennas N_{T}. Reprinted with permission from [23].

(MRT) beamforming is adopted. Figure 1.5 plots the coverage probability against the BS height h with $m = 1$, $\lambda = 10^{-2}$ BSs/m^2 and $\tilde{\lambda} = 10^{-1}$ buildings/m. In this setting, it is evident that increasing the number of transmit antennas N_{T} produces a substantial improvement in the network performance. Observe that the gain derived from MRT does not depend on the LOS/NLOS propagation conditions of the serving BS and, therefore, there is no additional advantage in increasing the BS height with respect to the case of single-antenna BSs.

1.3.2 The Effect of LOS Fading

In this section, we consider the effect of LOS propagation alone by extending the framework proposed in [1] and [6] to the case of Nakagami-m fading. In doing so, we fix $p_{\text{LOS}}(r) = 1$, $\forall r \in [0, \infty)$ so that all signals from both serving and interfering BSs are subject to Nakagami-m fading. Furthermore, we consider a single pathloss exponent β and neglect the BS height by fixing $h = 0$. Under this setting, we have $\ell_{\text{LOS}}(r, h) = \ell_{\text{NLOS}}(r, h) = r^{\beta}$. Hence, the results derived in this subsection implicitly assume non-elevated BSs; in turn, we make the dependence on the shape parameter m explicit in the resulting expressions of the Laplace transform of the interference and coverage probability.

Let us introduce the following preliminary definitions:

$$\eta(s, m, r) \triangleq {}_2F_1\left(m, -\frac{2}{\beta}, 1 - \frac{2}{\beta}, -\frac{s}{mr^{\beta}}\right), \quad \zeta(m) \triangleq -\frac{\Gamma\left(m + \frac{2}{\beta}\right)\Gamma\left(-\frac{2}{\beta}\right)}{\beta\Gamma(m)} \tag{1.24}$$

where

$$(z)_k \triangleq \frac{\Gamma(z+k)}{\Gamma(z)} = z(z+1)\cdots(z+k-1).$$

Proposition 1.2 [23] For LOS propagation conditions, the Laplace transforms of the interference for closest and strongest BS association are given by

$$\mathcal{L}_I^{\text{LOS,(C)}}(s,m) \triangleq \exp(-\pi\lambda r^2(\eta(s,m,r)-1)), \tag{1.25}$$

$$\mathcal{L}_I^{\text{LOS,(S)}}(s,m) \triangleq \exp\left(-2\pi\lambda\zeta(m)\left(\frac{s}{m}\right)^{2/\beta}\right) \tag{1.26}$$

respectively, with $\eta(s,m,r)$ and $\zeta(m)$ defined in Eq. (1.24). For NLOS propagation conditions, the Laplace transforms of the interference for closest and strongest BS association are given by

$$\mathcal{L}_I^{\text{NLOS,(C)}}(s) \triangleq \mathcal{L}_I^{\text{LOS,(C)}}(s,1), \tag{1.27}$$

$$\mathcal{L}_I^{\text{NLOS,(S)}}(s) \triangleq \mathcal{L}_I^{\text{LOS,(S)}}(s,1) \tag{1.28}$$

respectively, and the corresponding coverage probabilities can be written as

$$\mathrm{P}_\mathrm{c}^{\text{NLOS,(C)}}(\lambda,\theta) \triangleq \frac{1}{\eta(\theta,1,1)}, \tag{1.29}$$

$$\mathrm{P}_\mathrm{c}^{\text{NLOS,(S)}}(\lambda,\theta) \triangleq \frac{1}{2\zeta(1)\theta^{2/\beta}} \tag{1.30}$$

respectively.

Let $B_k(z_1,z_2,\dots,z_k) = \sum_{j=1}^{k} B_{k,j}(z_1,z_2,\dots,z_{k-j+1})$ denote the kth complete Bell polynomial, where $B_{k,j}(z_1,z_2,\dots,z_{k-j+1})$ is the incomplete Bell polynomial. The following theorem provides closed-form expressions of the coverage probabilities in the presence of LOS propagation, i.e. with Nakagami-m fading.

Theorem 1.2 [23] For LOS propagation conditions, the coverage probability is given as follows.

(a) For closest BS association, we have

$$\mathrm{P}_\mathrm{c}^{\text{LOS,(C)}}(\lambda,\theta,m) \triangleq \frac{1}{\eta(m\theta,m,1)}\left(1+\sum_{k=1}^{m-1}\sum_{j=1}^{k}\frac{j!}{k!}B_{k,j}\left(\frac{\psi_1(\theta,m)}{\eta(m\theta,m,1)},\frac{\psi_2(\theta,m)}{\eta(m\theta,m,1)},\dots,\frac{\psi_{k-j+1}(\theta,m)}{\eta(m\theta,m,1)}\right)\right), \tag{1.31}$$

with $\eta(s,m,r)$ defined in Eq. (1.24) and

$$\psi_k(\theta,m) \triangleq -\left(-\frac{2}{\beta}\right)_k\left(\eta(m\theta,m,1)-\sum_{j=1}^{k}\frac{(m)_{k-j}}{(1-2\beta)_{k-j}}\theta^{k-j}(1+\theta)^{-m-k+j}\right). \tag{1.32}$$

(b) For the strongest BS association, we have $\mathrm{P}_{\mathrm{c}}^{\mathrm{LOS,(S)}}(\lambda, \theta, m) = \mathrm{P}_{\mathrm{c}}^{\mathrm{NLOS,(S)}}(\lambda, \theta)$, with $\mathrm{P}_{\mathrm{c}}^{\mathrm{NLOS,(S)}}(\lambda, \theta)$ defined in Eq. (1.30).

For strongest BS association, the above result states that Nakagami-m fading does not affect the coverage probability with respect to Rayleigh fading: this stems from the fact that, under LOS fading, the desired signal power grows with the shape parameter m at the same rate as the interference power.

As the above expressions, although in closed form, are quite involved, we summarize the trend of the coverage probability for LOS propagation conditions and closest BS association as follows: $\mathrm{P}_{\mathrm{c}}^{\mathrm{LOS,(C)}}(\lambda, \theta, m+1) > \mathrm{P}_{\mathrm{c}}^{\mathrm{LOS,(C)}}(\lambda, \theta, m) > \mathrm{P}_{\mathrm{c}}^{\mathrm{NLOS,(C)}}(\lambda, \theta)$, $\forall m \geq 1$, and $\lim_{m\to\infty} \mathrm{P}_{\mathrm{c}}^{\mathrm{LOS,(C)}}(\lambda, \theta, m) = \mathrm{P}_{\mathrm{c}}^{\mathrm{NLOS,(S)}}(\lambda, \theta)$.

Note that in closest BS association, LOS (Nakagami-m) fading has a beneficial effect on the coverage probability: this stems from the fact that, under LOS fading, the desired signal power grows at a higher rate than the interference power. In addition, as the shape parameter m increases, the performance with the closest BS association converges toward that with the strongest BS association.

Consider that all signals from both serving and interfering BSs are subject to Nakagami-m fading, single pathloss exponent $\beta = 4$, non-elevated BSs ($h = 0$), and SIR threshold $\theta = 0$ dB. Figure 1.6 plots the coverage probability against the shape parameter of the Nakagami-m fading. We can see that the coverage probability for the closest BS association increases with m until it approaches the (constant) value of the coverage probability with the strongest BS association: in particular, the two values are approximately the same already for $m = 25$.

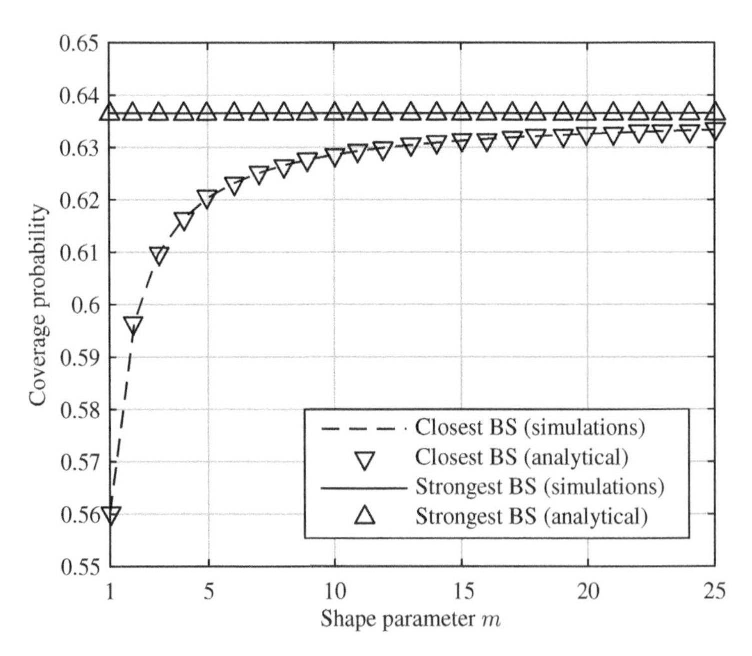

Figure 1.6 Coverage probability with Nakagami-m fading, $\beta = 4$, and non-elevated BSs against shape parameter m. Reprinted with permission from [23].

1.3.3 The Effect of BS Height

We now focus on the effect of BS height on the coverage probability. In doing so, we set the shape parameter $m = 1$ and, as in Section 3.2, we consider a single pathloss exponent, which yields $\ell_{\text{LOS}}(r, h) = \ell_{\text{NLOS}}(r, h) = (r^2 + h^2)^{\beta/2}$. Hence, the results derived in this subsection implicitly assume that all signals from both serving and interfering BSs are subject to Rayleigh fading. While the coverage probability for both closest and strongest BS association is independent of the BS density λ when $h = 0$, as can be observed from Eqs. (1.29) and (1.30) (see also [1] and [6], respectively, for details), the impact of BS height becomes visible as λ increases. Therefore, we make the dependence on λ explicit in the resulting expressions of the coverage probability and the potential throughput.

1. *Impact on Interference*

We begin by observing that $\ell(r_x, h)$ yields a bounded pathloss model for any BS height $h > 0$, since BS x cannot get closer than h to the typical UE (this occurs when $r_x = 0$). The following lemma expresses the Laplace transforms of the interference for a fixed BS height h.

Lemma 1.1 [23] For elevated BSs, the Laplace transforms of the interference for closest and strongest BS association can be written as

$$\mathcal{L}_I^{(C)}(s) \triangleq \mathcal{L}_I^{\text{NLOS},(C)}(s) \cdot \exp\left(2\pi\lambda \int_r^{\sqrt{r^2+h^2}} \left(1 - \frac{1}{1 + st^{-\beta}}\right) t \, \mathrm{d}t\right) \tag{1.33}$$

$$= \exp(-\pi\lambda(r^2 + h^2)(\eta(s, 1, \sqrt{r^2 + h^2}) - 1)), \tag{1.34}$$

$$\mathcal{L}_I^{(S)}(s) \triangleq \mathcal{L}_I^{\text{NLOS},(S)}(s) \exp\left(2\pi\lambda \int_0^h \left(1 - \frac{1}{1 + st^{-\beta}}\right) t \, \mathrm{d}t\right)$$

$$= \exp(-\pi\lambda h^2(\eta(s, 1, h) - 1)) \tag{1.35}$$

respectively, where $\mathcal{L}_I^{\text{NLOS},(C)}(s)$ and $\mathcal{L}_I^{\text{NLOS},(S)}(s)$ are the Laplace transforms of the interference with non-elevated BSs defined in Eqs. (1.27) and (1.28).

From the above expressions, it is straightforward to see that the interference is reduced when $h > 0$ with respect to when $h = 0$, since the Laplace transforms in Eqs. (1.27) and (1.28) are multiplied by exponential terms with positive arguments.

Recall that, for the strongest BS association and for $h = 0$, the expected interference power is infinite [2, Ch. 5.1]. Let $U(a, b, z) \triangleq 1/\Gamma(a) \int_0^\infty e^{-zt} t^{a-1}(1 + t)^{b-a-1} dt$ denote Tricomi's confluent hypergeometric function and let $E_n(z) \triangleq \int_1^\infty e^{-zt} t^{-n} dt$ be the exponential integral function. A consequence of the bounded pathloss model is given in the following lemma, which characterizes the expected interference with elevated BSs for the strongest BS association.

Lemma 1.2 [23] For elevated BSs, the expected interference power for strongest BS association is finite and is upper bounded by

$$\mathbb{E}\left[\sum_{y \in \Phi \setminus \{x\}} g_y \ell(r_y, h)\right] < \sum_{i=1}^{\infty} (\pi\lambda)^i h^{2i-\beta} U\left(i, i + 1 - \frac{\beta}{2}, \pi\lambda h^2\right) \tag{1.36}$$

where the expected interference power from the nearest interfering BS, whose location is denoted by x_1, corresponds to

$$\mathbb{E}[g_{x_1}\ell(r_{y_1}, h)] = \pi\lambda h^{2-\beta} e^{\pi\lambda h^2} E_{\beta/2}(\pi\lambda h^2). \tag{1.37}$$

2. *Impact on Coverage Probability and Potential Throughput*

We now focus on the effect of BS height on the coverage probability and on the potential throughput. The following result provides the coverage probabilities for a fixed BS height h.

Theorem 1.3 [23] For elevated BSs, recalling the definition of $\eta(s, m, r)$ in Eq. (1.24), the coverage probability is given as follows.

(a) For closest BS association, we have

$$\mathrm{P}_\mathrm{c}^{(\mathrm{C})}(\lambda, \theta) \triangleq \mathrm{P}_\mathrm{c}^{\mathrm{NLOS},(\mathrm{C})}(\theta) \exp(-\pi\lambda h^2(\eta(\theta, 1, 1) - 1)) \tag{1.38}$$

where $\mathrm{P}_\mathrm{c}^{\mathrm{NLOS},(\mathrm{C})}(\theta)$ is the coverage probability with non-elevated BSs defined in Eq. (1.29).

(b) For the strongest BS association, we have

$$\mathrm{P}_\mathrm{c}^{(\mathrm{S})}(\lambda, \theta) \triangleq 2\pi\lambda \int_h^\infty \exp(-\pi\lambda h^2(\eta(\theta r^\beta, 1, h) - 1))r \, \mathrm{d}r. \tag{1.39}$$

Notably, for the closest BS association, a closed-form expression is available; therefore, Corollary 1.1 gives the optimal BS density that maximizes the potential throughput.

Corollary 1.1 [23] For elevated BS and closest BS association, let $\mathcal{T}^{(\mathrm{C})}(\lambda, \theta) \triangleq \lambda \mathrm{P}_\mathrm{c}^{(\mathrm{C})}(\lambda, \theta) \log(1 + \theta)$ denote the potential throughput. Then the optimal BS density is given by

$$\lambda_{\mathrm{opt}}^{(\mathrm{C})} \triangleq \underset{\lambda}{\mathrm{argmax}}\, \mathcal{T}^{(\mathrm{C})}(\lambda, \theta) = \frac{1}{\pi h^2(\eta(\theta, 1, 1) - 1)} \tag{1.40}$$

and the maximum potential throughput is given by

$$\mathcal{T}_{\max}^{(\mathrm{C})}(\theta) \triangleq \mathcal{T}^{(\mathrm{C})}(\lambda_{\mathrm{opt}}^{(\mathrm{C})}, \theta) = \frac{e^{-1}}{\pi h^2 \eta(\theta, 1, 1)(\eta(\theta, 1, 1) - 1)} \log(1 + \theta). \tag{1.41}$$

Theorem 1.3 unveils the detrimental effect of BS height on the system performance. In particular, Eq. (1.38) quantifies the coverage degradation due to elevated BSs, while Eq. (1.40) confirms the existence of an optimal BS density from the perspective of the potential throughput. The degradation in coverage probability stems from the fact that the distance of the typical user equipment (UE) from its serving BS is more affected by the BS height than the distances from the interfering BSs and, therefore, the desired signal power and interference power do not grow at the same rate as in the case with $h = 0$. The following corollary strengthens this claim by showing the asymptotic performance for both the closest and strongest BS associations.

Corollary 1.2 [23] For elevated BSs, recalling the definitions of the coverage probabilities with non-elevated BSs $\mathrm{P}_\mathrm{c}^{\mathrm{NLOS,(C)}}(\theta)$ and $\mathrm{P}_\mathrm{c}^{\mathrm{NLOS,(S)}}(\theta)$ in Equations (1.29) and (1.30), the following holds:

(a) $\lim_{\lambda\to 0} \mathrm{P}_\mathrm{c}^{(\mathrm{C})}(\lambda,\theta) = \mathrm{P}_\mathrm{c}^{\mathrm{NLOS,(C)}}(\theta)$;

(b) $\lim_{\lambda\to 0} \mathrm{P}_\mathrm{c}^{(\mathrm{S})}(\lambda,\theta) = \mathrm{P}_\mathrm{c}^{\mathrm{NLOS,(S)}}(\theta)$;

(c) $\lim_{\lambda\to \infty} \mathrm{P}_\mathrm{c}^{(\mathrm{C})}(\lambda,\theta) = \lim_{\lambda\to \infty} \mathrm{P}_\mathrm{c}^{(\mathrm{S})}(\lambda,\theta) = 0$.

For a fixed BS height $h > 0$, the coverage probability monotonically decreases as the BS density λ increases, eventually leading to near-universal outage: as a consequence, the area spectral efficiency also decays to zero as $\lambda \to \infty$. On the other hand, the effect of BS height becomes negligible as $\lambda \to 0$. In practice, the coverage probability and the potential throughput decay to zero even for moderately low BS densities (i.e. for $\lambda \simeq 10^{-2}$ BSs/m^2).

For a fixed BS density λ, the coverage probability monotonically decreases as the BS height h increases. More specifically, from Corollary 1.1, we have that $\mathcal{T}_{\max}^{(\mathrm{C})}(\theta) \propto 1/h^2$. Therefore, the optimal BS height is $h = 0$.

When the serving and interfering BSs are characterized by the same propagation conditions, the optimal BS height is always $h = 0$. Moreover, numerical results show that the same usually holds in the more general case where the serving and interfering BSs are subject to the same distance-dependent LOS probability function. However, under a propagation model where LOS interference is nearly absent (which can be obtained, for instance, by assuming downtilted antennas), a non-zero optimal BS height is expected: in fact, in that case, the desired signal would be subject to a tradeoff between pathloss (for which a low BS is desirable) and probability of LOS propagation conditions (for which a high BS is desirable).

Consider now a single pathloss exponent $\beta = 4$, shape parameter $m = 1$, and SIR threshold $\theta = 0$ dB. Figure 1.7 plots the coverage probability with elevated BSs against the BS density λ; two BS heights are considered, i.e. $h = 10$ m and $h = 20$ m. Interestingly, it is shown that the coverage probability decays to zero even for moderately low BS densities, i.e. at $\lambda \simeq 10^{-2}$ BSs/m^2 with $h = 20$ m and at $\lambda \simeq 3 \times 10^{-2}$ BSs/m^2 with $h = 10$ m. For comparison, the coverage probabilities with $h = 0$, i.e. $\mathrm{P}_\mathrm{c}^{\mathrm{NLOS,(C)}}(\lambda,\theta)$ and $\mathrm{P}_\mathrm{c}^{\mathrm{NLOS,(S)}}(\lambda,\theta)$ in Eqs. (1.29) and (1.30), are also depicted. The coverage probability with elevated BSs converges to that with $h = 0$ as $\lambda \to 0$: this is already verified at $\lambda \simeq 10^{-5}$ BSs/m^2, when the BSs are so far away from the typical UE that the effect of BS height become negligible.

In Figure 1.8, we show the potential throughput with elevated BSs against the BS density λ; three BS heights are considered, i.e. $h = 10$ m, $h = 15$ m, and $h = 20$ m. Here, the detrimental effect of BS height on the system performance appears even more evident. As an example, considering the closest BS association, the maximum potential throughput is 0.84×10^{-3} bps/Hz/m^2 for $h = 10$ m and 0.21×10^{-3} bps/Hz/m^2 for $h = 20$ m; likewise, considering the strongest BS association, the maximum potential throughput is 1.04×10^{-3} bps/Hz/m^2 for $h = 10$ m and 0.26×10^{-3} bps/Hz/m^2 for $h = 20$ m. Hence, doubling the BS height reduces the maximum potential throughput by a factor of four. Furthermore, it is worth noting that the BS density that maximizes the potential

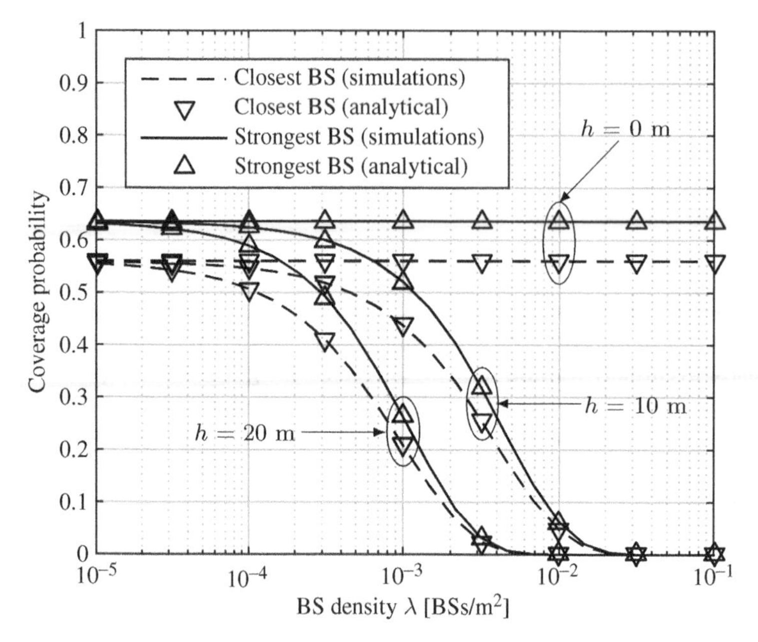

Figure 1.7 Coverage probability with elevated BSs, $m = 1$, and $\beta = 4$ against BS density λ. Reprinted with permission from [23].

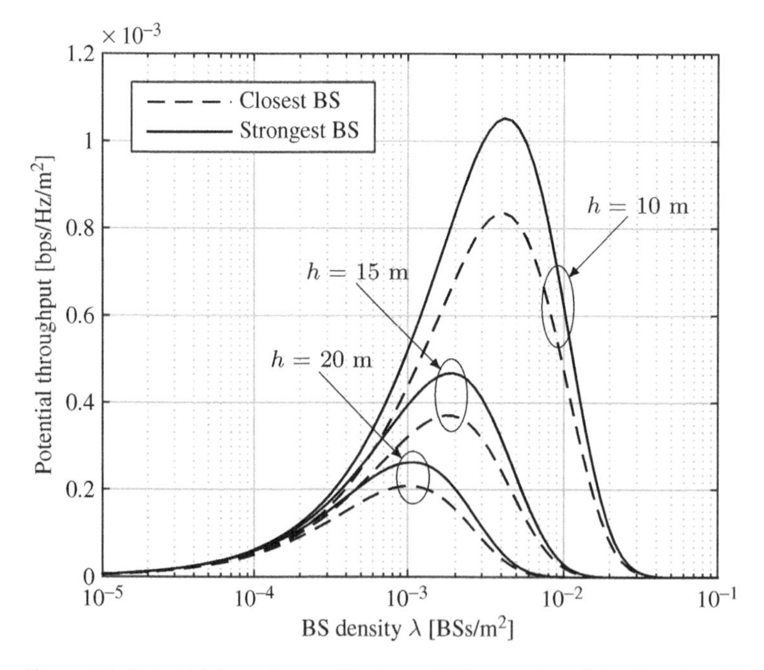

Figure 1.8 Potential throughput with elevated BSs, $m = 1$, and $\beta = 4$ against BS density λ. Reprinted with permission from [23].

throughput for the closest BS association, i.e. $\lambda_{\text{opt}}^{(C)}$, which is derived in closed form in Eq. (1.40) coincides with the optimal BS density for the case of the strongest BS association: this corresponds to $\lambda \simeq 4 \times 10^{-3}$ BSs/m^2 for $h = 10$ m and to $\lambda \simeq 10^{-3}$ BSs/m^2 for $h = 20$ m.

We now analyze the combined effect of LOS/NLOS fading, LOS/NLOS pathloss, and BS height. Considering elevated BSs with $h = 20$ m, Figure 1.9 illustrates the coverage

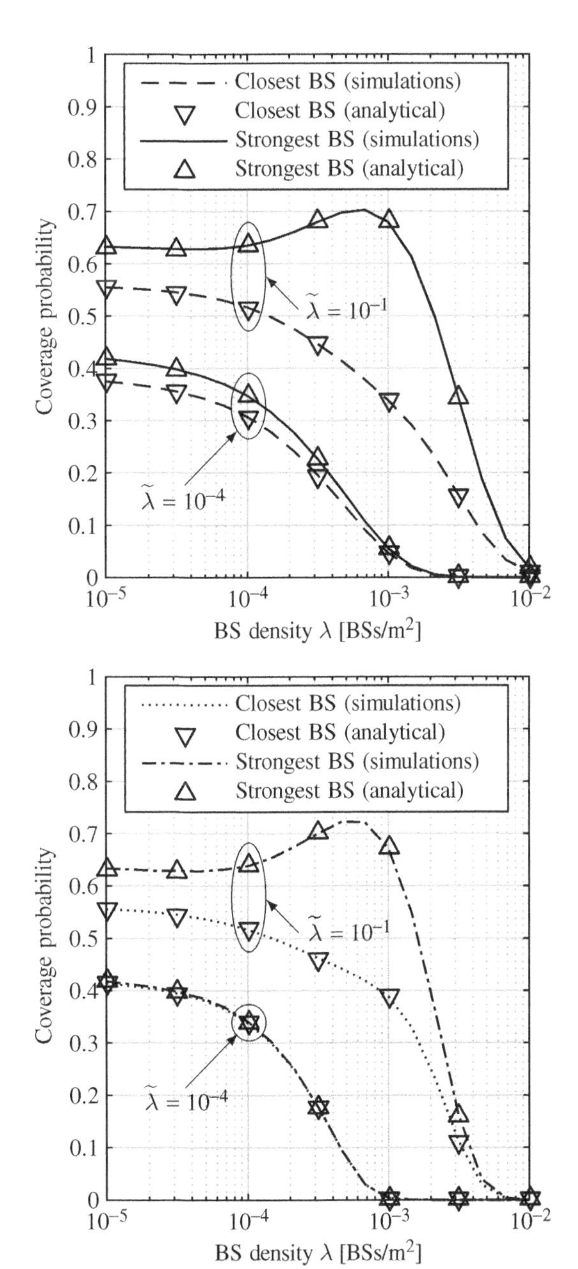

Figure 1.9 Coverage probability with the LOS/NLOS model with randomly placed buildings, $\beta_{\text{LOS}} = 3$, $\beta_{\text{NLOS}} = 4$, $h = 20$ m, and $\tilde{h} = 10$ m against BS density λ for $m = 1$ (top) and $m = 10$ (center); corresponding potential throughput (bottom). Reprinted with permission from [23].

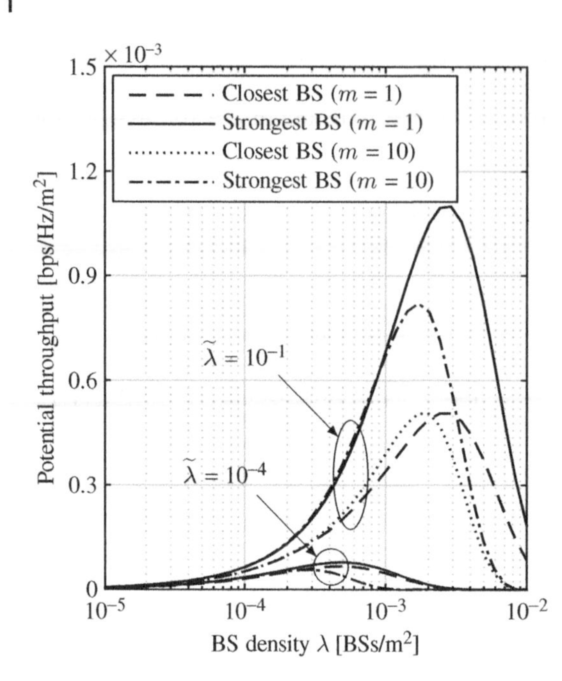

Figure 1.9 (*Continued*)

probability and the potential throughput against the BS density λ; two shape parameters of the Nakagami-m fading are considered, i.e. $m = 1$ and $m = 10$. Observing the curves in Figure 1.9, the detrimental effects on the system performance can be ranked in decreasing order of importance as follows:

(a) The BS height is evidently the dominant effect, since it eventually leads to near-universal outage regardless of the other parameters and even for moderately low BS densities.
(b) The LOS pathloss, which is determined by the building density, is the second most important effect: specifically, a high building density is beneficial in this setting since it creates nearly NLOS pathloss conditions (i.e. p_{LOS} is very low), whereas a low building density leads to a nearly LOS scenario (i.e. unconstrained propagation, p_{LOS} close to one in most cases).
(c) The LOS fading, also determined by the building density, has a minor impact as compared to the pathloss: in particular, for closest BS association, the coverage probability with $m = 10$ is slightly improved at low BS densities and marginally deteriorated at high BS densities as compared to the case with $m = 1$.

Moreover, for $\tilde{\lambda} = 10^{-1}$ buildings/m and strongest BS association, we observe a peak in the coverage probability around $\lambda = 10^{-3}$ BSs/m^2.

Lastly, we consider the case where each BS x is characterized by a different distance-dependent shape parameter $m(r_x) = (K(r_x) + 1)^2/(2K(r_x) + 1)$ (rounded to the nearest integer), where the Rician K-factor depends on the link distance. Building on the measurements for the urban microcell scenario in [27, Sec. 5.4.11], we express the Rician K-factor in dB as $K_{\text{dB}}(r) = 13 + 0.0142r$. The coverage probabilities with $m = 10$ and with the distance-dependent m are compared in Figure 1.10. The excellent matching between the two curves justifies our approach of using a fixed value of the

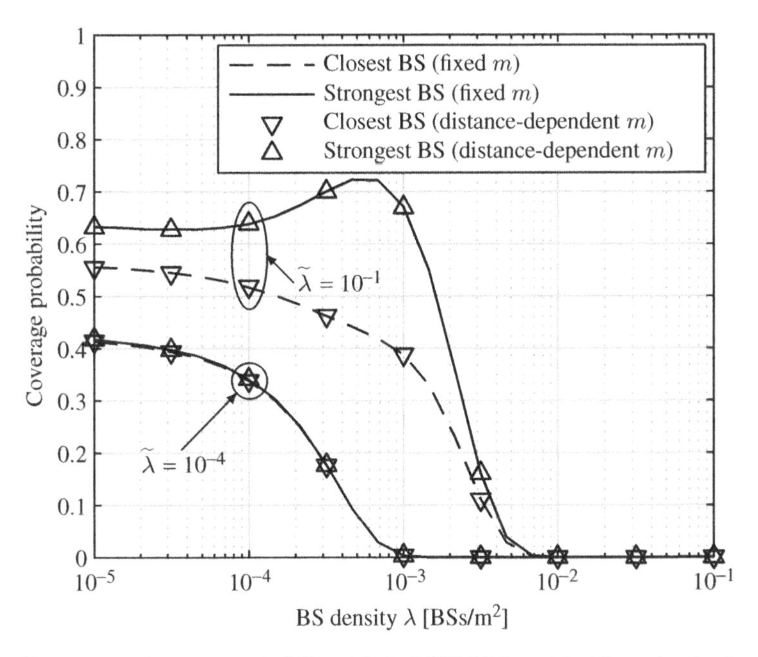

Figure 1.10 Coverage probability with the LOS/NLOS model with randomly placed buildings, $\beta_{\text{LOS}} = 3$, $\beta_{\text{NLOS}} = 4$, $h = 20$ m, and $\tilde{h} = 10$ m against BS density λ for fixed m ($m = 10$) and distance-dependent m (with $K_{\text{dB}}(r) = 13 + 0.0142r$). Reprinted with permission from [23].

shape parameter m: in fact, as the link distance increases, the LOS probability decreases and the impact of the increasing Rician K-factor becomes negligible.

1.4 The Quest for Scaling Laws

In previous sections, our main target was to derive exact results and analytical expressions for the downlink performance of UDNs as a means to characterize the performance limits of network densification and to get crisp insights into the coverage and throughput behavior in UDNs. However, the above expressions become involved quite easily and several approximations or simplifying assumptions are needed to proceed further. For that, here we take on a different approach and aim at characterizing the asymptotic scaling of key performance metrics as a means to obtain useful system design guidelines even in the absence of analytical closed-form expressions.

For assessing the maximum achievable performance in the asymptotic regime, it is both practically and theoretically relevant to consider: (i) a general multi-slope pathloss model, (ii) a general channel power model for g including various fast fading and shadowing distributions, and (iii) the strongest BS association, i.e. the SINR of the typical user can be expressed as

$$\text{SINR} = \frac{M_i}{I_{\text{tot}} + W - M_i}, \quad \text{with } M_i \triangleq \max_{i \in \Phi} P_i. \tag{1.42}$$

Note that the BSs are not elevated ($h = 0$), but the model can easily be extended to consider $h > 0$.

1.4.1 User Performance

The network performance precisely depends on the received SINR, which in turn depends on M and I_{tot} (cf. Eq. (1.42)). The behavior of the maximum M and the sum I_{tot} is totally determined by that of the received power P_i. The following theorem characterizes the signal power P_i, in particular its tail behavior using tools from regular variation theory.

Theorem 1.4 [28] The tail distribution of the received signal power $\overline{F}_P$ depends on the tail distribution of the channel power $\overline{F}_g$ and the pathloss function $\ell(r)$ as follows:

- If $\overline{F}_g \in \mathcal{R}_{-\alpha}$ with $\alpha \in [0, \infty]$, then $\overline{F}_P \in \mathcal{R}_{-\rho}$ where $\rho = \min(\delta_0, \alpha)$ with the convention that $\delta_0 = +\infty$ for $\beta_0 = 0$;
- If $\overline{F}_g(x) = o(\overline{H}(x))$ as $x \to \infty$ with $\overline{H} \in \mathcal{R}_{-\infty}$, then $\overline{F}_P(t)$ and $\overline{F}_g(A_0 t)$ are tail-equivalent for $\beta_0 = 0$ and $\overline{F}_P \in \mathcal{R}_{-\delta_0}$ for $\beta_0 > 0$.

Theorem 1.4 shows that the tail behavior of the wireless link depends not only on whether the pathloss function is bounded or not but also on the tail behavior of the channel power. More precisely, a key implication of Theorem 1.4 is that pathloss and channel power have interchangeable effects on the tail behavior of the wireless link. This can also be shown using Breiman's Theorem [29] and results from large deviation of product distributions. Specifically, if the channel power is regularly varying with index $-\rho$ and the ρ-moment of the pathloss is finite, then the tail behavior of the received signal is governed by the regularly varying tail, i.e., the wireless link is also regularly varying regardless of the pathloss singularity. For lighter-tailed channel power, the regular variation property of the wireless link is solely imposed by the pathloss singularity.

More importantly, Theorem 1.4 is a general result and covers all tail behaviors of the channel power. The first case covers the heaviest tails (i.e. $\mathcal{R}_{-\alpha}$ with $0 \leq \alpha < \infty$), such as Pareto distribution, as well as the moderately heavy tails (i.e. the class $\mathcal{R}_{-\infty}$), such as exponential, normal, lognormal, and gamma distributions. The second case covers all remaining tails (e.g. truncated distributions). Therefore, for any statistical distribution of the channel power, and in particular of fast fading and shadowing, Theorem 1.4 allows us to characterize the tail behavior of the wireless link, which is essential to understand the behavior of the interference, the maximum received power, and their asymptotic relationship.

In current wireless networks, the signal distribution $\overline{F}_g$ is governed by lognormal or gamma shadowing and Rayleigh fast fading. Since these fading distributions belong to the class $\mathcal{R}_{-\infty}$ and the pathloss is bounded, the tail distribution of the received signal follows $\overline{F}_P \in \mathcal{R}_{-\infty}$. Note also that in most relevant cases, it can be shown that $\overline{F}_P$ belongs to the maximum domain of attraction of a Gumbel distribution [30].

Based on the above result, we want now to better understand the signal power scaling and the interplay between the pathloss function and channel power distribution. The following result can be derived.

Corollary 1.3 [28] The tail distribution $\overline{F}_P$ is classified as follows:

- $\overline{F}_P \in \mathcal{R}_0$ if and only if $\overline{F}_g \in \mathcal{R}_0$;
- $\overline{F}_P \in \mathcal{R}_{-\alpha}$ with $\alpha \in (0, 1)$ if $\beta_0 > d$ or $\overline{F}_g \in \mathcal{R}_{-\alpha}$;

- $\overline{F}_P \in \mathcal{R}_{-\alpha}$ with $\alpha > 1$ if $0 < \beta_0 < d$ and $\overline{F}_g \notin \mathcal{R}_{-\rho}$ with $\rho \in [0, 1]$;
- $\overline{F}_P = o(\overline{H})$ with $\overline{H} \in \mathcal{R}_{-\infty}$ if $\beta_0 = 0$ and $\overline{F}_g = o(\overline{H})$.

The tail behavior of the wireless link as characterized in the previous section allows provision of the scaling laws for the received SINR in the high density regime.

Theorem 1.5 [28] In ultra-dense networks (i.e. $\lambda \to \infty$), the received SINR behaves as

1. $\text{SINR} \xrightarrow{p} \infty$ if $\overline{F}_P \in \mathcal{R}_0$;
2. $\text{SINR} \xrightarrow{d} G$ if $\overline{F}_P \in \mathcal{R}_{-\alpha}$ with $0 < \alpha < 1$, where G has a non-degenerate distribution;
3. $\text{SINR} \xrightarrow{a.s.} 0$ if $\overline{F}_P \notin \mathcal{R}_{-\alpha}$ with $\alpha \in [0, 1]$.

According to Corollary 1.3, $\overline{F}_P \in \mathcal{R}_0$ is due to the fact that $\overline{F}_g \in \mathcal{R}_0$. Since g is the channel power containing the transmit power and all potential gains and propagation phenomena (including fading, array gain, etc.), $\overline{F}_g \in \mathcal{R}_0$ means that the channel power is more likely to take large values. As a result, it compensates the pathloss and makes the desired signal power grow at the same rate as the aggregate interference (i.e. $M/I_{\text{tot}} \xrightarrow{p} 1$). This provides a theoretical justification for the fact that network densification always enhances the signal quality SINR.

When $\overline{F}_P \in \mathcal{R}_{-\alpha}$ with $0 < \alpha < 1$, $\text{SINR} \xrightarrow{d} G$ implies that the SINR distribution converges to a non-degenerate distribution. From Corollary 1.3, this is due to either a large near-field exponent or heavy-tailed channel power. In that case, for any SINR target θ, the coverage probability $\mathbb{P}(\text{SINR} > \theta)$ flattens out starting from some network density λ (ceiling effect). This means that further increasing the network density by installing more BSs does not improve the network performance. This saturation effect is confirmed by simulation results shown in Figure 1.11, where the tail distribution of SINR converges to a steady distribution for both cases: either $\beta_0 > d$ (left plot) or $\overline{F}_g \in \mathcal{R}_{-\alpha}$ with $\alpha \in (0, 1)$ (right plot). In Figure 1.11, by $F_g \sim$ Composite we mean that the channel power corresponds to the case with constant transmit power and with commonly known composite Rayleigh-lognormal fading, which belongs to the rapidly varying class $\mathcal{R}_{-\infty}$. Pareto(α) stands for channel power following a Pareto distribution of shape $1/\alpha$ and some scale $\sigma > 0$, i.e. Pareto(α) : $\overline{F}_g(x) = (1 + x/\sigma)^{-\alpha}$. Note that Pareto($\alpha$) $\in \mathcal{R}_{-\alpha}$.

In real-world networks, the pathloss attenuation is bounded (i.e. $\beta_0 = 0$) and channel power is normally less heavy-tailed or even truncated (i.e. $\overline{F}_g = o(\overline{H})$ for some $\overline{H} \in \mathcal{R}_{-\infty}$). In particular, the conventional case of lognormal shadowing and Rayleigh fading results in channel power belonging to the class $\mathcal{R}_{-\infty}$. As a result, based on Theorem 1.4, we have that $\overline{F}_P \in \mathcal{R}_{-\infty}$; hence $\text{SINR} \xrightarrow{a.s.} 0$. Therefore, the SINR is proven to be asymptotically decreasing with the network density. This means that there is a fundamental limit on network densification and the network should not operate in the ultra-dense regime since deploying excessively many BSs would decrease the network performance due to the fact that signal power boosting cannot compensate for the faster growing aggregate interference (i.e. $M/I_{\text{tot}} \xrightarrow{a.s.} 0$). In other words, there exists an optimal density value

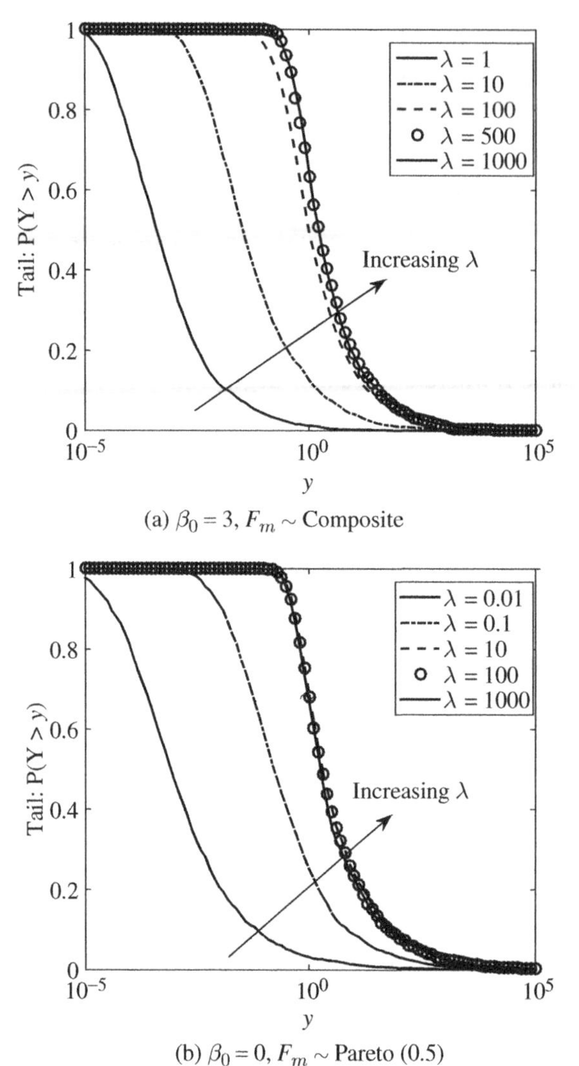

Figure 1.11 Validation of the convergence of SINR to a steady distribution in the case of $\overline{F}_P \in \mathcal{R}_{-\alpha}$ with $\alpha \in (0, 1)$. Simulated parameters include two-slope pathloss (i.e. $K = 2$) with $A_0 = 1$, $\beta_1 = 4$, $R_1 = 10$ m, $R_\infty = 40$ km, two-dimensional network domain (i.e. $d = 2$), and network density λ in BSs/km^2. Reprinted with permission from [28].

until which the SINR monotonically increases and after which the SINR monotonically decreases.

In Figure 1.12, we provide simulation results for $\overline{F}_P \notin \mathcal{R}_{-\alpha}$ with $\alpha \in [0, 1]$, with $\overline{F}_P \in \mathcal{R}_{-2}$ in Figure 1.12(a) and $\overline{F}_P \in \mathcal{R}_{-4}$ in Figure 1.12(b). The results in Figure 1.12 confirm that the tail of SINR distribution vanishes and converges to zero when λ increases. The convergence of SINR to zero in the high-density regime further emphasizes the cardinal importance of performing local scheduling among BSs, as well as signal processing mechanisms for interference mitigation.

To provide a complete characterization of the network performance under the densification effect, we investigate now the fundamental limits to the amount of densification, which depend not only on the pathloss but also on the channel power distribution. Based on previous analytical results, we obtain the following result.

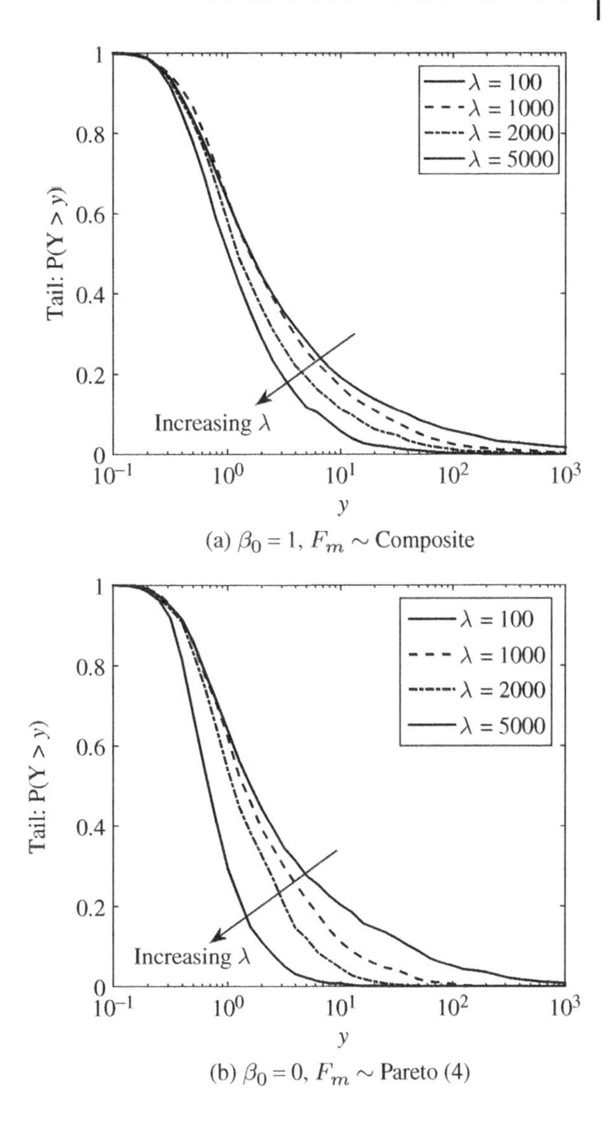

Figure 1.12 Validation of the convergence of SINR to zero when $\overline{F}_P \notin \mathcal{R}_{-\alpha}$ with $\alpha \in [0, 1]$. Simulated parameters include two-slope pathloss (i.e. $K = 2$) with $A_0 = 1$, $\beta_1 = 4$, $R_1 = 10$ m, $R_\infty = 40$ km, two-dimensional network domain (i.e. $d = 2$), and network density λ in BSs/km^2. Reprinted with permission from [28].

Theorem 1.6 [28] The coverage probability $\mathrm{P_c}(\lambda, \theta)$ for fixed θ and the mean rate $C(\lambda)$ scale as follows:

1. $\mathrm{P_c}(\lambda, \theta) \to 1$ and $C(\lambda) \to \infty$ as $\lambda \to \infty$ if $\overline{F}_P \in \mathcal{R}_0$;
2. $\mathrm{P_c}(u\lambda, \theta)/\mathrm{P_c}(\lambda, \theta) \to 1$ and $C(u\lambda)/C(\lambda) \to 1$ for $0 < u < \infty$ as $\lambda \to \infty$ if $\overline{F}_P \in \mathcal{R}_{-\alpha}$ with $0 < \alpha < 1$;
3. $\mathrm{P_c}(\lambda, \theta) \to 0$ and $C(\lambda) \to 0$ as $\lambda \to \infty$ if $\overline{F}_P \notin \mathcal{R}_{-\alpha}$ with $\alpha \in [0, 1]$; moreover there exist finite densities λ_p, λ_c such that $\mathrm{P_c}(\lambda_p, \theta) > \lim_{\lambda\to\infty} \mathrm{P_c}(\lambda, \theta)$ and $C(\lambda_c) > \lim_{\lambda\to\infty} C(\lambda)$.

Based on Theorem 1.6, we see that there exists a phase transition when the network density goes to infinity (ultra-densification). Specifically, depending on the pathloss attenuation (singularity and multi-slope) and the channel power distribution, there are

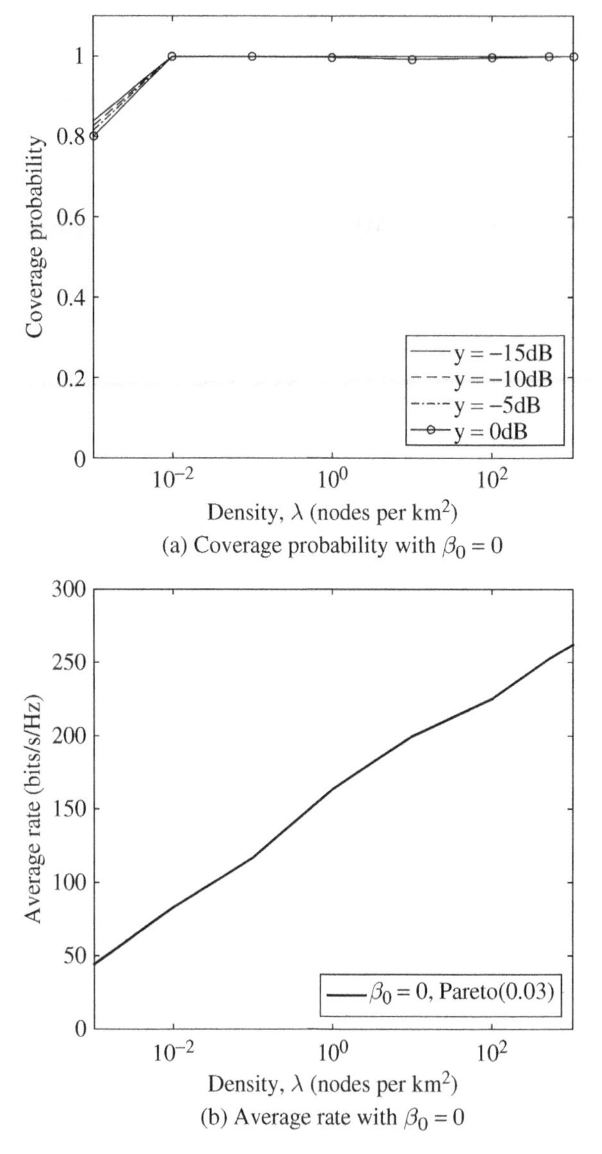

Figure 1.13 Experiment of *growth regime*. Approximation of $\overline{F}_P \in \mathcal{R}_0$ by $\overline{F}_g \sim$ Pareto(0.03) in Matlab simulation. Simulated parameters include two-slope pathloss (i.e. $K = 2$) with $A_0 = 1, \beta_1 = 4, R_1 = 10$ m, $R_\infty = 40$ km, two-dimensional network domain (i.e. $d = 2$), and network density λ in BSs/km^2. Reprinted with permission from [28].

three distinct regimes for the coverage and the average rate: monotonically increasing, saturation, and deficit:

- *Growth Regime* When $\overline{F}_P \in \mathcal{R}_0$, meaning that the channel power g is slowly varying with $\overline{F}_g \in \mathcal{R}_0$ (see Corollary 1.3), both the coverage probability and the average rate are monotonically increasing with λ. In particular, the average rate asymptotically grows with the network density. In Figure 1.13, we show simulation results with $\overline{F}_g \sim$ Pareto(0.03). We can see that, even though the pathloss is bounded, the coverage probability is almost one for all SINR thresholds, while throughput increases almost linearly with the logarithm of the network density in the evaluated range. This growth regime, revealed from Theorem 1.6, shows that the great expectations on the potential of network densification are theoretically possible. However, since slowly varying distributions are rather theoretical extremes and would be rarely observable in practice, the growth regime would be highly unlikely in real-world networks.

- *Saturation Regime* When the channel power is regularly varying with index within −1 and 0 or when the near-field pathloss exponent β_0 is larger than the network dimension d, the tail distribution of a wireless link behaves as $\overline{F}_P \in \mathcal{R}_{-\alpha}$ with $\alpha \in (0, 1)$. Consequently, both the coverage probability and the average rate saturate past a certain network density. This saturation behavior is also confirmed by simulation experiments as shown in Figures 1.14 and 1.15. In Figure 1.14, it is the pathloss function's singularity with $\beta_0 > d$ that creates performance saturation for any type of channel power distribution not belonging to the class $\mathcal{R}_0$. In Figure 1.15, the saturation is completely due to regularly varying channel power $\overline{F}_g \in \mathcal{R}_{-\alpha}$, $\alpha \in (0, 1)$, regardless of pathloss boundedness. Prior studies have shown that the network performance is invariant of the network density for unbounded pathloss and negligible background noise. Our results (Theorem 1.6) show that this performance saturation may happen in a much larger setting, including (i) with non-negligible thermal noise and (ii) even with bounded pathloss if channel power is in the class $\mathcal{R}_{-\alpha}$ with $\alpha \in (0, 1)$. More importantly, the *unbounded* property of the pathloss – widely used in the literature – is just a necessary condition to have saturated performance. A sufficient condition is that the near-field pathloss exponent has to be greater than the network dimension, i.e. $\beta_0 > d$. As we will see shortly, when channel power is less heavy-tailed than the class $\mathcal{R}_{-\alpha}$ with $\alpha \in [0, 1]$, then unbounded pathloss with $0 < \beta_0 < d$ results in the same scaling regime as that of bounded pathloss.
- *Deficit Regime* The third regime of network densification is determined first by a channel power distribution that is less heavy-tailed, precisely all remaining distributions not belonging to $\mathcal{R}_{-\alpha}$ with $\alpha \in [0, 1]$, and second by a near-field pathloss exponent smaller than the network dimension (i.e. $\beta_0 < d$). In this regime, both coverage probability and rate initially increase in the low-density regime, then achieve a maximum at a finite network density, after which they start decaying and go to zero in the ultra-dense regime. This behavior is also confirmed by simulations shown in Figure 1.16. More precisely, both coverage probability (left plot) and rate (right plot) exhibit '*inverse U*' curves with respect to the network density λ for different SINR thresholds y and different types of channel power distribution. This suggests that there is an optimal point of network density that maximize the rate gain of network densification. Particularly, this deficit regime can happen even with unbounded pathloss given that the near-field exponent is still smaller than the network dimension (i.e., $\beta_0 < d$); the bounded pathloss used in prior works ($\beta_0 = 0$) to obtain this deficit regime is a special case of this class.

Table 1.1 summarizes the behavior of user performance and shows the three different regimes. First, we see that the optimistic expectation of an ever-growing user's rate and full coverage in UDNs is theoretically possible, though unlikely in reality. Second, we shed light on the divergence between prior results in the literature on the fundamental limits of network densification. It is due to two different assumptions on the pathloss model, one with $\beta_0 > d$ that results in a saturation regime and the other with $\beta_0 = 0$ that results in a deficit regime.

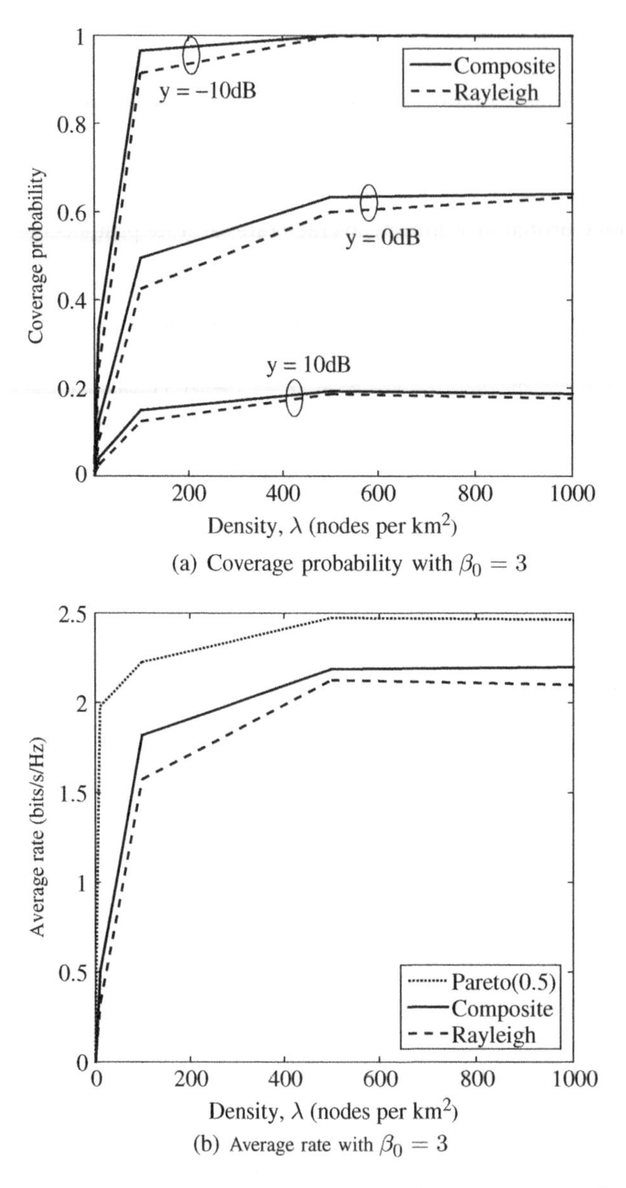

Figure 1.14 *Saturation regime* due to $\beta_0 > d$. Here, *saturation* of both coverage probability (left plot) and rate (right plot) is due to large near-field pathloss exponent $\beta_0 = 3$, which leads to $\overline{F}_P \in \mathcal{R}_{-\alpha}$ with $\alpha \in (0, 1)$ for all considered types of the channel power distribution including composite Rayleigh-lognormal, Rayleigh, and Pareto(0.5). Simulated parameters include a two-slope pathloss (i.e. $K = 2$) with $A_0 = 1$, $\beta_1 = 4$, $R_1 = 10$ m, $R_\infty = 40$ km, two-dimensional network domain (i.e. $d = 2$), and network density λ in BSs/km^2. Reprinted with permission from [28].

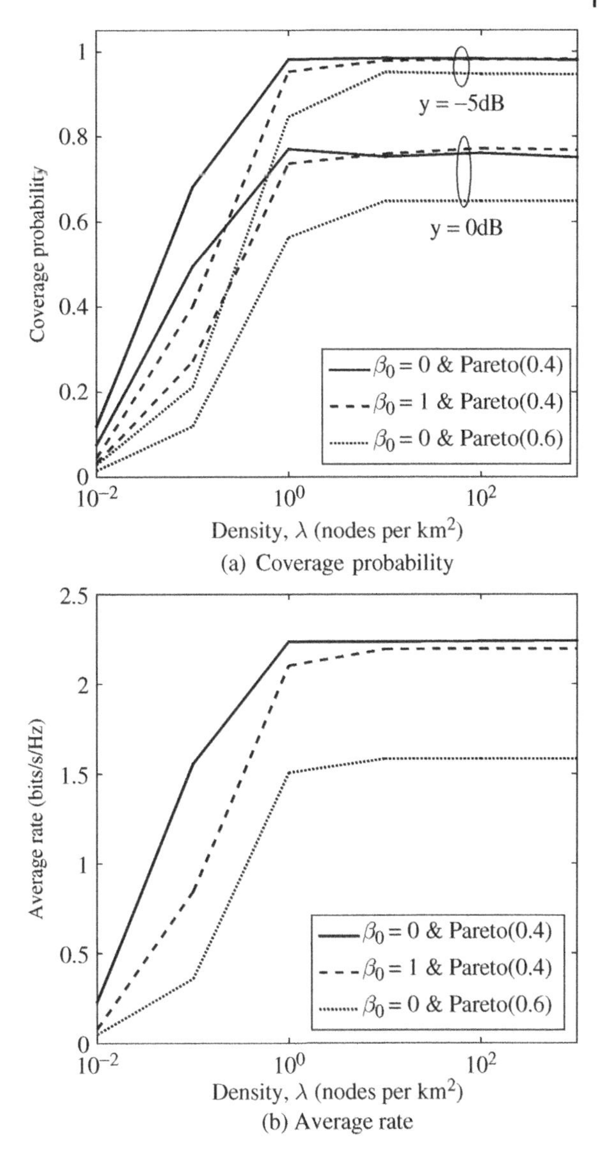

Figure 1.15 *Saturation regime* due to *regularly varying* channel power with index in $(-1, 0)$. Here, *saturation* of both coverage probability (left plot) and rate (right plot) is due to regularly varying channel power $\overline{F}_g \sim \text{Pareto}(\alpha)$ with $\alpha = 0.4$ and $\alpha = 0.6$. Simulated parameters include two-slope pathloss (i.e. $K = 2$) with $A_0 = 1$, $\beta_1 = 4$, $R_1 = 10$ m, $R_\infty = 40$ km, two-dimensional network domain (i.e. $d = 2$), and network density λ in BSs/km². Reprinted with permission from [28].

1.4.2 Network Performance

We focus now on the network-wide performance, i.e. on the sum average rate that can be obtained in UDNs. This can be summarized in the following corollary, which is based on Theorem 1.6:

Corollary 1.4 **[28]** The coverage density $\mathcal{D}(\lambda, \theta)$ for fixed θ and the area spectral efficiency $\mathcal{E}(\lambda)$ scale as follows:

1. $\mathcal{D}(\lambda, \theta)/\lambda \to 1$ and $\mathcal{E}(\lambda)/\lambda \to \infty$ as $\lambda \to \infty$ if $\overline{F}_P \in \mathcal{R}_0$;
2. $\mathcal{D}(\lambda, \theta)/\lambda \to c_{\theta,\alpha}$ and $\mathcal{E}(\lambda)/\lambda \to c_\alpha$ as $\lambda \to \infty$ if $\overline{F}_P \in \mathcal{R}_{-\alpha}$ with $0 < \alpha < 1$, where constant $c_{\theta,\alpha} \in [0, 1]$ depending on θ and α, and constant $c_\alpha > 0$ depending on α;
3. $\mathcal{D}(\lambda, \theta)/\lambda \to 0$ and $\mathcal{E}(\lambda)/\lambda \to 0$ as $\lambda \to \infty$ if $\overline{F}_P \notin \mathcal{R}_{-\alpha}$ with $\alpha \in [0, 1]$.

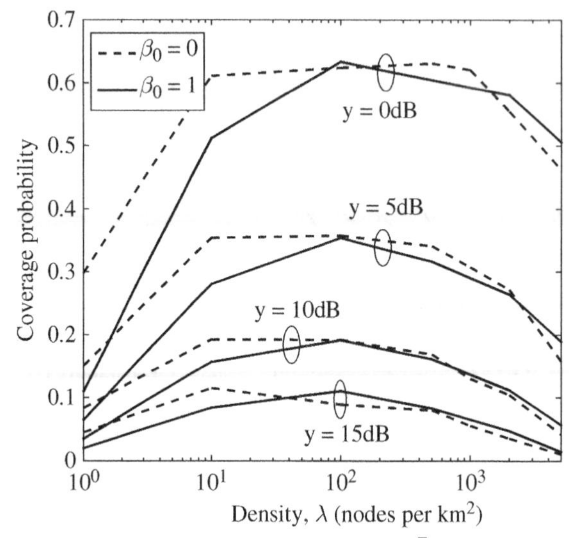

(a) Coverage probability, $\beta_0 < d$ and $\bar{F}_g \sim$ Composite

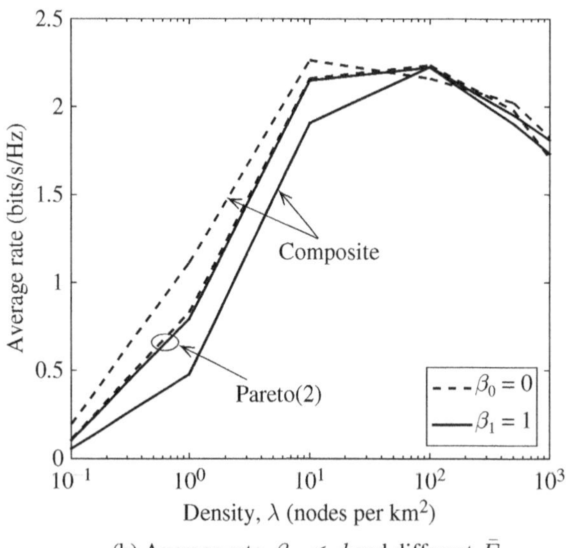

(b) Average rate, $\beta_0 < d$ and different $\bar{F}_g$

Figure 1.16 Validation of *deficit regime* when $\bar{F}_p \notin \mathcal{R}_{-\alpha}$ with $\alpha \in [0, 1]$. Simulated parameters include two-slope pathloss (i.e. $K = 2$) with $A_0 = 1$, $\beta_1 = 4$, $R_1 = 10$ m, $R_\infty = 40$ km, two-dimensional network domain (i.e. $d = 2$), and network density λ in BSs/km^2. Reprinted with permission from [28].

We see that the network-level performance scales with the network density more optimistically than the user performance does. In particular, both the *growth* and *saturation* regimes for the user performance lead to a *growth* regime for the network performance. In the *deficit* regime of user performance, the network performance scales sublinearly with the network density. Nevertheless, this does not necessarily mean that $\mathcal{D}(\lambda, \theta)$ and $\mathcal{E}(\lambda)$ always vanish in the same manner as $P_c(\lambda, \theta)$ and $C(\lambda)$ do when $\lambda \to \infty$. It depends

Table 1.1 Scaling regimes of user performance.

	$\overline{F}_g \in \mathcal{R}_{-\alpha}$			
Scaling regime	$\alpha = 0$	$0 < \alpha < 1$	$\alpha > 1$	$\overline{F}_g$ **is lighter-tailed**
$\beta_0 < d$	$\text{SINR} \xrightarrow{p} \infty$,	Saturation	Inverse U	
$\beta_0 > d$	$\text{P}_\text{c}(\lambda, \theta) \to 1$, $C(\lambda) \to \infty$		Saturation	

on the precise tail behavior of $\overline{F}_P$; in the case that the resulting $\mathbb{P}(\text{SINR} > \theta)$ does not vanish faster than $\lambda^{-\varepsilon}$ for some $0 < \varepsilon < 1$ as $\lambda \to \infty$, then $\mathcal{D}_\theta$ and $\mathcal{E}(\lambda)$ will be in the same order of $\lambda^{1-\varepsilon}$ as $\lambda \to \infty$. In such cases, $\mathcal{D}(\lambda, \theta)$ and $\mathcal{E}(\lambda)$ may still increase with the network density although they will increase with a much lower speed, i.e. $\mathcal{D}(\lambda, \theta) = o(\lambda)$ and $\mathcal{E}(\lambda) = o(\lambda)$ as $\lambda \to \infty$.

1.4.3 Network Ordering and Design Guidelines

The mathematical framework developed above characterizes the asymptotic behavior of coverage and throughput in ultra-dense networks. In short, knowing the tail behavior of fading and pathloss characteristics, we can see in which regime (growth, saturation, or deficit) the performance falls in. We want now to compare two networks with the same scaling regime but with different characteristics (e.g. different number of transmit antennas, different shadowing, etc.) and see which network would achieve the better network performance. One approach to answer this question would be to derive the performance metrics in closed form. Although the exact distribution of SINR is known for a simplified network model, e.g. [1] and [31], it is still analytically cumbersome for realistic and practically relevant system models. In the absence of handy analytical expressions, it is difficult to compare different transmission techniques (e.g. spatial multiplexing, beamforming, cooperative transmission, etc.) in general network settings. For that, in addition to the scaling laws, we develop ordering results for both coverage and average rate in order to facilitate a finer comparison between UDNs with different parameters. The ordering approach provides crisp insights and useful design guidelines into the relative performance of different transmission techniques, while circumventing the need to evaluate complicated coverage and rate expressions. As such, the scaling results provide an answer to the asymptotic performance behavior (increase, saturation, or inverse-U), while the ordering results aim at identifying which network provides superior performance in terms of coverage and rate among networks with the same asymptotic performance.

Theorem 1.7 **[28]** For two networks Ξ_1 and Ξ_2 with the same density λ and with distribution of wireless link F_1 and F_2, respively, if $\overline{F}_1 \in \mathcal{R}_{-\alpha_1}$, $\overline{F}_2 \in \mathcal{R}_{-\alpha_2}$, and $0 \leq \alpha_1 \leq \alpha_2 \leq \infty$, then

$$\mathbb{E}(\text{SINR}_1) \geq \mathbb{E}(\text{SINR}_2), \quad \text{and} \quad \mathbb{E}(\log(1 + \text{SINR}_1)) \geq \mathbb{E}(\log(1 + \text{SINR}_2)), \tag{1.43}$$

as $\lambda \to \infty$, where SINR_1 and SINR_2 are the received SINR of network Ξ_1 and Ξ_2, respectively.

Theorem 1.7 states that the heavier the tail of a wireless link distribution, the better the performance under ultra-densification. As shown by Theorem 1.4, a heavier-tailed wireless link distribution can be obtained through either a heavier-tailed channel power distribution g or a greater near-field pathloss exponent β_0. Physically, heavier-tailed channel power indicates a greater probability of high channel power, taking into account all relevant effects, such as transmit power, array gains, and beamforming gains. Since a higher channel power may also increase interference, a natural question is whether it is beneficial to have higher beamforming gains (or higher channel power in general) in the high-density regime. Note that large beamforming gains will increase the interference toward the users that are located in the beam direction of the intended user, but since large beamforming gains are achieved with narrow beams, this probability may be low. Theorem 1.7 states that achieving higher beamforming gains or using techniques that render the tail of wireless link distribution heavier are beneficial in terms of network performance. In short, directional transmissions can be beneficial for the network performance, as are envisioned for massive MIMO and millimeter wave systems.

1.5 Conclusions and Future Challenges

The results in this chapter, both in the asymptotic and non-asymptotic regime in terms of BS density, indicate that in many scenarios, there are fundamental (physical) limitations in the performance of UDNs. The SINR invariance does not hold at high densities, and the SINR will monotonically decrease as the network is densified past a certain point and it can even decrease to zero. Similar trends are observed for the throughput, although it is more robust to network densification than SINR. This is further exacerbated with elevated BSs as the BS height proves to be the most prominent factor in degrading the system performance. In that case the maximum area spectral efficiency is inversely proportional to the square of the BS height. Therefore, densifying the network leads to near-universal outage regardless of the other parameters and even at moderately low BS densities. Another key finding is that under the strongest cell association, the performance of UDNs exhibits three distinct regimes of the user performance, namely growth, saturation, and deficit regime. The tail behavior of the channel power and near-field pathloss exponent are the key parameters that determine the performance limits and the asymptotic scaling. Some particular implications include:

- Monotonically increasing per-user performance (coverage probability and average rate) by means of ultra-densification is theoretically possible, though highly unlikely in reality since it requires a slowly varying tail of the channel power distribution, which is a theoretical extreme.
- In practice, installing more BSs is beneficial to the user performance up to a certain density, after which further densification can become harmful to user performance due to the faster growth of interference compared to useful signal. This highlights the cardinal importance of interference mitigation, coordination among neighboring cells, and local spatial scheduling.
- The network performance in terms of coverage density and area spectral efficiency benefits from the network densification more than user performance does. In

particular, it scales linearly with the network density when the user performance is in the growth or saturation regime.

- Increasing the tail distribution of the channel power using advanced transmission techniques, such as massive MIMO, CoMP, and directional beamforming, is beneficial as it improves the performance scaling regime. Moreover, the effect of emerging technologies (e.g. D2D, mmWave) on near-field pathloss and channel power distribution needs to be studied.
- It is meaningful to determine the optimal network density beyond which further densification becomes destructive or cost-ineffective. This operating point will depend on properties of the channel power distribution, noise level, and pathloss in the near-field region, and is of cardinal importance for the successful deployment of 5G UDNs. It would also be interesting to consider the effects of advanced interference suppression techniques, such as joint (over multiple BSs) transmission or decoding, or successive interference cancellation, as the above conclusions have the caveats that interference is treated as noise and that the active user density scales with the base station density.

Bibliography

1 J.G. Andrews, F. Baccelli, and R.K. Ganti, "A tractable approach to coverage and rate in cellular networks," *IEEE Trans. Commun.*, vol. 59, no. 11, pp. 3122–3134, Nov. 2011.

2 M. Haenggi, *Stochastic Geometry for Wireless Networks*. New York, NY, USA: Cambridge University Press, 2012.

3 A. Guo and M. Haenggi, "Asymptotic deployment gain: A simple approach to characterize the SINR distribution in general cellular networks," *IEEE Trans. Commun.*, vol. 63, no. 3, pp. 962–976, Mar. 2015.

4 M. Haenggi and R. K. Ganti, *Interference in Large Wireless Networks*. Now Publishers Inc, 2009.

5 F. Baccelli and B. Blaszczyszyn, *Stochastic Geometry and Wireless Networks, Volume I - Theory*. Now Publishers, 2009.

6 H.S. Dhillon, R.K. Ganti, F. Baccelli, and J.G. Andrews, "Modeling and analysis of K-tier downlink heterogeneous cellular networks," *IEEE J. Sel. Areas Commun.*, vol. 30, no. 3, pp. 550–560, Apr. 2012.

7 X. Zhang and J. Andrews, "Downlink cellular network analysis with multi-slope path loss models," *IEEE Trans. Commun.*, vol. 63, no. 5, pp. 1881–1894, May 2015.

8 T. Bai and R.W. Heath, "Coverage and rate analysis for millimeter-wave cellular networks," *IEEE Trans. Wireless Commun.*, vol. 14, no. 2, pp. 1100–1114, Feb. 2015.

9 M. Di Renzo, "Stochastic geometry modeling and analysis of multi-tier millimeter wave cellular networks," *IEEE Trans. Wireless Commun.*, vol. 14, no. 9, pp. 5038–5057, Sept. 2015.

10 J. Kim, C. Jeong, H. Yu, and J. Park, "Areal capacity limit on the growth of small cell density in heterogeneous networks," in *Proc. IEEE Global Commun. Conf. (GLOBECOM)*, Austin, TX, USA, Dec. 2014, pp. 4263–4268.

11 C. Galiotto, N.K. Pratas, N. Marchetti, and L. Doyle, "A stochastic geometry framework for LOS–NLOS propagation in dense small cell networks," in *Proc. IEEE Int. Conf. Commun. (ICC)*, London, UK, June 2015, pp. 2851–2856.

12 A.K. Gupta, M.N. Kulkarni, E. Visotsky, F.W. Vook, A. Ghosh, J.G. Andrews, and R.W. Heath, "Rate analysis and feasibility of dynamic TDD in 5G cellular systems," in *Proc. IEEE Int. Conf. Commun. (ICC)*, Kuala Lumpur, Malaysia, May 2016.

13 C.S. Chen, V.M. Nguyen, and L. Thomas, "On small cell network deployment: A comparative study of random and grid topologies," in *Proc. IEEE Veh. Technol. Conf. (VTC Fall)* 2012, Sept. 2012.

14 J. Arnau, I. Atzeni, and M. Kountouris, "Impact of LOS/NLOS propagation and path loss in ultra-dense cellular networks," in *Proc. IEEE Int. Conf. Commun. (ICC)*, Kuala Lumpur, Malaysia, May 2016.

15 M. Ding, P. Wang, D. Lopez-Perez, G. Mao, and Z. Lin, "Performance impact of LoS and NLoS transmissions in dense cellular networks," *IEEE Trans. Wireless Commun.*, vol. 15, no. 3, pp. 2365–2380, Mar. 2016.

16 T. Samarasinghe, H. Inaltekin, and J.S. Evans, "On optimal downlink coverage in Poisson cellular networks with power density constraints," *IEEE Trans. Commun.*, vol. 62, no. 4, pp. 1382–1392, Apr. 2014.

17 F. Baccelli and A. Biswas, "On scaling limits of power law shot-noise fields," *Stochastic Models*, vol. 31, no. 2, pp. 187–207, 2015.

18 N. Lee, F. Baccelli, and R.W. Heath, "Spectral efficiency scaling laws in dense random wireless networks with multiple receive antennas," *IEEE Trans. Inf. Theory*, vol. 62, no. 3, pp. 1344–1359, Mar. 2016.

19 P. Embrechts, C. Klüppelberg, and T. Mikosch, *Modelling Extremal Events*. Springer, 1997.

20 N.H. Bingham, C.M. Goldie, and J.L. Teugels, *Regular Variation*. Cambridge University Press, 1989.

21 3GPP, "Technical specification group radio access network; evolved universal terrestrial radio access (E-UTRA); further advancements for E-UTRA physical layer aspects (Release 9). TR 36.814," 2010.

22 X. Li, T. Bai, and R.W. Heath, "Impact of 3D base station antenna in random heterogeneous cellular networks," in *Proc. IEEE Wireless Commun. and Netw. Conf. (WCNC)*, Istanbul, Turkey, Apr. 2014, pp. 2254–2259.

23 I. Atzeni, J. Arnau, and M. Kountouris, "Downlink cellular network analysis with LOS/NLOS propagation and elevated base stations," *IEEE Trans. Wireless Commun.*, vol. 17, no. 1, pp. 142–156, Jan. 2018.

24 J.G. Andrews, X. Zhang, G.D. Durgin, and A.K. Gupta, "Are we approaching the fundamental limits of wireless network densification?" *IEEE Commun. Mag.*, vol. 54, no. 10, pp. 184–190, Oct. 2016.

25 T. Bai, R. Vaze, and R.W. Heath, "Analysis of blockage effects on urban cellular networks," *IEEE Trans. Wireless Commun.*, vol. 13, no. 9, pp. 5070–5083, Sept. 2014.

26 J. Lee, X. Zhang, and F. Baccelli, "A 3-D spatial model for in-building wireless networks with correlated shadowing," *IEEE Trans. Wireless Commun.*, vol. 15, no. 11, pp. 7778–7793, Nov. 2016.

27 Information Society Technologies, "IST-2003-507581 WINNER: Final report on link level and system level channel models," Tech. Rep. D5.4 v. 1.4, Nov. 2005.

28 V.M. Nguyen and M. Kountouris, "Performance limits of network densification," *IEEE J. Sel. Areas Commun.*, vol. 35, no. 6, pp. 1294–1308, June 2017.

29 L. Breiman, "On some limit theorems similar to the arc-sin law," *Theory Probab. Appl.*, vol. 10, no. 2, pp. 323–331, 1965.

30 V.M. Nguyen, F. Baccelli, L. Thomas, and C.S. Chen, "Best signal quality in cellular networks: asymptotic properties and applications to mobility management in small cell networks," *EURASIP J. Wirel. Commun. Netw.*, vol. 2010, 2010.
31 V.M. Nguyen and F. Baccelli, "A stochastic geometry model for the best signal quality in a wireless network," in *Proc. 8th Int. Symp. Model. Optim. Mobile, Ad Hoc Wirel. Netw. (WiOpt)*, May 2010, pp. 465–471.

2

Performance Analysis of Dense Small Cell Networks with Line of Sight and Non-Line of Sight Transmissions under Rician Fading

Amir Hossein Jafari[1], Ming Ding[2] and David López-Pérez[3]

[1] *Samsung Research America, USA*
[2] *Data61, CSIRO, Australia*
[3] *Nokia Bell Labs, Dublin, Ireland*

2.1 Introduction

Ultra-dense networks (UDNs) hold the promise to rapidly increase the network capacity for the fifth generation (5G) of cellular communications [1] by deploying base stations (BSs) much closer to user equipment (UE) and reusing the spectrum intensively [2–4]. According to [7] and [8], a UDN refers to a cellular network where the BS density is larger than the UE density. However, in UDNs the small distances between transmitters and receivers also bring a change in the channel characteristics, which in turn may significantly impact the network performance. For example, the channel may become line-of-sight (LoS) dominated with the smaller distances between BSs and UEs, with the subsequent loss of channel diversity. However, some of the prior works for instance do not consider the probability of LoS propagations, and instead they use path loss models that do not differentiate between LoS and Non-LoS (NLoS) transmissions. These assumptions along with other simplifications can lead to misleading conclusions, or conclusions that do not apply to the full spectrum of BS densities, e.g. the coverage probability is independent of the density of deployed small cell BSs in interference-limited fully loaded cellular networks, and/or the area spectral efficiency (ASE) will linearly increase with network densification [9, 10].

A critical question now is whether such simplifications have a significant impact on the analysis of network performance, and whether conclusions regarding the impact of BS density on network performance still hold and when. To address the question with regard to the probability of LoS, the authors in [11] to [13] have studied the impact of the probability of LoS propagations using a piecewise path loss model that incorporates both LoS and NLoS transmissions in UDNs. The authors showed that, as the density of small cell BSs increases, the ASE will initially increase. However, as the density of small cell BSs exceeds a specific threshold, due to the transition of a large number of interfering signals from NLoS to LoS transmission, the coverage probability will start to decrease, and as a consequence the ASE may either continue to grow but at a much slower pace or even decrease [14]. In other words, due to such NLoS to LoS transition, the interference power will increase faster than the signal power for some BS densities, which negatively affects the UE's signal-to-interference-plus-noise-ratio (SINR),

Ultra-dense Networks for 5G and Beyond: Modelling, Analysis, and Applications,
First Edition. Edited by Trung Q. Duong, Xiaoli Chu and Himal A. Suraweera.

and verifies that the density of small cell BSs may have a key impact on network performance [15–18]. However, in terms of multi-path fading, it is important to note that the authors in [11] and [14] oversimplified the–NLoS link by assuming Rayleigh fading, which gives rise to the question of whether a more accurate multi-path fading channel model may change their conclusions and whether the multi-path fading can mitigate or exacerbate the SINR degradation caused by the NLoS to LoS transition. To answer this question, we consider a practical distance-dependent Rician fading channel with a variant Rician K-factor, and then derive the analytical results for the coverage probability as well as the ASE of a UDN.

The rest of this chapter is organized as follows. In Section 2.2, we propose a stochastic geometry framework for analysing the performance of a UDN, and derive the analytical results for the coverage probability as well as the ASE of a UDN under a practical distance-dependent Rician fading channel with a variant Rician K factor using a general path loss model that incorporates both NLoS and LoS transmissions. In Section 2.4, we obtain integral-form expressions for the coverage probability and the ASE using a 3GPP path loss model with a *linear* LoS probability function. Section 2.5 presents simulation results and derives the main insights of this chapter, which reveals an important finding, i.e. due to the dominance of path loss in UDNs, the impact of the multi-path fading is negligible, and thus when the density of small cell BSs exceeds a certain threshold, the network coverage probability will decrease as small cell BSs become denser. The multi-path fading is not able to mitigate this phenomenon. We finally draw the conclusions in Section 2.6.

2.2 System Model

In this section, we introduce our system model for evaluating the performance of a UDN using stochastic geometry, which is a useful tool to analyse the system performance in cellular networks [19]. We focus on the downlink (DL) coverage experienced by an outdoor user. We make the following assumptions in our system model.

2.2.1 BS Distribution

We assume that the small cell BSs form a homogeneous Poisson point process (HPPP) Φ of intensity λ BSs/km^2 on the plane consisting of small cell BSs. Note that all small cell BSs are assumed to have a constant transmit power P_t.

2.2.2 User Distribution

We assume that UEs form another stationary HPPP with an intensity of λ^{UE} UEs/km^2, which is independent from the small cell BSs one. Note that we consider a fully loaded network where λ^{UE} is considered to be sufficiently larger than λ so that each BS has at least one associated UE in its coverage. A typical UE is assumed to be located at the origin O, which is a standard approach in the analysis using stochastic geometry. According to Slivnyak's theorem, due to the stationarity and independence of all points in an HPPP, conditioning on a UE located at the origin or at a fixed distance from the origin does not change the distribution of the rest of the process, and therefore the downlink SINR and rate experienced by the typical UE have the same distributions as the aggregate ones in the network [20]. The typical UE is assumed to be an outdoor UE.

2.2.3 Path Loss

We denote the distance between an arbitrary BS and the typical UE by r in km. Considering practical LoS/NLoS transmissions, we propose to model the path loss with respect to distance r as shown below:

$$\zeta(r) = \begin{cases} \zeta_1(r) = \begin{cases} \zeta_1^{\mathrm{L}}(r), & \text{with probability } \Pr_1^{\mathrm{L}}(r) \\ \zeta_1^{\mathrm{NL}}(r), & \text{with probability } \left(1 - \Pr_1^{\mathrm{L}}(r)\right) \end{cases}, & \text{when } 0 \le r \le d_1 \\ \zeta_2(r) = \begin{cases} \zeta_2^{\mathrm{L}}(r), & \text{with probability } \Pr_2^{\mathrm{L}}(r) \\ \zeta_2^{\mathrm{NL}}(r), & \text{with probability } \left(1 - \Pr_2^{\mathrm{L}}(r)\right) \end{cases}, & \text{when } d_1 < r \le d_2 \\ \vdots & \vdots \\ \zeta_N(r) = \begin{cases} \zeta_N^{\mathrm{L}}(r), & \text{with probability } \Pr_N^{\mathrm{L}}(r) \\ \zeta_N^{\mathrm{NL}}(r), & \text{with probability } \left(1 - \Pr_N^{\mathrm{L}}(r)\right) \end{cases}, & \text{when } r > d_{N-1} \end{cases}. \tag{2.1}$$

As can be seen from Eq. (2.1), the path loss function $\zeta(r)$ is divided into N pieces, where each piece is represented by $\zeta_n(r)$, $n \in \{1, 2, \dots, N\}$. Moreover, $\zeta_n^{\mathrm{L}}(r)$, $\zeta_n^{\mathrm{NL}}(r)$, and $\Pr_n^{\mathrm{L}}(r)$ represent the n-th piece of path loss function for the LoS transmission, the n-th piece of the path loss function of the NLoS transmission, and the n-th piece of the LOS probability function, respectively. In addition, we model $\zeta_n^{\mathrm{L}}(r)$ and $\zeta_n^{\mathrm{NL}}(r)$ in Eq. (2.1) as $\zeta_n^{\mathrm{L}}(r) = A_n^{\mathrm{L}} r^{-\alpha_n^{\mathrm{L}}}$ and $\zeta_n^{\mathrm{NL}}(r) = A_n^{\mathrm{NL}} r^{-\alpha_n^{\mathrm{NL}}}$ where A_n^{L} and A_n^{NL}, $n \in \{1, 2, \dots, N\}$, are the path losses at a reference distance of $r = 1$ km for the LoS and the NLoS cases in $\zeta_n(r)$, respectively, and α_n^{L} and α_n^{NL} are the path loss exponents for the LoS and the NLoS cases in $\zeta_n(r)$, respectively. Typical values of reference path losses and path loss exponents for microurban scenarios include $A^{\mathrm{L}} = 10^{-10.38}$, $A^{\mathrm{NL}} = 10^{-14.54}$, $\alpha^{\mathrm{L}} = 2.09$, $\alpha^{\mathrm{NL}} = 3.75$, which have been obtained from field tests and can be found in [21], [22], and [23]. Furthermore, we stack $\{\zeta_n^{\mathrm{L}}(r)\}$ and $\{\zeta_n^{\mathrm{NL}}(r)\}$ into piece-wise functions as

$$\zeta^s(r) = \begin{cases} \zeta_1^s(r), & \text{when } 0 \le r \le d_1 \\ \zeta_2^s(r), & \text{when } d_1 < r \le d_2 \\ \vdots & \vdots \\ \zeta_N^s(r), & \text{when } r > d_{N-1} \end{cases} \tag{2.2}$$

where s is a string variable taking the values from $s \in \{\mathrm{L}, \mathrm{NL}\}$ for the LoS and the NLoS cases, respectively.

Note that in Eq. (2.1), $\Pr_n^{\mathrm{L}}(r)$, $n \in \{1, 2, \dots, N\}$, denotes the n-th piece LoS probability function for a BS and a UE that are separated by the distance r, and is typically a monotonically decreasing function with r. For notational convenience, $\{\Pr_n^{\mathrm{L}}(r)\}$ is further stacked into a piecewise LoS probability function as follows:

$$\mathrm{Pr}^{\mathrm{L}}(r) = \begin{cases} \mathrm{Pr}_1^{\mathrm{L}}(r), & \text{when } 0 \leq r \leq d_1 \\ \mathrm{Pr}_2^{\mathrm{L}}(r), & \text{when } d_1 < r \leq d_2 \\ \vdots & \vdots \\ \mathrm{Pr}_N^{\mathrm{L}}(r), & \text{when } r > d_{N-1} \end{cases}. \tag{2.3}$$

2.2.4 User Association Strategy (UAS)

Each UE is associated with the BS with the smallest downlink path loss, regardless of whether it is LoS or NLoS.

2.2.5 Antenna Radiation Pattern

Each BS as well as the typical UE are equipped with an omnidirectional antenna.

2.2.6 Multi-path Fading

The multi-path fading between an arbitrary BS and the typical UE is modeled as a distance-dependent Rician fading channel [24]. The Rician K-factor is defined as the ratio of the power in the specular LoS component to the power in all NLoS components. To capture the impact of the distance between a BS and a UE on the power of LoS and NLoS components, the distance-dependant Rician K-factor is computed as $K(r) = 13 - 0.03r$ (dB), where r denotes the distance the BS and the UE in meters [25]. Note that we denote $K(r)$ by K hereafter for notational simplicity, but it is critical not to interpret K as a constant value.

2.3 Coverage Probability Analysis Based on the Piecewise Path Loss Model

In this section, we analyse the coverage probability in the defined system model of a UDN. First, we provide some SINR ordering results, and then we derive expressions for the coverage probability in a UDN with a general LoS probability function.

One metric that can be used to evaluate the system's performance is the SINR coverage probability. It is defined as the probability that the SINR at the UE from its associated BS is larger than a predefined threshold γ, and is given by

$$p^{\mathrm{cov}}(\lambda, \gamma) = \Pr[\mathrm{SINR} > \gamma], \tag{2.4}$$

where the SINR is computed by

$$\mathrm{SINR} = \frac{P_t \zeta(r) h}{I_r + N_0}, \tag{2.5}$$

where h is the Rician distributed channel gain, and P_t and N_0 denote the transmission power of each BS and the additive white Gaussian noise (AWGN) power at the typical

UE, respectively. The aggregated interference power from all non-serving BSs is denoted by I_r, and is given by

$$I_r = \sum_{i:\, b_i \in \Phi \backslash b_o} P_t \beta_i g_i, \tag{2.6}$$

where b_o denotes the serving BS, and b_i, β_i and g_i refer to the i-th interfering BS, the corresponding path loss of b_i, and the corresponding Rician fading channel gain of b_i, respectively.

For a specific λ, the area spectral efficiency (ASE) in bps/Hz/km^2 can be expressed as

$$A^{\text{ASE}}(\lambda, \gamma_0) = \lambda \int_{\gamma_0}^{\infty} \log_2(1+\gamma) f_{\Gamma}(\lambda, \gamma) d\gamma, \tag{2.7}$$

where γ_0 denotes the minimum working SINR for the considered UDN and $f_{\Gamma}(\lambda, \gamma)$ represents the probability density function (PDF) of the SINR for a specific value of λ at the typical UE. The minimum working SINR denotes the smallest SINR that is required by a typical UE to properly receive signals, i.e. the minimum working SINR in long-term evolutionary (LTE) is around −6 dB, supported by the lowest code rate with QPSK modulation [26]. Note that the ASE defined in this chapter is different from that defined in [27], where regardless of the actual value of the SINR, a deterministic rate based on γ_0 was assumed for the typical UE. The ASE definition in Eq. (2.7) is more realistic due to the SINR-dependent rate, but it is more complex to analyse, since it requires one more integral than that in [27].

The PDF of SINR is then computed as

$$f_{\Gamma}(\lambda, \gamma) = \frac{\partial(1 - p^{\text{cov}}(\lambda, \gamma))}{\partial \gamma}. \tag{2.8}$$

In the following, we present Theorem 2.1 to obtain $p^{\text{cov}}(\lambda, \gamma)$ based on the path loss model in Eq. (2.1). Note that for analytical tractability in this chapter, we consider an interference-limited scenario, and thus in the coverage probability derivations, we concentrate on the signal-to-interference-ratio (SIR) rather than the SINR.

Theorem 2.1 Considering the path loss model in Eq. (2.1), $p^{\text{cov}}(\lambda, \gamma)$ is computed as

$$p^{cov}(\lambda, \gamma) = \sum_{n=1}^{N} (T_n^L + T_n^{NL}), \tag{2.9}$$

where $T_n^L = \int_{d_{n-1}}^{d_n} Pr[P_t \zeta_n^L(r) h / I_r > \gamma] f_{R,n}^L(r) dr$, $T_n^{NL} = \int_{d_{n-1}}^{d_n} Pr[P_t \zeta_n^{NL}(r) h / I_r > \gamma] f_{R,n}^{NL}(r) dr$, $d_0 = 0$, and $d_N = \infty$.

Moreover, $f_{R,n}^L(r)$ and $f_{R,n}^{NL}(r)$ refer to the piecewise PDF of the distance to LoS or NLoS serving BS, and are defined by

$$\begin{aligned} f_{R,n}^L(r) = {} & \exp\left(-\int_0^{r_1} (1 - Pr^L(u)) 2\pi u \lambda du\right) \times \exp\left(-\int_0^{r} Pr^L(u) 2\pi u \lambda du\right) \\ & \times Pr_n^L(r) \times 2\pi r \lambda \quad (d_{n-1} < r \le d_n) \end{aligned} \tag{2.10}$$

and

$$\begin{aligned} f_{R,n}^{NL}(r) = {} & \exp\left(-\int_0^{r_2} Pr^L(u) 2\pi u \lambda du\right) \times \exp\left(-\int_0^{r} (1 - Pr^L(u)) 2\pi u \lambda du\right) \\ & \times (1 - Pr_n^L(r)) \times 2\pi r \lambda \quad (d_{n-1} < r \le d_n) \end{aligned} \tag{2.11}$$

It is noteworthy that we can determine r_1 and r_2 as $\arg_{r_1}\{\zeta^{NL}(r_1) = \zeta_n^L(r)\}$ and $\arg_{r_2}\{\zeta^L(r_2) = \zeta_n^{NL}(r)\}$, respectively.

Furthermore, $Pr[P_t\zeta_n^L(r)h/I_r > \gamma]$ and $Pr[P_t\zeta_n^{NL}(r)h/I_r > \gamma]$ are respectively computed by

$$\Pr\left[\frac{P_t\zeta_n^L(r)h}{I_r} > \gamma\right] = \sum_{k=0}^{\infty}\sum_{m=0}^{k} J(m,k)\ \gamma^{k-m}(-1)^{k-m}\frac{\partial^{k-m}\mathcal{L}_{I_r}\left(\frac{\gamma}{P_t\zeta_n^L(r)}\right)}{\partial\gamma^{k-m}} \tag{2.12}$$

and

$$\Pr\left[\frac{P_t\zeta_n^{NL}(r)h}{I_r} > \gamma\right] = \sum_{k=0}^{\infty}\sum_{m=0}^{k} J(m,k)\,\gamma^{k-m}(-1)^{k-m}\frac{\partial^{k-m}\mathcal{L}_{I_r}\left(\frac{\gamma}{P_t\zeta_n^{NL}(r)}\right)}{\partial\gamma^{k-m}} \tag{2.13}$$

where $\mathcal{L}_{I_r}(s)$ is the Laplace transform of random variable (RV) I_r evaluated at s and $J(m,k) = e^{-K}K^k m!\left(\begin{smallmatrix}k\\m\end{smallmatrix}\right)/(k!)^2$ with K denoting the Rician K-factor.

Proof: See Appendix A.

In Theorem 2.1, for a certain r, $\Pr[P_t\zeta_n^L(r)h/I_r > \gamma]$ in Eq. (2.12) and $\Pr[P_t\zeta_n^{NL}(r)h/I_r > \gamma]$ in Eq. (2.13) compute the coverage probability for LoS signal transmission and that for NLoS transmission, respectively. The rational behind such expressions is that the h follows a Rician distribution, whose CDF is $-\sum_{k=0}^{\infty}\sum_{m=0}^{k} J(m,k)h^{k-m}e^{-h} + 1$.

2.4 Study of a 3GPP Special Case

In this section, as a special case for Theorem 2.1, we derive the coverage probability using a 3GPP adopted path loss function, $\zeta(r)$ [21], which is compatible with UDNs, because both the exponential path loss function in [21] and the LoS probability function in [23] are only valid for small cell BSs. The path loss function is given by

$$\zeta(r) = \begin{cases} A^{\mathrm{L}}r^{-\alpha^{\mathrm{L}}}, & \text{with probability } \mathrm{Pr}^{\mathrm{L}}(r) \\ A^{\mathrm{NL}}r^{-\alpha^{\mathrm{NL}}}, & \text{with probability } (1-\mathrm{Pr}^{\mathrm{L}}(r)) \end{cases} \tag{2.14}$$

where the linear LoS probability [23] function, $\mathrm{Pr}^{\mathrm{L}}(r)$, is given by

$$\mathrm{Pr}^{\mathrm{L}}(r) = \begin{cases} 1-\dfrac{r}{d_1}, & 0 < r \le d_1 \\ 0, & r > d_1 \end{cases} \tag{2.15}$$

where the steepness of $\mathrm{Pr}^{\mathrm{L}}(r)$ is defined by parameter d_1. According to Theorem 2.1, and Eqs. (2.14) and (2.15), the coverage probability can be computed as $p^{\mathrm{cov}}(\lambda,\gamma) = \sum_{n=1}^{2}(T_n^{\mathrm{L}} + T_n^{\mathrm{NL}})$, where $\zeta_1^{\mathrm{L}}(r) = \zeta_2^{\mathrm{L}}(r) = A^{\mathrm{L}}r^{-\alpha^{\mathrm{L}}}$, $\zeta_1^{\mathrm{NL}}(r) = \zeta_2^{\mathrm{NL}}(r) = A^{\mathrm{NL}}r^{-\alpha^{\mathrm{NL}}}$, $\mathrm{Pr}_1^{\mathrm{L}}(r) = 1-\frac{r}{d_1}$, and $\mathrm{Pr}_2^{\mathrm{L}}(r) = 0$. Note that studying the linear LoS probability function not only allows us to obtain more tractable results, but also helps us to deal with more complicated path loss models in practice, as they can be approximated by piecewise linear functions.

In the following, we present the computation of T_1^{L}, T_1^{NL}, T_2^{L}, and T_2^{NL}, respectively.

2.4.1 The Computation of T_1^{L}

From Theorem 2.1, T_1^{L} is computed as

$$\begin{aligned} T_1^{\mathrm{L}} &= \int_0^{d_1} \sum_{k=0}^{\infty}\sum_{m=0}^{k} J(m,k)\ \gamma^{k-m}(-1)^{k-m} \frac{\partial^{k-m}\mathscr{L}_{I_r}\left(\frac{\gamma}{P_t\zeta_n^{\mathrm{L}}(r)}\right)}{\partial \gamma^{k-m}} f_{R,1}^{\mathrm{L}}(r)dr \\ &\overset{(a)}{=} \int_0^{d_1} \sum_{k=0}^{\infty}\sum_{m=0}^{k} J(m,k)\ \gamma^{k-m}(-1)^{k-m} \frac{\partial^{k-m}\mathscr{L}_{I_r}\left(\frac{\gamma r^{\alpha^{\mathrm{L}}}}{P_tA^{\mathrm{L}}}\right)}{\partial \gamma^{k-m}} f_{R,1}^{\mathrm{L}}(r)dr, \end{aligned} \tag{2.16}$$

where $\zeta_1^{\mathrm{L}}(r) = A^{\mathrm{L}}r^{-\alpha^{\mathrm{L}}}$ from Eq. (2.14) is plugged into the step (a) of Eq. (2.16). Note that $\mathscr{L}_{I_r}(s)$ represents the Laplace transform of RV I_r evaluated at s.

In Eq. (2.16), according to Theorem 2.1 and Eq. (2.15), $f_{R,1}^{\mathrm{L}}(r)$ is computed as

$$\begin{aligned} f_{R,1}^{\mathrm{L}}(r) &= \exp\left(-\int_0^{r_1} \lambda\frac{u}{d_1}2\pi u du\right) \\ &\times \exp\left(-\int_0^{r} \lambda\left(1-\frac{u}{d_1}\right)2\pi u du\right)\left(1-\frac{r}{d_1}\right)2\pi r\lambda \\ &= \exp\left(-\pi\lambda r^2 + 2\pi\lambda\left(\frac{r^3}{3d_1}-\frac{r_1^3}{3d_1}\right)\right)\left(1-\frac{r}{d_1}\right)2\pi r\lambda \quad (0 < r \le d_1), \end{aligned} \tag{2.17}$$

where $r_1 = (A^{\mathrm{NL}}/A^{\mathrm{L}})^{1/\alpha^{\mathrm{NL}}} r^{\alpha^{\mathrm{L}}/\alpha^{\mathrm{NL}}}$. Moreover, in order to compute $\mathscr{L}_{I_r}(\gamma r^{\alpha^{\mathrm{L}}}/P_tA^{\mathrm{L}})$ in Eq. (2.16) for the range of $0 < r \le d_1$, we propose Lemma 2.1.

Lemma 2.1 $\mathscr{L}_{I_r}(\gamma r^{\alpha^{\mathrm{L}}}/P_tA^{\mathrm{L}})$ in the range of $0 < r \le d_1$ can be computed as

$$\begin{aligned} \mathscr{L}_{I_r}\left(\frac{\gamma r^{\alpha^{\mathrm{L}}}}{P_tA^{\mathrm{L}}}\right) &= \exp\Big(-2\pi\lambda(\rho_1(\alpha^{\mathrm{L}},1,(1-K)(\gamma r^{\alpha^{\mathrm{L}}})^{-1},d_1) \\ &- \rho_1(\alpha^{\mathrm{L}},1,(1-K)(\gamma r^{\alpha^{\mathrm{L}}})^{-1},r))) \exp(-2\pi\lambda(\rho_1(\alpha^{\mathrm{L}},\alpha^{\mathrm{L}}+1,(1-K)(\gamma r^{\alpha^{\mathrm{L}}})^{-1},d_1) \\ &- \rho_1(\alpha^{\mathrm{L}},\alpha^{\mathrm{L}}+1,(1-K)(\gamma r^{\alpha^{\mathrm{L}}})^{-1},r))) \exp\Big(\frac{2\pi\lambda}{d_0}(\gamma r^{\alpha^{\mathrm{L}}})^{-1}(1-K-e^{-K}) \\ &\times(\rho_1(\alpha^{\mathrm{L}},2,(1-K)(\gamma r^{\alpha^{\mathrm{L}}})^{-1},d_1) - \rho_1(\alpha^{\mathrm{L}},2,(1-K)(\gamma r^{\alpha^{\mathrm{L}}})^{-1},r))\Big) \\ &\times \exp\Big(\frac{2\pi\lambda}{d_0}(\gamma r^{\alpha^{\mathrm{L}}})^{-1}(1-K-e^{-K})(\rho_1(\alpha^{\mathrm{L}},\alpha^{\mathrm{L}}+2,(1-K)(\gamma r^{\alpha^{\mathrm{L}}})^{-1},d_1) \\ &-\rho_1(\alpha^{\mathrm{L}},\alpha^{\mathrm{L}}+2,(1-K)(\gamma r^{\alpha^{\mathrm{L}}})^{-1},r_1))\Big) \\ &\times \exp\left(\frac{-2\pi\lambda}{d_0}\left(\rho_1\left(\alpha^{\mathrm{NL}},2,(1-K)\left(\frac{\gamma A^{\mathrm{NL}}}{A^{\mathrm{L}}}r^{\alpha^{\mathrm{L}}}\right)^{-1},d_1\right)\right.\right. \\ &\left.\left.-\rho_1\left(\alpha^{\mathrm{NL}},2,(1-K)\left(\frac{\gamma A^{\mathrm{NL}}}{A^{\mathrm{L}}}r^{\alpha^{\mathrm{L}}}\right)^{-1},r_1\right)\right)\right)\exp\left(\frac{-2\pi\lambda}{d_0}\left(\frac{\gamma A^{\mathrm{NL}}}{A^{\mathrm{L}}}r^{\alpha^{\mathrm{L}}}\right)^{-1}(1-K-e^{-K})\right. \\ &\times\left(\rho_1\left(\alpha^{\mathrm{NL}},\alpha^{\mathrm{NL}}+2,(1-K)\left(\frac{\gamma A^{\mathrm{NL}}}{A^{\mathrm{L}}}r^{\alpha^{\mathrm{L}}}\right)^{-1},d_1\right)\right. \end{aligned}$$

$$-\rho_1\left(\alpha^{\mathrm{NL}},\alpha^{\mathrm{NL}}+2,(1-K)\left(\frac{\gamma A^{\mathrm{NL}}}{A^{\mathrm{L}}}r^{\alpha^{\mathrm{L}}}\right)^{-1},r_1\right)\Bigg)\Bigg)$$

$$\times\exp\left(-2\pi\lambda\rho_2\left(\alpha^{\mathrm{NL}},1,(1-K)\left(\frac{\gamma A^{\mathrm{NL}}}{A^{\mathrm{L}}}r^{\alpha^{\mathrm{L}}}\right)^{-1},d_1\right)\right)$$

$$\times\exp\left(-2\pi\lambda\left(\frac{\gamma A^{\mathrm{NL}}}{A^{\mathrm{L}}}r^{\alpha^{\mathrm{L}}}\right)^{-1}(1-K-e^{-K})\right.$$

$$\left.\rho_2\left(\alpha^{\mathrm{NL}},\alpha^{\mathrm{NL}}+1,(1-K)\left(\frac{\gamma A^{\mathrm{NL}}}{A^{\mathrm{L}}}r^{\alpha^{\mathrm{L}}}\right)^{-1},d_1\right)\right)\quad(0<r\leq d_1),\tag{2.18}$$

where

$$\rho_1(\alpha,\beta,t,d)=\left[\frac{d^{(\beta+1)}}{\beta+1}\right]{}_2F_1\left[1,\frac{\beta+1}{\alpha};1+\frac{\beta+1}{\alpha};-td^{\alpha}\right]\tag{2.19}$$

and

$$\rho_2(\alpha,\beta,t,d)=\left[\frac{d^{-(\alpha-\beta-1)}}{t(\alpha-\beta-1)}\right]{}_2F_1\left[1,1-\frac{\beta+1}{\alpha};2-\frac{\beta+1}{\alpha};-\frac{1}{td^{\alpha}}\right]$$
$$\times(\alpha>\beta+1),\tag{2.20}$$

where ${}_2F_1[\cdot,\cdot;\cdot;\cdot]$ is the hyper-geometric function [28].

Proof: See Appendix B.

It is important to note that the Laplace term computed by Lemma 2.2 gives the probability that the first-piece LoS signal power exceeds the aggregate interference power by a factor of at least γ.

Overall, we can evaluate T_1^{L} as

$$T_1^{\mathrm{L}}=\int_0^{d_1}\sum_{k=0}^{\infty}\sum_{m=0}^{k}J(m,k)\,\gamma^{k-m}(-1)^{k-m}\frac{\partial^{k-m}\mathscr{L}_{I_r}\left(\frac{\gamma r^{\alpha^{\mathrm{L}}}}{P_tA^{\mathrm{L}}}\right)}{\partial\gamma^{k-m}}f_{R,1}^{\mathrm{L}}(r)dr,\tag{2.21}$$

where $f_{R,1}^{\mathrm{L}}(r)$ and $\mathscr{L}_{I_r}(\gamma r^{\alpha^{\mathrm{L}}}/P_tA^{\mathrm{L}})$ are determined by Eqs. (2.17) and (2.18), respectively.

2.4.2 The Computation of T_1^{NL}

From Theorem 2.1, T_1^{NL} is computed as

$$T_1^{\mathrm{NL}}=\int_0^{d_1}\sum_{k=0}^{\infty}\sum_{m=0}^{k}J(m,k)\,\gamma^{k-m}(-1)^{k-m}\frac{\partial^{k-m}\mathscr{L}_{I_r}\left(\frac{\gamma}{P_t\zeta_n^{\mathrm{NL}}(r)}\right)}{\partial\gamma^{k-m}}f_{R,1}^{\mathrm{NL}}(r)dr$$

$$\stackrel{(a)}{=}\int_0^{d_1}\sum_{k=0}^{\infty}\sum_{m=0}^{k}J(m,k)\,\gamma^{k-m}(-1)^{k-m}\frac{\partial^{k-m}\mathscr{L}_{I_r}\left(\frac{\gamma r^{\alpha^{\mathrm{NL}}}}{P_tA^{\mathrm{NL}}}\right)}{\partial\gamma^{k-m}}f_{R,1}^{\mathrm{NL}}(r)dr,\tag{2.22}$$

where $\zeta_1^{\mathrm{NL}}(r)=A^{\mathrm{NL}}r^{-\alpha^{\mathrm{NL}}}$ from Eq. (2.14) is plugged into step (a) of Eq. (2.22).

In Eq. (2.22), according to Theorem 2.1 and Eq. (2.15), $f_{R,1}^{\mathrm{NL}}(r)$ can be written as

$$f_{R,1}^{\mathrm{NL}}(r) = \exp\left(-\int_0^{r_2} \lambda \mathrm{Pr}^{\mathrm{L}}(u) 2\pi u du\right) \times \exp\left(-\int_0^{r} \lambda (1-\mathrm{Pr}^{\mathrm{L}}(u)) 2\pi u du\right)\left(\frac{r}{d_1}\right) 2\pi r\lambda \quad (0 < r \leq d_1), \tag{2.23}$$

where $r_2 = (A^{\mathrm{L}}/A^{\mathrm{NL}})^{1/\alpha^{\mathrm{L}}} r^{\alpha^{\mathrm{NL}}/\alpha^{\mathrm{L}}}$. In the following, we discuss the cases of $0 < r_2 \leq d_1$ and $r_2 > d_1$ separately.

If $0 < r_2 \leq d_1$, i.e. $0 < r \leq y_1 = d_1^{\alpha^{\mathrm{L}}/\alpha^{\mathrm{NL}}} (A^{\mathrm{NL}}/A^{\mathrm{L}})^{1/\alpha^{\mathrm{NL}}}$, the $f_{R,1}^{\mathrm{NL}}(r)$ is calculated as

$$\begin{aligned} f_{R,1}^{\mathrm{NL}}(r) &= \exp\left(-\int_0^{r_2} \lambda \left(1-\frac{u}{d_1}\right) 2\pi u du\right) \exp\left(-\int_0^{r} \lambda \frac{u}{d_1} 2\pi u du\right)\left(\frac{r}{d_1}\right) 2\pi r\lambda \\ &= \exp\left(-\pi\lambda r_2^2 + 2\pi\lambda\left(\frac{r_2^3}{3d_1} - \frac{r^3}{3d_1}\right)\right)\left(\frac{r}{d_1}\right) 2\pi r\lambda \quad (0 < r \leq y_1). \end{aligned} \tag{2.24}$$

Otherwise, if $r_2 > d_1$, i.e. $y_1 < r \leq d_1$, the $f_{R,1}^{\mathrm{NL}}(r)$ is calculated as

$$\begin{aligned} f_{R,1}^{\mathrm{NL}}(r) &= \exp\left(-\int_0^{d_1} \lambda \left(1-\frac{u}{d_1}\right) 2\pi u du\right) \exp\left(-\int_0^{r} \lambda \frac{u}{d_1} 2\pi u du\right)\left(\frac{r}{d_1}\right) 2\pi r\lambda \\ &= \exp\left(-\frac{\pi\lambda d_1^2}{3} - \frac{2\pi\lambda r^3}{3d_1}\right)\left(\frac{r}{d_1}\right) 2\pi r\lambda \quad (y_1 < r \leq d_1). \end{aligned} \tag{2.25}$$

In the following, Lemma 2.2 is proposed to compute $\mathscr{L}_{I_r}(\gamma r^{\alpha^{\mathrm{NL}}}/P_t A^{\mathrm{NL}})$ in Eq. (2.22) for the range of $0 < r \leq d_1$. Note that the computation of $\mathscr{L}_{I_r}(\gamma r^{\alpha^{\mathrm{NL}}}/P_t A^{\mathrm{NL}})$ will also be performed separately in the two ranges of $0 < r_2 \leq d_1$ and $r_2 > d_1$.

Lemma 2.2 $\mathscr{L}_{I_r}(\gamma r^{\alpha^{\mathrm{NL}}}/P_t A^{\mathrm{NL}})$ in the range of $0 < r \leq d_1$ is considered separately for two different cases, i.e. $0 < r \leq y_1$ and $y_1 < r \leq d_1$:

$$\begin{aligned} \mathscr{L}_{I_r}\left(\frac{\gamma r^{\alpha^{\mathrm{NL}}}}{P_t A^{\mathrm{NL}}}\right) &= \exp\left(-2\pi\lambda\left(\rho_1\left(\alpha^{\mathrm{L}}, 1, (1-K)\left(\frac{\gamma A^{\mathrm{L}}}{A^{\mathrm{NL}}} r^{\alpha^{\mathrm{NL}}}\right)^{-1}, d_1\right)\right.\right. \\ &\left.\left. - \rho_1\left(\alpha^{\mathrm{L}}, 1, (1-K)\left(\frac{\gamma A^{\mathrm{L}}}{A^{\mathrm{NL}}} r^{\alpha^{\mathrm{NL}}}\right)^{-1}, r_2\right)\right)\right) \\ &\times \exp\left(-2\pi\lambda\left(\rho_1\left(\alpha^{\mathrm{L}}, \alpha^{\mathrm{L}}+1, (1-K)\left(\frac{\gamma A^{\mathrm{L}}}{A^{\mathrm{NL}}} r^{\alpha^{\mathrm{NL}}}\right)^{-1}, d_1\right)\right.\right. \\ &\left.\left. - \rho_1\left(\alpha^{\mathrm{L}}, \alpha^{\mathrm{L}}+1, (1-K)\left(\frac{\gamma A^{\mathrm{L}}}{A^{\mathrm{NL}}} r^{\alpha^{\mathrm{NL}}}\right)^{-1}, r_2\right)\right)\right) \exp\left(\frac{2\pi\lambda}{d_0}\left(\frac{\gamma A^{\mathrm{L}}}{A^{\mathrm{NL}}} r^{\alpha^{\mathrm{NL}}}\right)^{-1}(1-K-e^{-K})\right. \\ &\left.\times\left(\rho_1\left(\alpha^{\mathrm{L}}, 2, (1-K)\left(\frac{\gamma A^{\mathrm{L}}}{A^{\mathrm{NL}}} r^{\alpha^{\mathrm{NL}}}\right)^{-1}, d_1\right) - \rho_1\left(\alpha^{\mathrm{L}}, 2, (1-K)\left(\frac{\gamma A^{\mathrm{L}}}{A^{\mathrm{NL}}} r^{\alpha^{\mathrm{NL}}}\right)^{-1}, r_2\right)\right)\right) \\ &\times \exp\left(\frac{2\pi\lambda}{d_0}\left(\frac{\gamma A^{\mathrm{L}}}{A^{\mathrm{NL}}} r^{\alpha^{\mathrm{NL}}}\right)^{-1}(1-K-e^{-K})\left(\rho_1\left(\alpha^{\mathrm{L}}, \alpha^{\mathrm{L}}+2, (1-K)\left(\frac{\gamma A^{\mathrm{L}}}{A^{\mathrm{NL}}} r^{\alpha^{\mathrm{NL}}}\right)^{-1}, d_1\right)\right.\right. \end{aligned}$$

$$
\begin{aligned}
&-\rho_1\left(\alpha^{\mathrm{L}},\alpha^{\mathrm{L}}+2,(1-K)\left(\frac{\gamma A^{\mathrm{L}}}{A^{\mathrm{NL}}}r^{\alpha^{\mathrm{NL}}}\right)^{-1},r_2\right)\Bigg)\Bigg)\Bigg)\\
&\times\exp\left(\frac{-2\pi\lambda}{d_0}(\rho_1(\alpha^{\mathrm{NL}},2,(1-K)(\gamma r^{\alpha^{\mathrm{NL}}})^{-1},d_1)-\rho_1(\alpha^{\mathrm{NL}},2,(1-K)(\gamma r^{\alpha^{\mathrm{NL}}})^{-1},r))\right)\\
&\times\exp\left(\frac{-2\pi\lambda}{d_0}(\gamma r^{\alpha^{\mathrm{NL}}})^{-1}(1-K-e^{-K})(\rho_1(\alpha^{\mathrm{NL}},\alpha^{\mathrm{NL}}+2,(1-K)(\gamma r^{\alpha^{\mathrm{NL}}})^{-1},d_1)\right.\\
&\left.-\rho_1(\alpha^{\mathrm{NL}},\alpha^{\mathrm{NL}}+2,(1-K)(\gamma r^{\alpha^{\mathrm{NL}}})^{-1},r))\right)\\
&\times\exp(-2\pi\lambda\rho_2(\alpha^{\mathrm{NL}},1,(1-K)(\gamma r^{\alpha^{\mathrm{NL}}})^{-1},d_1))\\
&\times\exp(-2\pi\lambda(\gamma r^{\alpha^{\mathrm{NL}}})^{-1}(1-K-e^{-K})\rho_2(\alpha^{\mathrm{NL}},\alpha^{\mathrm{NL}}+1,(1-K)(\gamma r^{\alpha^{\mathrm{NL}}})^{-1},d_1))\\
&\quad(0<r\le y_1)
\end{aligned}
\tag{2.26}
$$

and

$$
\begin{aligned}
\mathscr{L}_{I_r}\left(\frac{\gamma r^{\alpha^{\mathrm{NL}}}}{P_t A^{\mathrm{NL}}}\right)&=\exp\left(\frac{-2\pi\lambda}{d_0}(\rho_1(\alpha^{\mathrm{NL}},2,(1-K)(\gamma r^{\alpha^{\mathrm{NL}}})^{-1},d_1)\right.\\
&\left.-\rho_1(\alpha^{\mathrm{NL}},2,(1-K)(\gamma r^{\alpha^{\mathrm{NL}}})^{-1},r))\right)\exp\left(\frac{-2\pi\lambda}{d_0}(\gamma r^{\alpha^{\mathrm{NL}}})^{-1}(1-K-e^{-K})\right.\\
&\times(\rho_1(\alpha^{\mathrm{NL}},\alpha^{\mathrm{NL}}+2,(1-K)(\gamma r^{\alpha^{\mathrm{NL}}})^{-1},d_1)\\
&\left.-\rho_1(\alpha^{\mathrm{NL}},\alpha^{\mathrm{NL}}+2,(1-K)(\gamma r^{\alpha^{\mathrm{NL}}})^{-1},r))\right)\\
&\times\exp(-2\pi\lambda\ \rho_2(\alpha^{\mathrm{NL}},1,(1-K)(\gamma r^{\alpha^{\mathrm{NL}}})^{-1},d_1))\exp(-2\pi\lambda(\gamma r^{\alpha^{\mathrm{NL}}})^{-1}\\
&\times(1-K-e^{-K})\rho_2(\alpha^{\mathrm{NL}},\alpha^{\mathrm{NL}}+1,(1-K)(\gamma r^{\alpha^{\mathrm{NL}}})^{-1},d_1))\quad(y_1<r\le d_1),
\end{aligned}
\tag{2.27}
$$

where $\rho_1(\alpha,\beta,t,d)$ and $\rho_2(\alpha,\beta,t,d)$ are defined in Eqs. (2.19) and (2.20), respectively.

Proof: See Appendix C.

It is important to note that the Laplace term computed by Lemma 2.2 gives the probability that the first-piece NLoS signal power exceeds the aggregate interference power by a factor of at least γ.

Overall, we evaluate T_1^{NL} as

$$
\begin{aligned}
T_1^{\mathrm{NL}}&=\int_0^{y_1}\sum_{k=0}^{\infty}\sum_{m=0}^{k}J(m,k)\,\gamma^{k-m}(-1)^{k-m}\frac{\partial^{k-m}[\mathscr{L}_{I_r}\left(\frac{\gamma r^{\alpha^{\mathrm{NL}}}}{P_t A^{\mathrm{NL}}}\right)}{\partial\gamma^{k-m}}f_{R,1}^{\mathrm{NL}}(r)|0<r\le y_1]dr\\
&+\int_{y_1}^{d_1}\sum_{k=0}^{\infty}\sum_{m=0}^{k}J(m,k)\,\gamma^{k-m}(-1)^{k-m}\frac{\partial^{k-m}[\mathscr{L}_{I_r}\left(\frac{\gamma r^{\alpha^{\mathrm{NL}}}}{P_t A^{\mathrm{NL}}}\right)}{\partial\gamma^{k-m}}f_{R,1}^{\mathrm{NL}}(r)|y_1<r\le d_1]dr,
\end{aligned}
\tag{2.28}
$$

where $f_{R,1}^{\mathrm{NL}}(r)$ is computed using Eqs. (2.24) and (2.25), and $\mathscr{L}_{I_r}(\gamma r^{\alpha^{\mathrm{NL}}}/P_t A^{\mathrm{NL}})$ is given by Eqs. (2.26) and (2.27).

2.4.3 The Computation of T_2^{L}

From Theorem 2.1, T_2^{L} is computed as

$$T_2^{\mathrm{L}} = \int_{d_1}^{\infty} \sum_{k=0}^{\infty} \sum_{m=0}^{k} J(m,k)\ \gamma^{k-m}(-1)^{k-m} \frac{\partial^{k-m} \mathscr{L}_{I_r}\left(\frac{\gamma}{P_t \zeta_n^{\mathrm{L}}(r)}\right)}{\partial \gamma^{k-m}} f_{R,2}^{\mathrm{L}}(r) dr. \tag{2.29}$$

It is important to note that the Laplace term in Eq. (2.29), gives the probability that the second-piece LoS signal power exceeds the aggregate interference power by a factor of at least γ. According to Theorem 2.1 and Eq. (2.15), the $f_{R,2}^{\mathrm{L}}(r)$ can be written as

$$\begin{aligned} f_{R,2}^{\mathrm{L}}(r) &= \exp\left(-\int_0^{r_1} \lambda(1-\mathrm{Pr}^{\mathrm{L}}(u))2\pi u du\right) \\ &\times \exp\left(-\int_0^{r} \lambda \mathrm{Pr}^{\mathrm{L}}(u) 2\pi u du\right) \times 0 \times 2\pi r \lambda = 0 \quad (r > d_1). \end{aligned} \tag{2.30}$$

2.4.4 The Computation of T_2^{NL}

From Theorem 2.1, T_2^{NL} is computed as

$$\begin{aligned} T_2^{\mathrm{NL}} &= \int_{d_1}^{\infty} \sum_{k=0}^{\infty} \sum_{m=0}^{k} J(m,k)\ \gamma^{k-m}(-1)^{k-m} \frac{\partial^{k-m} \mathscr{L}_{I_r}\left(\frac{\gamma}{P_t \zeta_n^{\mathrm{NL}}(r)}\right)}{\partial \gamma^{k-m}} f_{R,2}^{\mathrm{NL}}(r) dr \\ &\overset{(a)}{=} \int_{d_1}^{\infty} \sum_{k=0}^{\infty} \sum_{m=0}^{k} J(m,k)\ \gamma^{k-m}(-1)^{k-m} \frac{\partial^{k-m} \mathscr{L}_{I_r}\left(\frac{\gamma r^{\alpha^{\mathrm{NL}}}}{P_t A^{\mathrm{NL}}}\right)}{\partial \gamma^{k-m}} f_{R,2}^{\mathrm{NL}}(r) dr, \end{aligned} \tag{2.31}$$

where $\zeta_2^{\mathrm{NL}}(r) = A^{\mathrm{NL}} r^{-\alpha^{\mathrm{NL}}}$ from Eq. (2.14) is plugged into step (a) of Eq. (2.31).

Moreover, based on Theorem 1.1 and Eq. (2.15), the $f_{R,2}^{\mathrm{NL}}(r)$ can be written as

$$\begin{aligned} f_{R,2}^{\mathrm{NL}}(r) &= \exp\left(-\int_0^{d_1} \lambda\left(1-\frac{u}{d_1}\right) 2\pi u du\right) \\ &= \exp\left(-\int_0^{d_1} \lambda \frac{u}{d_1} 2\pi u du - \int_{d_1}^{r} \lambda 2\pi u du\right) 2\pi r \lambda \\ &= \exp(-\pi \lambda r^2) 2\pi r \lambda \quad (r > d_1). \end{aligned} \tag{2.32}$$

In the following, Lemma 2.3 is proposed to calculate $\mathscr{L}_{I_r}(\gamma r^{\alpha^{\mathrm{NL}}}/P_t A^{\mathrm{NL}})$ in Eq. (2.31) for the range of $r > d_1$.

Lemma 2.3 $\mathscr{L}_{I_r}(\gamma r^{\alpha^{\mathrm{NL}}}/P_t A^{\mathrm{NL}})$ in the range of $r > d_1$ can be computed as

$$\begin{aligned} \mathscr{L}_{I_r}\left(\frac{\gamma r^{\alpha^{\mathrm{NL}}}}{P_t A^{\mathrm{NL}}}\right) &= \exp(-2\pi\lambda \rho_2(\alpha^{\mathrm{NL}}, 1, (1-K)(\gamma r^{\alpha^{\mathrm{NL}}})^{-1}, d_1)) \\ &\times \exp(-2\pi\lambda(\gamma r^{\alpha^{\mathrm{NL}}})^{-1}(1-K-e^{-K}) \\ &\rho_2(\alpha^{\mathrm{NL}}, \alpha^{\mathrm{NL}}+1, (1-K)(\gamma r^{\alpha^{\mathrm{NL}}})^{-1}, d_1)) \quad (r > d_1), \end{aligned} \tag{2.33}$$

where $\rho_2(\alpha, \beta, t, d)$ is defined in Eq. (2.20).

Proof: See Appendix D.

It is important to note that the Laplace term computed by Lemma 2.3 gives the probability that the second-piece NLoS signal power exceeds the aggregate interference power by a factor of at least γ.

Overall, we evaluate T_2^{NL} as

$$T_2^{\mathrm{NL}} = \int_{d_1}^{\infty} \sum_{k=0}^{\infty} \sum_{m=0}^{k} J(m,k) \quad \gamma^{k-m}(-1)^{k-m} \frac{\partial^{k-m} \mathscr{L}_{I_r}\left(\frac{\gamma r^{\alpha^{\mathrm{NL}}}}{P_t A^{\mathrm{NL}}}\right)}{\partial \gamma^{k-m}} f_{R,2}^{\mathrm{NL}}(r) dr. \tag{2.34}$$

where $f_{R,2}^{\mathrm{NL}}(r)$ and $\mathscr{L}_{I_r}(\gamma r^{\alpha^{\mathrm{NL}}}/P_t A^{\mathrm{NL}})$ are computed by Eqs, (2.32) and (2.33), respectively.

2.4.5 The Results of $p^{\mathrm{cov}}(\lambda, \gamma)$ and $A^{\mathrm{ASE}}(\lambda, \gamma_0)$

Based on the above derivations given in Eqs. (2.21), (2.28), and (2.34), the coverage probability can be written as

$$p^{\mathrm{cov}}(\lambda, \gamma) = T_1^{\mathrm{L}} + T_1^{\mathrm{NL}} + T_2^{\mathrm{NL}}, \tag{2.35}$$

Plugging $p^{\mathrm{cov}}(\lambda, \gamma)$ into Eq. (2.8), the ASE $A^{\mathrm{ASE}}(\lambda, \gamma_0)$ can also be obtained.

2.5 Simulation and Discussion

In this section, we use simulation results to study the performance of UDNs under the Rician fading channel with the path loss model in Section 2.1, and validate the accuracy of our analysis. Table 2.1 lists the simulation parameters.

2.5.1 Validation of the Analytical Results of $p^{\mathrm{cov}}(\lambda, \gamma)$ for the 3GPP Case

Figure 2.1 shows the analytical and simulated results of $p^{\mathrm{cov}}(\lambda, \gamma)$ versus the density of small cell BSs for two different SINR thresholds of $\gamma = 0$ dB and $\gamma = 3$ dB. Figure 2.1 shows that the coverage probability initially increases as the BS density increases. However, once the BS density exceeds a certain threshold, i.e. $\lambda > \lambda_1$ (e.g. $\lambda_1 = 100$ BSs/km^2 in Figure 2.1), the coverage probability starts to decline. This can be explained as follows.

Table 2.1 Simulation settings.

Parameter	Values [21]
α^{L}	2.09
α^{NL}	3.75
A^{L}	$10^{-10.38}$
A^{NL}	$10^{-14.54}$
d_1	0.3 km
P_t	24 dBm
N_0	−95 dBm

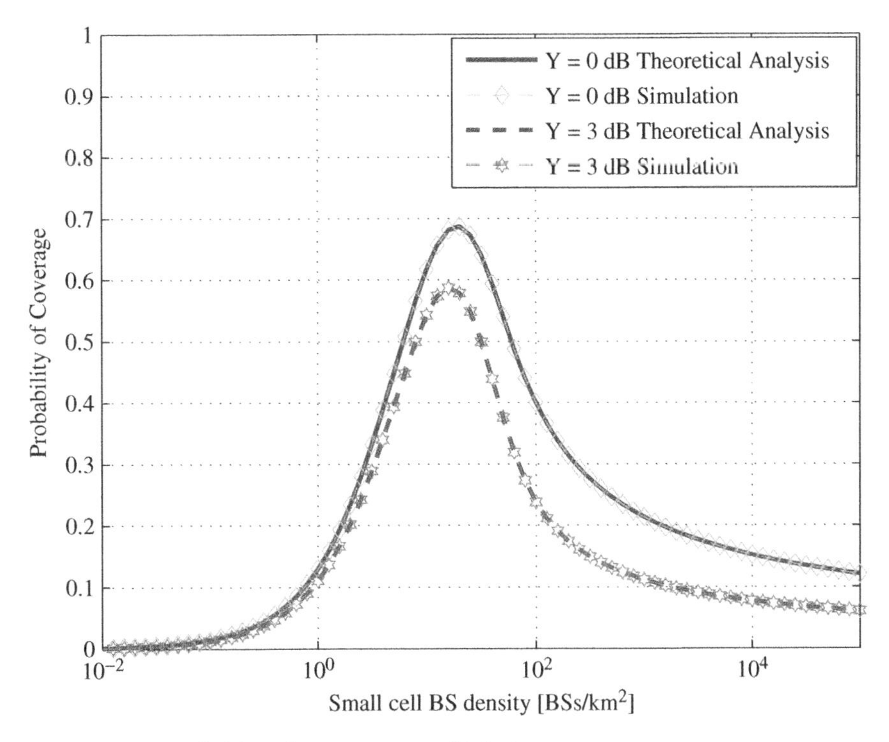

Figure 2.1 The probability of coverage versus BS denisty.

When the BS density is lower than λ_1, the network performance is noise-limited and thus there is a rapid increase in coverage probability with the BS density because more BSs provide better coverage in noise-limited networks. However, once the network becomes denser and the density of BSs is larger than λ_1, then a large number of interfering links transit from NLoS to LoS, and thus the increase in interference power cannot be counterbalanced by the increase in signal power, which is already LoS. Note that further densification beyond λ_1 results in a slower decline rate in coverage probability, since both signals corresponding to interfering and serving BSs are LoS dominated. It is perceived that the coverage probability in the case of Ricean fading follows the same trend as in the case of Rayleigh fading presented in [11].

Note that the theoretical analysis results match well with the simulation results, and thus we only show theoretical results in the sequel. Comparing these probability of coverage results with Rician fading with those in [11] with Rayleigh fading, as shown in Figure 2.2, it can be concluded that the impact of Rician fading on the coverage probability is negligible. The difference in coverage probability is less than 0.02 for all BS densities. This is because the impact of the transition of many interfering links from NLoS to LoS is in the order of 15–20 dB according to the 3GPP path loss functions [21], while the impact of Rayleigh to Rician fading transition is in the order of $\sim$ 3 dB. Thus, Rayleigh or Rician fading in the LoS components makes little difference against the abrupt change of interference strength caused by the transition of many NLoS interfering links to LoS ones.

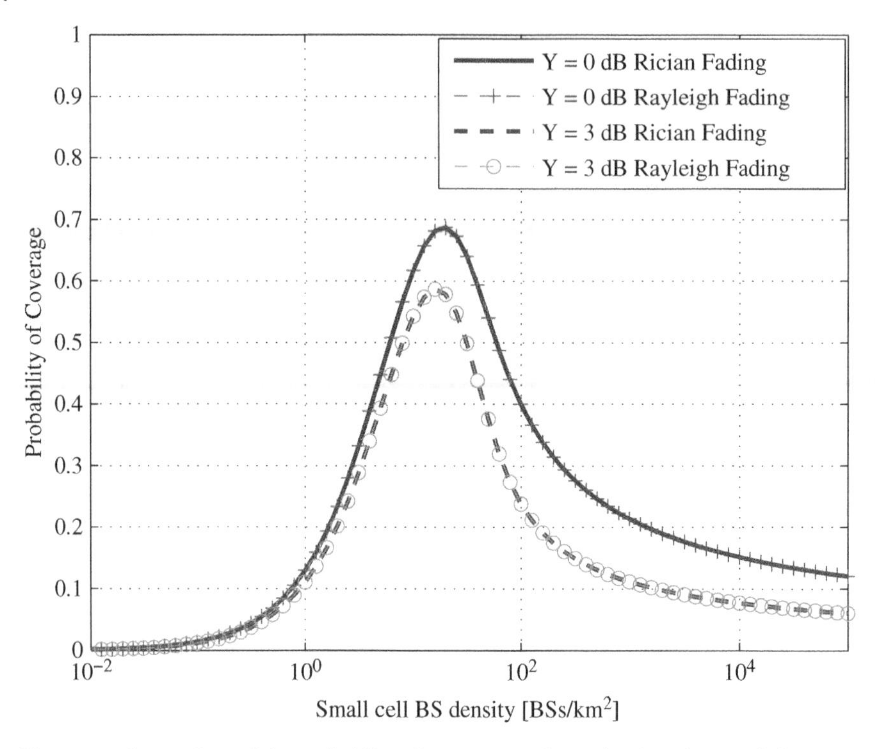

Figure 2.2 Comparison of the probability of coverage under Rayleigh and Rician fadings.

2.5.2 Discussion on the Analytical Results of $A^{ASE}(\lambda, \gamma_0)$ for the 3GPP Case

Figure 2.3 shows the ASE for the different SINR thresholds of $\gamma = 0$ dB and $\gamma = 3$ dB. Note that the ASE results are derived based on the results from the probability of coverage presented in Eq. (2.7). Similar to the observed trend for the coverage probability, the ASE trend also shows three phases. In the first phase, when the BS density is lower than λ_1, the ASE increases with the BS density as coverage holes are mitigated. In the second phase, when the BS density exceeds λ_1, the ASE suffers from a slower growth pace or even a decrease due to the decline in the coverage probability originated by the transition of a large number of interfering links from NLoS to LoS. In the third phase, when all interfering signals have transited from NLoS to LoS, the ASE starts to linearly increase with the BS density since the network has become statistically stable with all interfering and serving downlinks being LoS dominated.

Comparing the ASE results in this paper with Rician fading with those in [11] with Rayleigh fading, as shown in Figure 2.3, it can be concluded that the impact of Rician fading on the ASE is negligible with a peak Rician to Rayleigh gain of about 1.02× at a BS density of 15.85 BSs/km^2. The reason for this conclusion is the same as that explained before, i.e. the power variation of the NLoS to LoS transition is in the order of 15–20 dB according to the 3GPP path loss functions [21], while that of Rayleigh to Rician is in the order of ~ 3 dB. Thus, Rayleigh or Rician fading makes little difference against this abrupt change of interference strength.

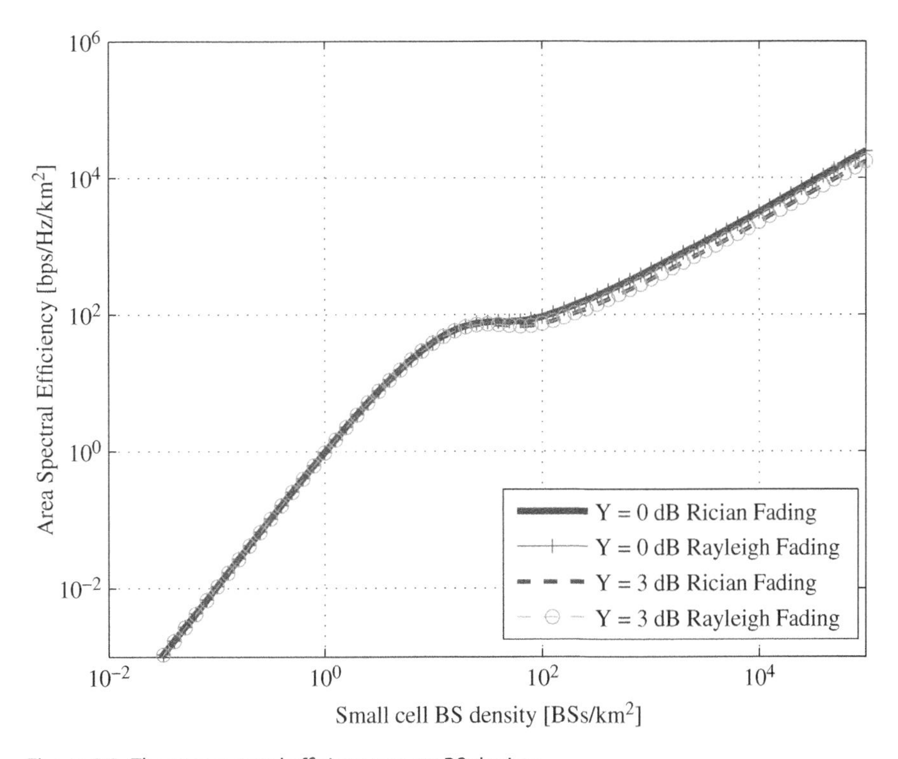

Figure 2.3 The area spectral efficiency versus BS denisty.

2.6 Conclusion

In this chapter, we have discussed the impact of a multi-path fading channel in UDNs, focusing on a path loss model that incorporates both LoS and NLoS transmissions. The analytical and simulation results show that Rician fading has a negligible impact compared to Rayleigh fading on the system performance, indicating that it is the LoS and NLoS path loss characteristics and not the multi-path fading that dominate the UDN performance in a single-input single-output (SISO) scenario. Our results show that, when the density of BSs exceeds a threshold, the ASE starts to suffer a slow growth or even a decrease for a given BS density range. The intuition is as follows. Network densification causes a transition from NLoS to LoS for a large number of interference signals as well as a channel diversity loss as LoS dominates. Due to the dominance of the path loss characteristics over the multi-path fading in UDNs, the interference power increases faster than the signal power, degrading the user SINR.

Appendix A: Proof of Theorem 1.1

To compute $p^{\mathrm{cov}}(\lambda, \gamma)$, we first need to calculate the distance PDFs for the corresponding events of the typical UE being associated with a BS with either a LoS or NLoS path.

Recalling from Eqs. (2.4) and (2.5), the $p^{\rm cov}(\lambda,\gamma)$ can be computed as

$$\begin{aligned}
p^{\rm cov}(\lambda,\gamma) &\stackrel{(a)}{=} \int_{r>0} \Pr[\mathrm{SINR} > \gamma | r]\, f_R(r) dr \\
&= \int_{r>0} \Pr\left[\frac{P_t\zeta(r)h}{I_r} > \gamma\right] f_R(r) dr \\
&= \int_0^{d_1} \Pr\left[\frac{P_t\zeta_1^{\rm L}(r)h}{I_r} > \gamma\right] f_{R,1}^{\rm L}(r) dr + \int_0^{d_1} \Pr\left[\frac{P_t\zeta_1^{\rm NL}(r)h}{I_r} > \gamma\right] f_{R,1}^{\rm NL}(r) dr \\
&+ \cdots \\
&+ \int_{d_{N-1}}^{\infty} \Pr\left[\frac{P_t\zeta_N^{\rm L}(r)h}{I_r} > \gamma\right] f_{R,N}^{\rm L}(r) dr + \int_{d_{N-1}}^{\infty} \Pr\left[\frac{P_t\zeta_N^{\rm NL}(r)h}{I_r} > \gamma\right] f_{R,N}^{\rm NL}(r) dr \\
&\stackrel{\triangle}{=} \sum_{n=1}^{N} (T_n^{\rm L} + T_n^{\rm NL}),
\end{aligned} \tag{2.36}$$

where $f_{R,n}^{\rm L}(r)$ and $f_{R,n}^{\rm NL}(r)$ refer to the piecewise PDFs of the RVs $R_n^{\rm L}$ and $R_n^{\rm NL}$, respectively. $R_n^{\rm L}$ and $R_n^{\rm NL}$ denote the distances that the UE is associated with its serving BS in the LoS case and in the NLoS case, respectively, where the corresponding events are assumed to be disjoint.

In the following, we present two events in order to calculate $f_{R,n}^{\rm L}(r)$ in Eq. (2.36).

- Event $B^{\rm L}$. The nearest BS with a LoS path to the UE is placed at distance $X^{\rm L}$. According to [27], the complementary cumulative distribution function (CCDF) of $X^{\rm L}$ can be written as $\bar{F}_X^{\rm L}(x) = \exp\left(-\int_0^x \Pr^{\rm L}(u) 2\pi u\lambda du\right)$.

  The PDF of $X^{\rm L}$ can then be obtained by taking the derivative of $(1-\bar{F}_X^{\rm L}(x))$ with regard to x as

$$f_X^{\rm L}(x) = \exp\left(-\int_0^x \Pr^{\rm L}(u) 2\pi u\lambda du\right) \Pr^{\rm L}(x) 2\pi x\lambda. \tag{2.37}$$

- Event $C^{\rm NL}$ conditioned on the value of $X^{\rm L}$. Given that $X^{\rm L} = x$, the UE is associated with the nearest BS with a LoS path placed at distance $X^{\rm L}$, giving the smallest path loss (i.e. the largest $\zeta(r)$) from such BS to the UE. To ensure that the UE is associated with such LoS BS at distance $X^{\rm L} = x$, there must be no BS with a NLoS path inside the disk centered on the UE with a radius of $x_1 < x$ to outperform such LoS BS at distance $X^{\rm L} = x$, where x_1 satisfies $x_1 = \underset{x_1}{\arg}\{\zeta^{\rm NL}(x_1) = \zeta^{\rm L}(x)\}$. According to [27], such conditional probability of $C^{\rm NL}$ on condition of $X^{\rm L} = x$ can be written as

$$\Pr[C^{\rm NL} | X^{\rm L} = x] = \exp\left(-\int_0^{x_1} (1 - \Pr^{\rm L}(u)) 2\pi u\lambda du\right). \tag{2.38}$$

Note that Event $B^{\rm L}$ guarantees that the path loss value $\zeta^{\rm L}(x)$ associated with *an arbitrary LoS BS* is always smaller than that associated with *the considered LoS BS* at distance $X^{\rm L} = x$. Moreover, conditioned on $X^{\rm L} = x$, Event $C^{\rm NL}$ guarantees that the path loss value $\zeta^{\rm NL}(x)$ associated with *an arbitrary NLoS BS* must always be smaller than the one associated with *the considered LoS BS* at distance x.

Another Event that has to be taken into account is the one that the UE is associated with a BS with a LoS path where the BS is placed at distance R^{L}. The CCDF of R^{L}, denoted by $\overline{F}_R^{\mathrm{L}}(r)$, is derived as

$$\begin{aligned}
\overline{F}_R^{\mathrm{L}}(r) &- \Pr[R^{\mathrm{L}} > r] \\
&\overset{(a)}{=} \mathrm{E}_{[X^{\mathrm{L}}]}\{\Pr[R^{\mathrm{L}} > r|X^{\mathrm{L}}]\} \\
&= \int_0^{+\infty} \Pr[R^{\mathrm{L}} > r|X^{\mathrm{L}} = x] f_X^{\mathrm{L}}(x)dx \\
&\overset{(b)}{=} \int_0^{r} 0 \times f_X^{\mathrm{L}}(x)dx + \int_r^{+\infty} \Pr[C^{\mathrm{NL}}|X^{\mathrm{L}} = x] f_X^{\mathrm{L}}(x)dx \\
&= \int_r^{+\infty} \Pr[C^{\mathrm{NL}}|X^{\mathrm{L}} = x] f_X^{\mathrm{L}}(x)dx,
\end{aligned} \tag{2.39}$$

where $\mathbb{E}_{[X]}\{\cdot\}$ in step (a) of Eq. (2.39) represents the expectation operation taking the expectation over the variable X and the step (b) of Eq. (2.39) is valid since $\Pr[R^{\mathrm{L}} > r|X^{\mathrm{L}} = x] = 0$ when $0 < x \leq r$ and the conditional event $[R^{\mathrm{L}} > r|X^{\mathrm{L}} = x]$ is equivalent to the conditional event $[C^{\mathrm{NL}}|X^{\mathrm{L}} = x]$ when $x > r$. In order to obtain the PDF of R^{L}, we can take the derivative of $(1 - \overline{F}_R^{\mathrm{L}}(r))$ with regard to r which results in

$$f_R^{\mathrm{L}}(r) = \Pr[C^{\mathrm{NL}}|X^{\mathrm{L}} = r] f_X^{\mathrm{L}}(r). \tag{2.40}$$

Considering the distance range of $(d_{n-1} < r \leq d_n)$, the segment of $f_{R,n}^{\mathrm{L}}(r)$ from $f_R^{\mathrm{L}}(r)$ can be derived as

$$\begin{aligned}
f_{R,n}^{\mathrm{L}}(r) = \exp &\left(- \int_0^{r_1} (1 - \Pr^{\mathrm{L}}(u)) 2\pi u \lambda du \right) \\
&\times \exp \left(- \int_0^{r} \Pr^{\mathrm{L}}(u) 2\pi u \lambda du \right) \Pr_n^{\mathrm{L}}(r) 2\pi r \lambda \quad (d_{n-1} < r \leq d_n),
\end{aligned} \tag{2.41}$$

where $r_1 = \underset{r_1}{\arg}\{\zeta^{\mathrm{NL}}(r_1) = \zeta_n^{\mathrm{L}}(r)\}$.

Having obtained $f_{R,n}^{\mathrm{L}}(r)$, we move on to evaluate $\Pr[P\zeta_n^{\mathrm{L}}(r)h/I_r > \gamma]$ in Eq. (2.36) as

$$\Pr\left[\frac{P_t \zeta_n^{\mathrm{L}}(r)h}{I_r} > \gamma\right] = 1 - \Pr\left[\frac{P_t \zeta_n^{\mathrm{L}}(r)h}{I_r} < \gamma\right] \tag{2.42}$$

where $\Pr[P_t\zeta_n^{\mathrm{L}}(r)h/I_r > \gamma]$ and $\Pr[P_t\zeta_n^{\mathrm{L}}(r)h/I_r < \gamma]$ refer to the CCDF and CDF of SINR, respectively. It is worth reminding that for tractability of analysis, we have considered an interference limited scenario.

The interference is normalized with respect to $P_t\zeta_n^{\mathrm{L}}(r)$ and therefore the normalized interference is defined as $I_{rn} = I_r/P_t\zeta_n^{\mathrm{L}}(r)$. Hence, Eq. (2.42) can be expressed as

$$\Pr\left[\frac{h}{I_{rn}} > \gamma\right] = 1 - \Pr\left[\frac{h}{I_{rn}} < \gamma\right] \tag{2.43}$$

Subsequently, the coverage probability can be computed as

$$\Pr\left[\frac{h}{I_{rn}} > \gamma\right] = 1 - \iint_{x/y<\gamma} f_h(x) f_{I_{rn}}(y)\, dx\, dy = 1 - \int_0^{\infty} F_h(\gamma y) f_{I_{rn}}(y)\, dy \tag{2.44}$$

where $f_h(x)$ and $F_h(x)$ denote the PDF and CDF of random variable h, respectively. Assuming that random variable h is Rician distributed, its PDF is given by

$$f_h(x) = \frac{(K+1)e^{-K}}{\overline{x}} \exp\left(-\frac{(K+1)x}{\overline{x}}\right) I_0\left(\sqrt{\frac{4K(K+1)x}{\overline{x}}}\right) \tag{2.45}$$

where K refers to the Rician K-factor, I_0 is the zeroth-order first-kind modified Bessel function and $\overline{x}$ refers to the expectation of h. Applying the series expansion from [29], the $f_h(x)$ can be expressed as

$$f_h(x) = \exp(-K-x) \sum_{k=0}^{\infty} \frac{(Kx)^k}{(k!)^2} \tag{2.46}$$

and therefore the CDF of h can be derived from its PDF as

$$\begin{aligned} F_h(x) &= e^{-K} \sum_{k=0}^{\infty} \frac{K^k}{(k!)^2} \left(e^{-x} \sum_{m=0}^{k} (-1)^{2m+1}\, m! \binom{k}{m} x^{k-m} + k! \right) \\ &= -\sum_{k=0}^{\infty} \sum_{m=0}^{k} J(m,k)\; x^{k-m} e^{-x} + \sum_{k=0}^{\infty} \frac{K^k}{k!} e^{-K} \\ &= -\sum_{k=0}^{\infty} \sum_{m=0}^{k} J(m,k)\; x^{k-m} e^{-x} + 1 \end{aligned} \tag{2.47}$$

where $J(m,k) = e^{-K} K^k m! \binom{k}{m} / (k!)^2$ and $\sum_{k=0}^{\infty} K^k / k! = e^K$ based on the combination of the Taylor series.

By replacing Eq. (2.47) in Eq. (2.44), the coverage probability can be derived as

$$\begin{aligned} \Pr\left[\frac{h}{I_{rn}} > \gamma\right] &= \sum_{k=0}^{\infty} \sum_{m=0}^{k} J(m,k) \int_0^{\infty} (y\gamma)^{k-m} e^{-y\gamma} f_{I_{rn}}(y)\, dy\} \\ &= \sum_{k=0}^{\infty} \sum_{m=0}^{k} J(m,k)\; \gamma^{k-m} Q(\gamma, k-m) \end{aligned} \tag{2.48}$$

where $Q(\tau, n) = \int_0^{\infty} y^n e^{-y\tau} f_{I_{rn}}(y)\, dy = (-1)^n \partial^n \mathscr{L}_{I_{rn}}(\tau)/\partial \tau^n$ and $n = 0, 1, .., \infty$ [30, 31]. Also, note that $\int_0^{\infty} f_{I_{rn}}(y)\, dy = 1$. Finally, the coverage probability can be presented as

$$\Pr\left[\frac{h}{I_{rn}} > \gamma\right] = \sum_{k=0}^{\infty} \sum_{m=0}^{k} J(m,k)\; \gamma^{k-m} (-1)^{k-m} \frac{\partial^{k-m} \mathscr{L}_{I_{rn}}(\gamma)}{\partial \gamma^{k-m}} \tag{2.49}$$

Plugging $I_r = I_{rn}\, P_t \zeta_n^{\mathrm{L}}(r)$ into Eq. (2.49), we can derive

$$\Pr\left[\frac{P_t \zeta_n^{\mathrm{L}}(r) h}{I_r} > \gamma\right] = \sum_{k=0}^{\infty} \sum_{m=0}^{k} J(m,k)\; \gamma^{k-m} (-1)^{k-m} \frac{\partial^{k-m} \mathscr{L}_{I_r}\left(\frac{\gamma}{P_t \zeta_n^{\mathrm{L}}(r)}\right)}{\partial \gamma^{k-m}} \tag{2.50}$$

where $\mathscr{L}_{I_r}(s)$ is the Laplace transform of RV I_r evaluated at s.

Similarly, $f_{R,n}^{\mathrm{NL}}(r)$ can also be computed. In this regard, we define the following two events.

- Event B^{NL}. The nearest BS with a NLoS path to the UE is placed at distance X^{NL}. Similar to Eq. (2.37), the PDF of X^{NL} is given by

$$f_X^{\mathrm{NL}}(x) = \exp\left(-\int_0^x (1-\mathrm{Pr}^{\mathrm{L}}(u))2\pi u\lambda du\right)(1-\mathrm{Pr}^{\mathrm{L}}(x))2\pi x\lambda. \tag{2.51}$$

- Event C^{L} conditioned on the value of X^{NL}. Given that $X^{\mathrm{NL}} = x$, the UE is associated with the nearest BS with a NLoS path placed at distance X^{NL}, which gives the smallest path loss (i.e. the largest $\zeta(r)$) from such BS to the UE. Consequently, there should be no BS with a LoS path inside the disk centred on the UE with a radius of $x_2 < x$, where x_2 satisfies $x_2 = \arg_{x_2}\{\zeta^{\mathrm{L}}(x_2) = \zeta^{\mathrm{NL}}(x)\}$. Similar to Eq. (2.38), such conditional probability of C^{L} on condition of $X^{\mathrm{NL}} = x$ can be expressed as

$$\mathrm{Pr}[C^{\mathrm{L}}|X^{\mathrm{NL}} = x] = \exp\left(-\int_0^{x_2} \mathrm{Pr}^{\mathrm{L}}(u)2\pi u\lambda du\right). \tag{2.52}$$

Another Event that must be taken into account is the one where the UE is associated with a BS with a NLoS path and such BS is placed at distance R^{NL}. Similar to Eq. (2.39), the CCDF of R^{NL}, denoted by $\overline{F}_R^{\mathrm{NL}}(r)$, can be computed as

$$\begin{aligned}\overline{F}_R^{\mathrm{NL}}(r) &= \mathrm{Pr}[R^{\mathrm{NL}} > r] \\ &= \int_r^{+\infty} \mathrm{Pr}[C^{\mathrm{L}}|X^{\mathrm{NL}} = x] f_X^{\mathrm{NL}}(x)dx.\end{aligned} \tag{2.53}$$

The PDF of R^{NL} can be obtained by taking the derivative of $(1-\overline{F}_R^{\mathrm{NL}}(r))$ with regard to r, which results in

$$f_R^{\mathrm{NL}}(r) = \mathrm{Pr}[C^{\mathrm{L}}|X^{\mathrm{NL}} = r] f_X^{\mathrm{NL}}(x). \tag{2.54}$$

Considering the distance range of $(d_{n-1} < r \leq d_n)$, the segment of $f_{R,n}^{\mathrm{NL}}(r)$ from $f_R^{\mathrm{NL}}(r)$ can be derived as

$$\begin{aligned}f_{R,n}^{\mathrm{NL}}(r) = {} & \exp\left(-\int_0^{r_2} \mathrm{Pr}^{\mathrm{L}}(u)2\pi u\lambda du\right) \times \exp\left(-\int_0^{r} (1-\mathrm{Pr}^{\mathrm{L}}(u))2\pi u\lambda du\right) \\ & \times(1-\mathrm{Pr}_n^{\mathrm{L}}(r))2\pi r\lambda, (d_{n-1} < r \leq d_n)\end{aligned} \tag{2.55}$$

where $r_2 = \arg_{r_2}\{\zeta^{\mathrm{L}}(r_2) = \zeta_n^{\mathrm{NL}}(r)\}$.

Similarly, $\mathrm{Pr}[P_t\zeta_n^{\mathrm{NL}}(r)h/I_r > \gamma]$ can be calculated as

$$\mathrm{Pr}\left[\frac{P_t\zeta_n^{\mathrm{NL}}(r)h}{I_r} > \gamma\right] = \sum_{k=0}^{\infty}\sum_{m=0}^{k} J(m,k)\ \gamma^{k-m}(-1)^{k-m}\frac{\partial^{k-m}\mathscr{L}_{I_r}\left(\frac{\gamma}{P_t\zeta_n^{\mathrm{NL}}(r)}\right)}{\partial\gamma^{k-m}}. \tag{2.56}$$

Appendix B: Proof of Lemma 2.2

In the following, we derive $\mathscr{L}_{I_r}(s)$ in the range of $0 < r \leq d_1$ as

$$\begin{aligned}\mathscr{L}_{I_r}(s) &= \mathbb{E}_{[I_r]}\{\exp(-sI_r)|0 < r \leq d_1\}\\ &= \mathbb{E}_{[\Phi,\{\beta_i\},\{g_i\}]}\left\{\exp\left(-s\sum_{i\in\Phi/b_o} P_t\beta_i g_i\right)\middle| 0 < r \leq d_1\right\}\\ &\overset{(a)}{=} \exp\left(-2\pi\lambda\int_r^{\infty}(1-\mathbb{E}_{[g]}\{\exp(-sP_t\beta(u)g)\})udu\middle| 0 < r \leq d_1\right),\end{aligned} \tag{2.57}$$

where step (a) of Eq. (2.57) is obtained from [27].

Considering that $0 < r \leq d_1$, $\mathrm{E}_{[g]}\{\exp(-sP\beta(u)g)\}$ in Eq. (2.57) must take into account the interference from both the LoS and NLoS paths and note that the random variable g follows a Rician distribution. $\mathscr{L}_{I_r}(s)$ can be expressed as

$$\begin{aligned}\mathscr{L}_{I_r}(s) &= \exp\left(-2\pi\lambda\int_r^{d_1}\left(1-\frac{u}{d_1}\right)[1-\mathrm{E}_{[g]}\exp(-sP_tA^{\mathrm{L}}u^{-\alpha^{\mathrm{L}}}g)]\,udu\right)\\ &\times\exp\left(-2\pi\lambda\int_{r_1}^{d_1}\frac{u}{d_1}[1-\mathrm{E}_{[g]}\exp(-sP_tA^{\mathrm{NL}}u^{-\alpha^{\mathrm{NL}}}g)]\,udu\right)\\ &\times\exp\left(-2\pi\lambda\int_{d_1}^{\infty}[1-\mathrm{E}_{[g]}\exp(-sP_tA^{\mathrm{NL}}u^{-\alpha^{\mathrm{NL}}}g)]udu\right)\end{aligned} \tag{2.58}$$

For the sake of presentation, $sP_tA^{\mathrm{L}}u^{-\alpha^{\mathrm{L}}}$ is denoted by M and hence $\mathrm{E}_{[g]}\{\exp(-Mg)\}$ is computed as

$$\mathrm{E}_{[g]}\exp(-Mg) = \int_0^{\infty}\exp(-Mg)\,\exp(-K-g)\sum_{k=0}^{\infty}\frac{(Kg)^k}{(k!)^2}\,dg \tag{2.59}$$

where $\exp(-K-g)\sum_{k=0}^{\infty}(Kg)^k/(k!)^2$ denotes the PDF of random variable g. According to the Taylor series, it is realized that $\sum_{k=0}^{\infty}K^k/k! = e^K$ and hence Eq. (2.59) can be written as

$$\begin{aligned}\mathrm{E}_{[g]}\{\exp(-Mg)\} &= \int_0^{\infty}\exp(-Mg)\exp(-K-g)\exp(Kg)\,dg\\ &= \exp(-K)\int_0^{\infty}\exp(-g(1+M-K))\,dg = \frac{\exp(-K)}{1+M-K}\end{aligned} \tag{2.60}$$

Plugging $M = sP_tA^{\mathrm{L}}u^{-\alpha^{\mathrm{L}}}$ into Eq. (2.60), the term $1-\mathrm{E}_{[g]}\exp(-sP_tA^{\mathrm{L}}u^{-\alpha^{\mathrm{L}}}g)$ is derived as

$$\begin{aligned}&1-\mathrm{E}_{[g]}\exp(-sP_tA^{\mathrm{L}}u^{-\alpha^{\mathrm{L}}}g)\\ &= \frac{1+(sP_tA^{\mathrm{L}})^{-1}u^{\alpha^{\mathrm{L}}}-K(sP_tA^{\mathrm{L}})^{-1}u^{\alpha^{\mathrm{L}}}-(e^KsP_tA^{\mathrm{L}})^{-1}u^{\alpha^{\mathrm{L}}}}{1+(sP_tA^{\mathrm{L}})^{-1}u^{\alpha^{\mathrm{L}}}-K(sP_tA^{\mathrm{L}})^{-1}u^{\alpha^{\mathrm{L}}}}.\end{aligned} \tag{2.61}$$

Similarly, the term $1 - \mathrm{E}_{[g]}\{\exp(-sP_tA^{\mathrm{NL}}u^{-\alpha^{\mathrm{NL}}}g)\}$ is computed, and therefore, Eq. (2.58) is written as

$$\begin{aligned}
\mathscr{L}_{I_r}(s) &= \exp\left(-2\pi\lambda\int_r^{d_1}\left(1-\frac{u}{d_1}\right)\right.\\
&\times\left.\left(\frac{1+(sP_tA^{\mathrm{L}})^{-1}u^{\alpha^{\mathrm{L}}}-K(sP_tA^{\mathrm{L}})^{-1}u^{\alpha^{\mathrm{L}}}-(e^{K}sP_tA^{\mathrm{L}})^{-1}u^{\alpha^{\mathrm{L}}}}{1+(sP_tA^{\mathrm{L}})^{-1}u^{\alpha^{\mathrm{L}}}-K(sP_tA^{\mathrm{L}})^{-1}u^{\alpha^{\mathrm{L}}}}\right)udu\right)\\
&\times\exp\left(-2\pi\lambda\int_{r_1}^{d_1}\frac{u}{d_1}\right.\\
&\times\left.\left(\frac{1+(sP_tA^{\mathrm{NL}})^{-1}u^{\alpha^{\mathrm{NL}}}-K(sP_tA^{\mathrm{NL}})^{-1}u^{\alpha^{\mathrm{NL}}}-(e^{K}sP_tA^{\mathrm{NL}})^{-1}u^{\alpha^{\mathrm{NL}}}}{1+(sP_tA^{\mathrm{NL}})^{-1}u^{\alpha^{\mathrm{NL}}}-K(sP_tA^{\mathrm{NL}})^{-1}u^{\alpha^{\mathrm{NL}}}}\right)udu\right)\\
&\times\exp\left(-2\pi\lambda\int_{d_1}^{\infty}\right.\\
&\times\left.\left(\frac{1+(sP_tA^{\mathrm{NL}})^{-1}u^{\alpha^{\mathrm{NL}}}-K(sP_tA^{\mathrm{NL}})^{-1}u^{\alpha^{\mathrm{NL}}}-(e^{K}sP_tA^{\mathrm{NL}})^{-1}u^{\alpha^{\mathrm{NL}}}}{1+(sP_tA^{\mathrm{NL}})^{-1}u^{\alpha^{\mathrm{NL}}}-K(sP_tA^{\mathrm{NL}})^{-1}u^{\alpha^{\mathrm{NL}}}}\right)udu\right)
\end{aligned}\tag{2.62}$$

Plugging $s = \gamma r^{\alpha^{\mathrm{L}}}/P_tA^{\mathrm{L}}$ into Eq. (2.62), and considering the definition of $\rho_1(\alpha,\beta,t,d)$ and $\rho_2(\alpha,\beta,t,d)$ in Eqs. (2.19) and (2.20), we can obtain $\mathscr{L}_{I_r}(\gamma r^{\alpha^{\mathrm{L}}}/P_tA^{\mathrm{L}})$, as shown in Eq. (2.18).

Appendix C: Proof of Lemma 2.3

Similar to Appendix B, we derive $\mathscr{L}_{I_r}(\gamma r^{\alpha^{\mathrm{NL}}}/P_tA^{\mathrm{NL}})$ in the range of $0 < r \leq y_1$ as

$$\begin{aligned}
\mathscr{L}_{I_r}\left(\frac{\gamma r^{\alpha^{\mathrm{NL}}}}{P_tA^{\mathrm{NL}}}\right) &= \exp\left(-2\pi\lambda\int_{r_2}^{d_1}\left(1-\frac{u}{d_1}\right)\right.\\
&\times\left.\frac{1+\left(\frac{\gamma r^{\alpha^{\mathrm{NL}}}}{P_tA^{\mathrm{NL}}}P_tA^{\mathrm{L}}\right)^{-1}u^{\alpha^{\mathrm{L}}}-K\left(\frac{\gamma r^{\alpha^{\mathrm{NL}}}}{P_tA^{\mathrm{NL}}}P_tA^{\mathrm{L}}\right)^{-1}u^{\alpha^{\mathrm{L}}}-\left(e^{K}\frac{\gamma r^{\alpha^{\mathrm{NL}}}}{P_tA^{\mathrm{NL}}}P_tA^{\mathrm{L}}\right)^{-1}u^{\alpha^{\mathrm{L}}}}{1+\left(\frac{\gamma r^{\alpha^{\mathrm{NL}}}}{P_tA^{\mathrm{NL}}}P_tA^{\mathrm{L}}\right)^{-1}u^{\alpha^{\mathrm{L}}}-K\left(\frac{\gamma r^{\alpha^{\mathrm{NL}}}}{P_tA^{\mathrm{NL}}}P_tA^{\mathrm{L}}\right)^{-1}u^{\alpha^{\mathrm{L}}}})udu\right)\\
&\times\exp\left(-2\pi\lambda\int_{r}^{d_1}\frac{u}{d_1}\right.\\
&\times\left.\frac{1+\left(\frac{\gamma r^{\alpha^{\mathrm{NL}}}}{P_tA^{\mathrm{NL}}}P_tA^{\mathrm{NL}}\right)^{-1}u^{\alpha^{\mathrm{NL}}}-K\left(\frac{\gamma r^{\alpha^{\mathrm{NL}}}}{P_tA^{\mathrm{NL}}}P_tA^{\mathrm{NL}}\right)^{-1}u^{\alpha^{\mathrm{NL}}}-\left(e^{K}\frac{\gamma r^{\alpha^{\mathrm{NL}}}}{P_tA^{\mathrm{NL}}}P_tA^{\mathrm{NL}}\right)^{-1}u^{\alpha^{\mathrm{NL}}}}{1+\left(\frac{\gamma r^{\alpha^{\mathrm{NL}}}}{P_tA^{\mathrm{NL}}}P_tA^{\mathrm{NL}}\right)^{-1}u^{\alpha^{\mathrm{NL}}}-K\left(\frac{\gamma r^{\alpha^{\mathrm{NL}}}}{P_tA^{\mathrm{NL}}}P_tA^{\mathrm{NL}}\right)^{-1}u^{\alpha^{\mathrm{NL}}}}udu\right)\\
&\times\exp\left(-2\pi\lambda\int_{d_1}^{\infty}\right.
\end{aligned}$$

$$\times \frac{1+\left(\frac{\gamma r^{\alpha^{\mathrm{NL}}}}{P_tA^{\mathrm{NL}}}P_tA^{\mathrm{NL}}\right)^{-1}u^{\alpha^{\mathrm{NL}}}-K\left(\frac{\gamma r^{\alpha^{\mathrm{NL}}}}{P_tA^{\mathrm{NL}}}P_tA^{\mathrm{NL}}\right)^{-1}u^{\alpha^{\mathrm{NL}}}-\left(e^K\frac{\gamma r^{\alpha^{\mathrm{NL}}}}{P_tA^{\mathrm{NL}}}P_tA^{\mathrm{NL}}\right)^{-1}u^{\alpha^{\mathrm{NL}}}}{1+\left(\frac{\gamma r^{\alpha^{\mathrm{NL}}}}{P_tA^{\mathrm{NL}}}P_tA^{\mathrm{NL}}\right)^{-1}u^{\alpha^{\mathrm{NL}}}-K\left(\frac{\gamma r^{\alpha^{\mathrm{NL}}}}{P_tA^{\mathrm{NL}}}P_tA^{\mathrm{NL}}\right)^{-1}u^{\alpha^{\mathrm{NL}}}}udu\Bigg) \tag{2.63}$$

Similarly, $\mathscr{L}_{I_r}(\gamma r^{\alpha^{\mathrm{NL}}}/P_tA^{\mathrm{NL}})$ in the range of $y_1 < r \leq d_1$ can be calculated by

$$\begin{aligned}\mathscr{L}_{I_r}\left(\frac{\gamma r^{\alpha^{\mathrm{NL}}}}{P_tA^{\mathrm{NL}}}\right) &= \exp\left(-2\pi\lambda\int_r^{d_1}\frac{u}{d_1}\right.\\ &\times \left.\frac{1+\left(\frac{\gamma r^{\alpha^{\mathrm{NL}}}}{P_tA^{\mathrm{NL}}}P_tA^{\mathrm{NL}}\right)^{-1}u^{\alpha^{\mathrm{NL}}}-K\left(\frac{\gamma r^{\alpha^{\mathrm{NL}}}}{P_tA^{\mathrm{NL}}}P_tA^{\mathrm{NL}}\right)^{-1}u^{\alpha^{\mathrm{NL}}}-\left(e^K\frac{\gamma r^{\alpha^{\mathrm{NL}}}}{P_tA^{\mathrm{NL}}}P_tA^{\mathrm{NL}}\right)^{-1}u^{\alpha^{\mathrm{NL}}}}{1+\left(\frac{\gamma r^{\alpha^{\mathrm{NL}}}}{P_tA^{\mathrm{NL}}}P_tA^{\mathrm{NL}}\right)^{-1}u^{\alpha^{\mathrm{NL}}}-K\left(\frac{\gamma r^{\alpha^{\mathrm{NL}}}}{P_tA^{\mathrm{NL}}}P_tA^{\mathrm{NL}}\right)^{-1}u^{\alpha^{\mathrm{NL}}}}udu\right)\\ &\times \exp\left(-2\pi\lambda\int_{d_1}^{\infty}\right.\\ &\times \left.\frac{1+\left(\frac{\gamma r^{\alpha^{\mathrm{NL}}}}{P_tA^{\mathrm{NL}}}P_tA^{\mathrm{NL}}\right)^{-1}u^{\alpha^{\mathrm{NL}}}-K\left(\frac{\gamma r^{\alpha^{\mathrm{NL}}}}{P_tA^{\mathrm{NL}}}P_tA^{\mathrm{NL}}\right)^{-1}u^{\alpha^{\mathrm{NL}}}-\left(e^K\frac{\gamma r^{\alpha^{\mathrm{NL}}}}{P_tA^{\mathrm{NL}}}P_tA^{\mathrm{NL}}\right)^{-1}u^{\alpha^{\mathrm{NL}}}}{1+\left(\frac{\gamma r^{\alpha^{\mathrm{NL}}}}{P_tA^{\mathrm{NL}}}P_tA^{\mathrm{NL}}\right)^{-1}u^{\alpha^{\mathrm{NL}}}-K\left(\frac{\gamma r^{\alpha^{\mathrm{NL}}}}{P_tA^{\mathrm{NL}}}P_tA^{\mathrm{NL}}\right)^{-1}u^{\alpha^{\mathrm{NL}}}}udu\right)\end{aligned} \tag{2.64}$$

We conclude our proof by plugging Eqs. (2.19) and (2.20) into Eqs. (2.63) and (2.64).

Appendix D: Proof of Lemma 2.4

Considering only NLoS interference, $\mathscr{L}_{I_r}(\gamma r^{\alpha^{\mathrm{NL}}}/P_tA^{\mathrm{NL}})$ in the range of $r > d_1$ can be derived as

$$\begin{aligned}\mathscr{L}_{I_r}\left(\frac{\gamma r^{\alpha^{\mathrm{NL}}}}{P_tA^{\mathrm{NL}}}\right) &= \exp\left(-2\pi\lambda\int_{d_1}^{\infty}\right.\\ &\times \left.\left(\frac{1+\left(\frac{\gamma r^{\alpha^{\mathrm{NL}}}}{P_tA^{\mathrm{NL}}}P_tA^{\mathrm{NL}}\right)^{-1}u^{\alpha^{\mathrm{NL}}}-K\left(\frac{\gamma r^{\alpha^{\mathrm{NL}}}}{P_tA^{\mathrm{NL}}}P_tA^{\mathrm{NL}}\right)^{-1}u^{\alpha^{\mathrm{NL}}}-\left(e^K\frac{\gamma r^{\alpha^{\mathrm{NL}}}}{P_tA^{\mathrm{NL}}}P_tA^{\mathrm{NL}}\right)^{-1}u^{\alpha^{\mathrm{NL}}}}{1+\left(\frac{\gamma r^{\alpha^{\mathrm{NL}}}}{P_tA^{\mathrm{NL}}}P_tA^{\mathrm{NL}}\right)^{-1}u^{\alpha^{\mathrm{NL}}}-K\left(\frac{\gamma r^{\alpha^{\mathrm{NL}}}}{P_tA^{\mathrm{NL}}}P_tA^{\mathrm{NL}}\right)^{-1}u^{\alpha^{\mathrm{NL}}}}\right)udu\right)\end{aligned} \tag{2.65}$$

where ($r > d_1$). We conclude our proof by plugging Eq. (2.20) into Eq. (2.65).

Bibliography

1 CISCO, "Cisco visual networking index: global mobile data traffic forecast update (2013-2018)," Feb. 2014.

2 D. López-Pérez, M. Ding, H. Claussen, and A.H. Jafari, "Towards 1 Gbps/UE in cellular systems: understanding ultra-dense small cell deployments," *IEEE Commun. Surveys Tuts.*, vol. 17, no. 4, pp. 2078–2101, Fourth quarter 2015.

3 J. Ling and D. Chizhik, "Capacity scaling of indoor pico-cellular networks via reuse", *IEEE Commun. Letters*, vol. 16, no. 2, pp. 231–233, Feb. 2012.
4 H. Holma and A. Toskala, *LTE for UMTS – OFDMA and SC-FDMA Based Radio Access*, John Wiley & Sons Ltd., 2009.
5 A.H. Jafari, D. López-Pérez, H. Song, H. Claussen, L. Ho, and J. Zhang, "Small cell backhaul: challenges and prospective solutions," *EURASIP J. on Wireless Commun. and Netw.*, vol. 2015, no. 1, pp. 1–18, 2015.
6 H. Zhang, Y. Dong, J. Cheng, M.J. Hossain, and V.C.M. Leung, "Fronthauling for 5G LTE-U ultra dense cloud small cell networks," *IEEE Wireless Commun. Mag.*, vol. 23, no. 6, pp. 48–53, Dec. 2016.
7 A.H. Jafari, V. Venkateswaran, D. López-Pérez, and J. Zhang, "Diversity pulse shaped transmission in ultra-dense small cell networks," *IEEE Trans. Veh. Technol.*, vol. 66, no. 7, pp. 5866–5878, Jul. 2017.
8 A.H. Jafari, V. Venkateswaran, D. López-Pérez, and J. Zhang, "Pulse shaping diversity to enhance throughput in ultra-dense small cell networks," *Proc. of IEEE SPAWC*, Jul. 2016.
9 H.S. Dhillon, R. Ganti, F. Baccelli, and J.G. Andrews, "Modeling and analysis of K-tier downlink heterogeneous cellular networks," *IEEE J. Sel. Areas Commun.*, vol. 30, no. 3, pp. 550–560, Apr. 2012.
10 S. Singh, H.S. Dhillon, and J.G. Andrews, "Offloading in heterogeneous networks: modeling, analysis, and design insights," *IEEE Trans. on Wireless Commun.*, vol. 12, no. 5, pp. 2484–2497, May 2013.
11 M. Ding, P. Wang, D. López-Pérez, M. Ding, G. Mao, and Z. Lin, "Performance impact of LoS and NLoS transmissions in dense cellular networks," *IEEE Trans. on Commun.*, vol. 15, no. 3, pp. 2365–2380, Mar. 2016.
12 A.H. Jafari, D. López-Pérez, M. Ding, and J. Zhang, "Performance analysis of dense small cell networks with practical antenna heights under Rician fading," *IEEE Access*, no. 99, Oct. 2017.
13 A.H. Jafari, J. Park, and R.W. Heath Jr., "Analysis of interference mitigation in mmwave communications," *Proc. of IEEE ICC*, May 2017.
14 M. Ding, D. López-Pérez, G. Mao, P. Wang, and Z. Lin, "Will the area spectral efficiency monotonically grow as small cells go dense?," *Proc. of IEEE Globecom*, Dec. 2015.
15 T. Bai, R. Vaze, and R.W. Heath, "Analysis of blockage effects on urban cellular networks," *IEEE Trans. on Wireless Commun.*, vol. 13, no. 9, pp. 5070–5083, Sep. 2014.
16 X. Zhang and J.G. Andrews, "Downlink cellular network analysis with multi-slope path loss models," *IEEE Trans. on Commun.*, vol. 63, no. 5, pp. 1881–1894, Mar. 2015.
17 T. Bai and R.W. Heath Jr., "Coverage and rate analysis for millimeter wave cellular networks," *IEEE Trans. on Wireless Commun.*, vol. 14, no. 2, pp. 1100–1114, Oct. 2014.
18 C. Galiotto, N.K. Pratas, N. Marchetti, and L. Doyle, "A stochastic geometry framework for LOS/NLOS propagation in dense small cell networks," *Proc. of IEEE ICC*, May 2015.
19 M. Haenggi, *Stochastic Geometry for Wireless Networks*. Cambridge University Press, 2012.

20 J.G. Andrews, A.K. Gupta, and H.S. Dhillon, "A Primer on Cellular Network Analysis Using Stochastic Geometry, " Oct. 2016 (Online: ftp://arxiv.org/abs/1604.03183).

21 3GPP, "TR 36.828 (V11.0.0): Further enhancements to LTE Time Division Duplex (TDD) for Downlink-Uplink (DL-UL) interference management and traffic adaptation," Jun. 2012.

22 3GPP, "TR 36.872, Small cell enhancements for E-UTRA and E-UTRAN – physical layer aspects," Dec. 2013.

23 Spatial Channel Model AHG, "Subsection 3.5.3, Spatial Channel Model Text Description V6.0," Apr. 2003 (Online: ftp://www.3gpp.org/tsg_ran/WG1_RL1/3GPP_3GPP2_SCM/ConfCall-16-20030417).

24 A.H. Jafari, D. López-Pérez, M. Ding, and J. Zhang, "Study on scheduling techniques for ultra dense small cell networks," *Proc. of IEEE VTC*, Sep. 2015.

25 3GPP, "TR 25.996, Spatial channel model for Multiple Input Multiple Output (MIMO) simulations," Sep. 2012.

26 3GPP, "TR 36.213, Evolved Universal Terrestrial Radio Access (E-UTRA); physical layer procedures," May. 2016.

27 J.G. Andrews, F. Baccelli, and R.K. Ganti, "A tractable approach to coverage and rate in cellular networks," *IEEE Trans. on Commun.*, vol. 59, no.11, pp. 3122–3134, Nov. 2011.

28 I.S. Gradshteyn and I.M. Ryzhik, *Table of Integrals, Series, and Products* (7th Ed.), Academic Press, 2007.

29 R.L. Burden and J.D. Faires, *Numerical Analysis* (3rd Ed.), PWS Publishers, 1985.

30 M. Peng, Y. Li, T.Q.S. Quek, and C. Wang, "Device-to-device underlaid cellular networks under rician fading channels," *IEEE Trans. on Wireless Commun.*, vol. 13, no. 8, pp. 4247–4259, Aug. 2014.

31 Y. Li, J. Li, J. Jiang, and M. Peng, "Performance analysis of device-to-device underlay communication in Rician fading channels," *Proc. of IEEE Globecom*, Dec. 2013.

3

Mean Field Games for 5G Ultra-dense Networks: A Resource Management Perspective

Mbazingwa E. Mkiramweni[1], *Chungang Yang*[2] *and Zhu Han*[3]

[1] *Xidian University, China*
[2] *School of Telecommunications Engineering, Xidian University, China*
[3] *Department of Electrical and Computer Engineering, University of Houston, USA*

In this chapter, the network densification in 5G from the lower to the upper layer perspective, i.e. terminals, access nodes, and cloud edge layers, is explained, followed by the introduction to the mean field game, which is more suitable for designing and analyzing the ultra-dense networks with spatial-temporal dynamics and densification. Then a survey of the current applications of the mean field game to device-to-device, 5G ultra-dense radio access, and cloud-edge networks, with the main concentration on the featured technical problems of interference mitigation, energy management, and caching, is presented. This can help in understanding the typical characteristics of mean field games better. Finally, several open problems are discussed for further development of mean field games in the 5G era.

3.1 Introduction

Nowadays, wireless communication technology has become a necessary component of our everyday life, showing a core socioeconomic impact and inspiring our daily activity. Communication networks are used in various areas such as education, military, stock market, travel, medical emergency care, disaster recovery, and package delivery. The wide range of applications and substantial improvement of wireless technology have made wireless devices affordable and convenient. It is expected that billions of devices including machine-type communication, the Internet of Vehicles (IoV), and the Internet of Things (IoT) will lead to smart cities where machines, sensors, vehicles, smart homes, and health devices will be connected [1].

As the connected devices increase in number, the traffic volume in wireless communication grows dramatically. The Cisco visual networking index estimated that the volume of mobile data traffic would increase annually at a compound growth rate of 47% from 2016 through 2021 [2]. Also, requirements towards mobile networks are further placed by the introduction of many new services that are supported by smart mobile devices, including ultra-high definition video (UHDV), virtual reality, 3D video, and augmented reality. The emerging data-hungry applications motivated by the proliferation of smart devices will result in the 1000× data challenge in 2020 [3].

Ultra-dense Networks for 5G and Beyond: Modelling, Analysis, and Applications,
First Edition. Edited by Trung Q. Duong, Xiaoli Chu and Himal A. Suraweera.

To be able to support such traffic demands, the future wireless networks have to be extremely dense and multi-layered. The densification of the network access nodes (ANs) and communication links result in the formation of ultra-dense networks (UDNs) [4]. In UDNs, a massive number of devices are located in physical proximity, generating massive amounts of independent traffic with different requirements, and at the same time are to share the same pool of radio resources. The IMT 2020 (5G) promotion group [5] identified stadiums, train stations, shopping malls, open-air assemblies, dense urban areas, and airports as typical UDN scenarios. Nokia [6] expects most urban outdoor and indoor areas to be covered with the UDNs by 2025 or 2030, where small cells will be providing data at the rate of 100 Mbps to all users.

UDNs are expected to significantly improve the system performance regarding throughput of the future 5G networks. However, the deployment of 5G UDN in practical cases will be faced with many challenges [7]. In the UDNs, typical problems include interference, limited energy, backhaul, high costs in small cell deployment and management, handover, spectrum reuse, and limited infrastructure resources. The essential reasons for almost all the problems are the scarcity and uneven distribution of the resources with respect to the traffic. For instance, the increased demands of mobile traffic results in a heavily uneven load between cells in heterogeneous ultra-dense networks (H-UDNs), which therefore demands a load balancing technique to shift appropriate traffic from heavy loaded cells to light loaded neighbors in order to optimize system performances, such as throughput, user fairness, resource utilization, and processing delays [8]. In addition, because of the increasing complexity in coordination and management among multiple network tiers in UDNs, the nodes will need to have a self-organization capability. Therefore, special techniques such as game theory [9] and machine learning [10] that allow network self-organization for addressing resource management problems are needed.

The rational behaviors, environmental dynamics, and the various preferences of the nodes in wireless networks have allowed game theory to have extensive applications in modeling, analyzing, and designing the distributed schemes for solving various problems in complex networks. Game theory is widely used for deriving distributed resource management techniques in different fields. The mean field game (MFG) is a novel game with unique characteristics suitable for UDNs. The MFG can model the interaction of an individual with the effect of the players' collective behavior, a property that makes the game very appropriate for solving problems in UDNs.

In this chapter, an extensive literature review on 5G UDN is given, where the network key enablers, services that 5G is expected to support, and the three different layers of the network architecture, i.e. terminals, radio access, and cloud-edge, are discussed. Moreover, the appropriateness and advantages of MFG over the conventional game theory for designing and analyzing UDN are reviewed. The basics of MFG are presented, where the definition of the mean field and the two differential partial equations of Hamilton-Jacobi-Bellman (HJB) and Fokker-Planck-Kolmogorov (FPK) are introduced. A survey of the current applications of MFG in dealing with various resource management challenges in ultra-dense D2D, radio access networks (RANs), and cloud-edge communication networks is given. Finally, to help understand and be able to apply the MFG to different networking problems, examples in which MFGs are used for managing interferences and energy in 5G UDN are presented.

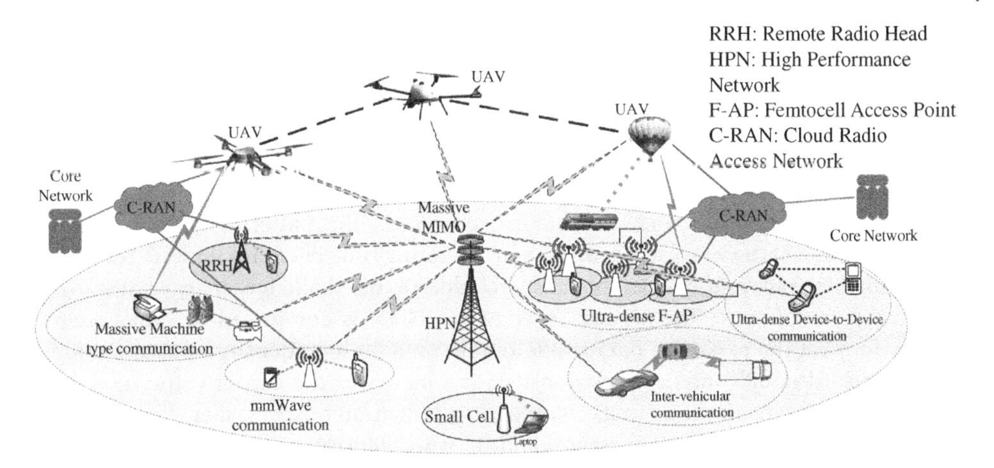

Figure 3.1 5G ultra-dense network architecture scenario.

3.2 Literature Review

The future 5G networks are expected to provide a new level of efficiency and performance that will enhance a user's experience [11]. The 5G networks are visioned to support and provide services that meet the communication requirements of ultra-reliable low-latency communications (uRLLC), enhanced mobile broadband (eMBB), and massive machine-type communication (mMTC) [12]. The 5G will consist of nodes and cells with heterogeneous characteristics and capacities including D2D user equipment, femtocells, picocells, macrocells, and cloudlets, which form a multi-tier network architecture [3, 13]. Figure 3.1 presents a multi-tier architecture scenario of the 5G mobile network with different enablers. The network is comprised of D2D networks, macro and small cells, cloud-enabled networks and different enabling technologies.

The primary technologies that will enable 5G to meet the expected requirements are classified as densification of heterogeneous cellular networks, big data and mobile cloud computing, massive multiple-input multiple-output (massive MIMO), simultaneous transmission and reception, the use of millimeter-wave (mmWave), direct D2D communication, scalable Internet of Things, cloud-RAN, and virtualization of resources [14–16]. Splitting and densification of cells are considered to be new frontiers and dominant themes for the realization of 5G [17, 18]. To provide seamless coverage a very large number of cells needs to be densely deployed in 5G wireless networks, which will result in forming a 5G ultra-dense network [19]. UDN is one of the core features of the 5G networks.

3.2.1 5G Ultra-dense Networks

The 5G UDN can be divided into three layers, the terminal, network access, and the cloud-edge layers. Terminals are considered to be the lower layer of the network, comprised of billions of devices that either communicate directly to each other or link cellular networks through base stations. If a device communicates directly with another device or is supported by other devices to transmit, its information is referred to as Device-to-Device (D2D) communication. D2D can support new application fields including Vehicle-to-Vehicle (V2V) and Machine-to-Machine (M2M) communication [20]. D2D connections can be used to establish M2M communication in the IoT [21] and can also be applied to V2V communication to share information between neighboring vehicles quickly and offload traffic efficiently. Typically, terminals are many in number and move or locate randomly in the network area; therefore special technologies are required for management resources. Also, many IoT devices are equipped with sensors, which are battery powered. The battery-powered devices require periodic battery replacement as their operating time becomes limited, resulting in high maintenance and operational cost. To address this challenge, energy harvesting can provide a solution. The second layer of 5G UDN is comprised of radio access points. To meet the explosive data traffic requirement, a high density of access nodes is deployed. Although the network densification increases the capacity, the inter-cell interference level increases as well, resulting in a reduction of spectral efficiency of the cell. Therefore, interference, spectrum, and network density management techniques in 5G ultra-dense networks will be needed. The third layer is composed of cloud-based platforms to simplify network operation and management and minimize inter-cell interference in wireless networks. Introduction of cloud-based architectures brings network functions such as mobile edge computing and contents caching capabilities to the cellular network edge [22]. To efficiently utilize the existing resources, content caching is enabled at the edge of the network such as base stations (BSs). The network popular contents are cached at the BSs. When BSs receive content requests from users, they can provide the cached contents instead of downloading them from the original servers. In this way, the need for massive content distribution can be addressed. Table 3.1 summarizes the typical features and technologies that are needed to optimize the networks of the considered three layers of the 5G ultra-dense communication network. The ultra-dense D2D, radio access networks (RANs), and cloud-edge networks are the main components of 5G UDN. These networks and their scenarios in which MFG will be applicable are discussed as follows.

1. *Ultra-dense D2D*: It is evident that 5G UDN will be characterized by direct D2D communication where devices transmit and receive data between each other on the user plane without having to use BSs or access points (APs) [23]. D2D communication plays an important role in supporting the ever-increasing machine-to-machine (M2M) and context-aware applications. D2D communication is now being specified

Table 3.1 5G ultra-dense communication networks.

Network layer	Communication networks	Typical features	Technology
Terminal	Internet of Things	Sensor energy	Energy harvesting
	Device-to-Device, Machine-to-Machine, Vehicle-to-Vehicle	Random/Number	Typical technology
RAN/AP	Radio access networks	Spectrum	Interference management
	Ultra-dense deployment of various small cells		Network density management
Cloud-edge	Cloudlet, Cloud-RAN, personal cloud	Traffic distribution	Caching

by 3GPP in LTE Rel-12 [24]. In addition, D2D is regarded as one of the key enablers of the evolving 5G network architecture [25]. Bluetooth and WiFi-Direct are the two most used D2D techniques. D2D can provide a flexible, efficient, dynamic, and secure decentralized approach to proximity discovery and device-to-device communication, thus contributing significantly toward meeting the goals of 5G. D2D can also improve radio resource management, network performance, reduce network load, enhance spectrum reuse, energy efficiency, boost network security, and expand communication applications. Moreover, D2D can be an important feature of mmWave 5G wireless networks that will enhance network capacity and establish communication between two devices [26]. mmWave can be used on D2D-enabled devices to get direct short-range communications among users as well as machines within close proximity. In case line-of-sight (LOS) links between mmWave BSs and wireless devices are blocked, D2D communication can be used to establish a path between them as a relay.

Applications of D2D in 5G include emergency communication, public safety, cellular offloading, content distribution, relaying, traffic control, and provisioning of local services such as advertisements for by-passer and gaming. Also, D2D-based communication can be applied to enhance IoT services [27]. An example of D2D-based IoT enhancement is V2V communication on the Internet of Vehicles (IoVs) where vehicles can send and receive warnings from other vehicles in the D2D mode before slowing down or changing lanes.

Large shopping malls, airports, train stations, and heavy-traffic roads are example scenarios where hundreds or thousands of D2D links can be established. Through D2D, local services such as advertisements, news, and security alerts to ensure safety can be provided. In such scenarios, D2D communication challenges such as security and energy management arise. Since most devices are battery powered, optimizing energy becomes crucial. Due to the massive number of D2D links and network dynamics, the mean field game (MFG) will be an appropriate method for network energy management. Details about MFG are provided in Section 3.3.

2. *5G radio access networks*: A radio access network (RAN) is another integral part of the 5G ultra-dense network that connects individual devices to other parts of a network through radio connections [28, 29]. In UDNs, RANs comprise of a massive number of base stations (BSs) and multiple radio access technologies (RATs). The 5G RAN system will ubiquitously connect billions of wireless devices including computers, mobile phones, remotely controlled machines, and IoT devices.

   To optimize 5G RANs, different methods and technologies such as dense small cell deployment, massive MIMO, and mmWave aggregation have been proposed [30–32]. To improve spectral efficiency, densification of the radio access networks (RANs) and various small cells will be necessary [30]. A massive MIMO technique is considered as one of the major driving elements in 5G radio access technologies (RATs) [31]. Massive MIMO technology, where BSs are equipped with many antennas, is expected to improve throughput, coverage, and capacity of 5G. Millimeter-wave (mmWave) [32] is another RAT that has the potential to significantly enhance the capacity by extending the transmission bandwidth of 5G cellular wireless systems. However, different types of technologies will be required to manage interference in these networks.

   A typical scenario is the dense urban area where many small cell BSs can be densely deployed to meet the high capacity demands. Dense deployment of network access points will result in severe inter-cell interference. The inter-cell interference management problem of the complex large-scale network can be addressed by exploiting MFG.

3. *5G edge cloud computing*: To efficiently and securely process the massive IoT data and provide services with faster response and greater quality, edge computing has been introduced [33]. The processing of the data generated by IoT devices at the network edge instead of the centralized cloud minimize bandwidth, end-to-end latency, and energy consumption. Software-defined networking (SDN) and the network function virtualization (NFV) are considered to be the key technologies that drive the computing and communication landscape from centralized cloud computing into the network edge devices [34]. The two technologies are developed to cater for flexibility, programming, agility, and scalability requirements of 5G cellular networks.

   There are three typical edge computing technologies, namely, the cloudlets, fog computing, and mobile-edge computing (MEC) [33]. Cloudlets represent the middle layer, which is between the mobile device layer and cloud layer architecture. Cloudlets are trusted, resource-rich computers connected to the Internet, which can be utilized by nearby low-latency devices to support interactive and resource-intensive applications by providing computing resources. Fog computing is an architecture in which an edge is considered to work with the cloud; it distributes services and resources for storage, computing, control and networking across networks and between edge devices. MEC is a technology that provides cloud-computing functions to the mobile network edge, within the RAN and nearby mobile subscribers.

   In 5G networks, cloud-based radio access networks (C-RANs) are needed for the system throughput optimization, mobility management, and energy efficiency enhancement [35]. In C-RAN, all the BS computational resources are pooled in servers deployed at the BSs to bring cloud-computing capabilities to the RAN edge [36]. C-RAN uses low-cost Dense or Coarse Wavelength Division Multiplexing (DWDM/CWDM) technology and mmWave to facilitate the transmission of a baseband signal over long distances, thus obtaining large-scale centralized BS deployment. Moving the data access to the cloud will allow the network to be accessed everywhere by providing better communication and cooperation among BSs and also improves caching and processing of data traffic demands in 5G networks [37, 38].

   Cloud-based radio networks support virtualization on distributed or centralized cloud platforms. NFV can make network functions run on a cloud computing infrastructure, which results in the higher reuse of network infrastructure compared to the present network system [39]. For instance, in [40] a cloud-based architecture was proposed for heterogeneous cellular networks. The proposed architecture uses the cloud as the network control and management plan. Cloud also helps network operators to efficiently coordinate small cells and manage network resources on demand when providing services.

   Deployment of the small cloud remote radio heads (RRHs) in areas such as offices or floor of a building can greatly improve coverage and capacity. The small cloud RRHs can also be deployed in hotspot scenarios with ultra-dense devices, such as stadiums or busy streets. MFG will be an appropriate tool for optimizing caching is such scenarios when hundreds of the cloud RRHs are deployed.

3.2.2 Resource Management Challenges in 5G

The deployment of 5G UDN in the practical cases will be faced with many challenges [38]. In the UDNs, typical problems include interference, limited energy, backhaul, cost, and handover, spectrum reuse, and limited infrastructure resources. The essential reasons for almost all the problems are the scarcity and uneven distribution of the resource with respect to the traffic. Different alternatives have been proposed to solve these challenges, for example in [41], network slicing is introduced as a method for optimizing resources in 5G networks. However, due to the increasing network coordination and and management complexity among multiple network tiers, the nodes of the network will have to be capable of self-organization, for example autonomous interference minimization, load balancing, power adaptation, and spectrum allocation. Therefore, there is still a need for a better mechanism for addressing challenges to be designed.

3.2.3 Game Theory for Resource Management in 5G

The rational behaviors, environmental dynamics, and the various preferences of the nodes in wireless networks have allowed game theory to have extensive applications in modeling, analyzing, and designing of the distributed schemes for solving various problems in complex networks such as 5G [42]. Game theory has widely been utilized to derive distributed resource allocation and management techniques. However, the dynamic nature and size of ever-growing communication systems require game models that are based on evolutionary algorithms [43]. The traditional games model the interaction of each player with every other player. Thus analyzing a system with a huge number of players using traditional games can be complicated. In addition, the conventional games demonstrated slow convergence to equilibrium because of the immense signaling and communication overhead caused by information exchange among players in the network. Moreover, the conventional games take longer to reach equilibrium for, as the number of players grows, the size of the payoff matrix grows as well. In particular, for an ultra-dense network, it is very difficult and sometimes impossible to solve problems such as the interference management using conventional games due to the massive number of interacting players.

As an appropriate method, the mean field game (MFG) can be used for modeling and analyzing resource management schemes in 5G UDNs. Different from the conventional game theory, which models the interaction of each player with all the other players, MFG models the interaction of a player with the effect of the collective behavior of the players in the system [43]. The collective behavior of players is reflected in the mean field. The interaction of an individual player with the mean field is modeled by the Hamilton-Jacobi-Bellman (HJB) equation. The dynamics of the mass according to action of the players is described by a Fokker-Planck-Kolmogorov (FPK) equation. The HJB and FPK equations are also known as forward and backward equations, respectively [44]. The MFG solution is obtained by solving the FPK and HJB equations.

3.3 Basics of Mean field game

In this section, mean field game theory and mean field approximation concepts are introduced. The origin, applications, limitations, and the distinguishing features of MFG are presented. The mean field is defined and the two partial differential equations, the HJB and FPK, are discussed.

3.3.1 Background

Mean field game (MFG) theory was introduced by Pierre-Louis Lions and Jean-Michel Lasry in 2006 [45], and further developed and implemented in the following years [44]. MFG was introduced to the engineering community by M. Huang and co-workers to solve optimization problems for a large number of agents [46]. MFG theory originates from three avenues, particle physics, game theory, and economics [47]. By replacing the particles with interacting agents, MFG theory adapts a methodology for dealing with a wide range of situations where the number of particles is too big to allow the dynamics or equilibrium to be described by modeling all the inter-particle interactions in particle physics. The departure of MFG from game theory involves studying the limit of a large class of N-player games when the number of players goes to infinity. MFG also utilized the general economic equilibrium theory in which agents look only at their own interests and to the market prices and have little concern for each other.

MFG models have found many applications in different fields, such as physics, economics, finance, and engineering [47]. For instance, MFG has been applied to model economic growth of the population of rational but non-cooperative agents [48]. Other areas of application include crowd models and segregation models in social sciences, data networks and energy systems in engineering, and formation of volatility in finance. Figure 3.2, presents various applications of MFG with emphasis on the field of wireless communication. In wireless communication, MFG has been used in sensor networks,

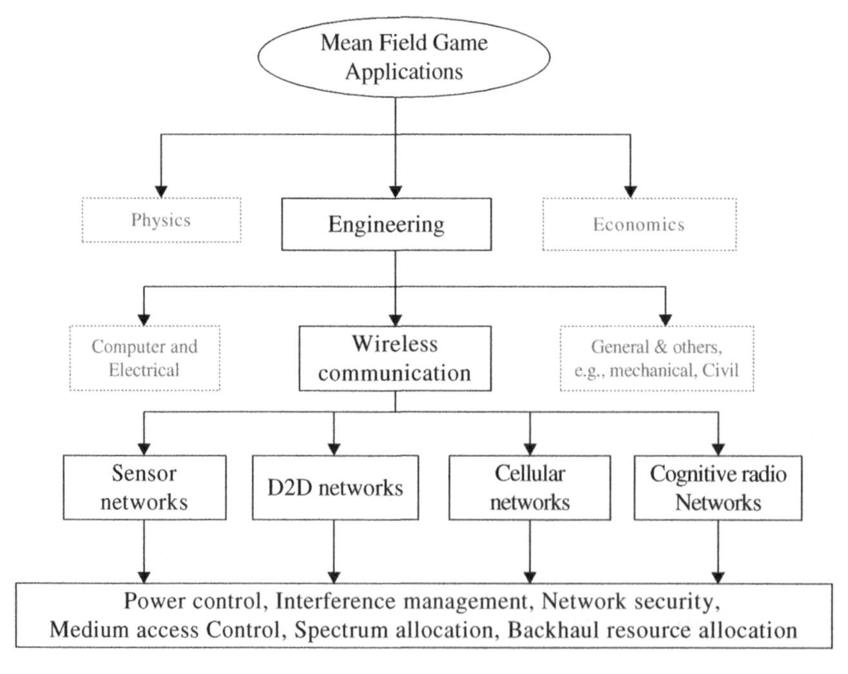

Figure 3.2 Applications of MFG in different fields of studies, with the emphasis on the field of wireless communications.

D2D networks, cellular networks, and cognitive radio networks [49]. MFG is used to address various problems such as power control, interference management, network security, medium access control, spectrum allocation, and backhaul resource allocations.

MFG models are considered to fit inherent characteristics of large-scale multi-agent dynamic systems, and therefore can be applied to address the challenges in such systems [43]. MFG theory can analyze the dynamics of a large number of interacting rational agents in which there is a heavy exchange of information; this is because it can model the interaction of an individual player with the effect of the mass of others (i.e. the mean field) instead of modeling the interaction between each player and every other player. The aim of each player in MFG is to find a strategy that gives maximum utility over a predefined period of time by considering the collective behavior of other players. With MFGs, the resource allocation problems for ultra-dense communication systems can be modeled by describing the behavior of the system using only two differential equations. As a special form of differential games, MFG is the right tool to develop mechanisms for solving the resource allocation problem, because it allows the stochastic nature of the system such as channel dynamics to be taken into consideration. In addition, MFG models allow the derivation of offline algorithms that do not depend on any exchange of information among entities while executing the algorithm. This is possible because the entities are required to collect the needed information only at the initialization stage before algorithm execution. This element makes MFG significantly attractive because it needs limited backhaul/fronthaul connectivity. Various features of MFG for large dynamic systems in comparison to the conventional game's models are summarized in Table 3.2.

3.3.2 Mean Field Games

MFG is a special form of the differential game. A differential game framework adopts the tools, methods, and models of optimal control theory. Optimal control theory has been

Table 3.2 Comparison of various features of MFG and conventional games.

Feature	Mean field games	Conventional games
Scalability	Allow network scalability	Limited network scalability
Convergence to equilibrium	Fast convergence	Slow convergence due to signal overhead and large size of payoff matrix
Computational complexity	Low complexity	Complexity increases with the number of players
Signal overhead	Low	Immerse due to information exchange
Equilibrium efficiency	High	Inefficiency in terms of welfare
Player rationality	Full rationality	Full rationality
Information requirement	Initial distribution of the players' states	Constant exchange of information

developed to obtain the optimal solution of planning problems that involve dynamic systems, where the state evolves over time under the influence of control input. Differential games can be viewed as an extension of optimal control problems in two directions: (i) the evolution of the state is controlled by not one input but multiple inputs with each under the control of a different player and (ii) the objective function is no longer single, for each player has a possible different objective function (payoff or cost), which is dependent over time intervals of interest and relevance to the problem.

MFGs can be expressed as a coupled system of two partial differential equations, the Fokker-Planck-Kolmogorov (FPK) and Hamilton-Jacobi-Bellman (HJB) [44]. Evolving forward in time, the FPK type of equation governs the evolution of the density function of the agents, and the HJB type of equation evolves backward in time governing the computation of the optimal path for each agent.

MFGs have found extensive applications in different fields to formulate the rational and interactive problems with the following characteristics.

(1) Rationality of the players, which is applied in any type of game to ensure that the players can take logical decisions.
(2) The existence of a continuum of the players (i.e. continuity of the mean field) and is generally applied in any type of game to ensure that players can take logical decisions.
(3) Interchangeability of the states among the players (i.e. permutation of the states among the players would not affect the outcome of the game) and each player is infinitesimal. Consequently, the contribution of each player on the mean-field game is infinitesimally small.
(4) Interaction of the players with the mean field. Each player selects its action/strategy according to its own interests, condition/state, and the condition/state of the infinite mass of other players (i.e. mean-field value) who simultaneously select theirs in the same way. In analyzing the game and finding the equilibrium, studying one typical player is sufficient.

1. *Mean field*: Given the state space $s_i(t) = [\mu_i(t), E_i(t)]$, the mean field $m(t, s)$ is defined as

$$m(t,s) = \lim_{N\to\infty} \frac{1}{N} \sum_{i=1}^{N} \mathbb{1}_{\{s_i(t)=s\}}, \tag{3.1}$$

   where $\mathbb{1}$ denotes an indicator function that returns 1 if the given condition is true and zero otherwise. For a given time instant, the mean field is the probability distribution of the states over the set of players.

   The mean field $m = m(t, s)$ is a familiar concept in physics and economics that describes the mass behaviors of many particles. The mean field is a middle term relating the investigated generic player with other players. In fact, the generic player does not care about interactive impacts from others but the mass behavior.
2. *HJB and FPK equations*: If $c(t, x, p)$ is the instantaneous cost function of the generic player at time t using policy p with state x, then the optimal policy designed to minimize the cost function over a period of $[0, T]$ is given as

$$p(t)^{\star} = \arg\min_{p(t)} \int_0^T c(t,x,p)dt + c(x(T), p(T)). \tag{3.2}$$

According to Bellman's optimality principle, an optimal control policy should have the property that for any initial state or initial decision, the remaining decisions must form an optimal policy with regard to the states resulting from the first decision. This value function should satisfy a partial differential equation, which is an HJB equation. The solution of the HJB equation is the value function, which gives the minimum cost for a given dynamic system with an associated cost function. The HJB and FPK equations are derived using the linear function of the cost function and mean field. It is assumed that a Brownian stochastic is involved in the linear dynamics function. The obtained HJB and FPK partial differential equations are given in (3.3) and (3.4), respectively.

$$\partial_t u - \kappa \Delta u + H(s, \nabla u) = V(m), \tag{3.3}$$

$$\partial_t m + \kappa \Delta m + \text{div}\left(m \frac{\partial H(s, \nabla u)}{\partial p} \right) = 0. \tag{3.4}$$

The HJB and FPK partial differential equations interact with each other as shown in Figure 3.3. The classical notion of solution in MFG is given by a pair of maps (u, m), where $u = u(t, x)$ is the value function of a typical small player while $m = m(t, x)$ denotes the density at time t and at position x of the population. The value function u satisfies an HJB, in which m enters as a parameter and describes the influence of the population on the cost of each agent, while the density m evolves in time according to an FPK equation in which u enters as a drift. More precisely, the pair (u, m) is a solution of the MFG system. The HJB equation governs the computation of the optimal path of control of the player, while the FPK equation governs the evolution of the mean field function of players. The HJB and FPK equations are termed as the backward and forward functions, respectively. The combination of both forward propagation and backward propagation in time creates some unusual phenomena in the time variable that is not seen in more conventional games. Backward means that the final value function is known and then we determine what is the value at time $[0, T]$ with the control vector of $u(t)$. Therefore, the HJB equation is always solved backward in time, starting from $t = T$ and ending at $t = 0$. When solved over the whole of the state space, the HJB equation is a necessary and sufficient condition for an optimum. The FPK equation evolves forward with time. The interactive evolution finally leads to the mean field equilibrium.

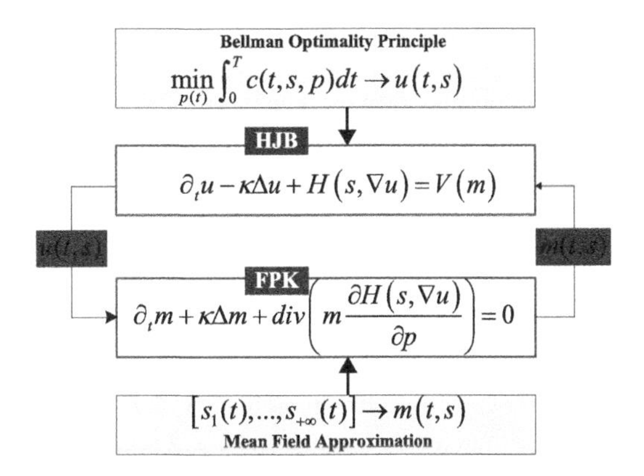

Figure 3.3 The interaction between FPK and HJB partial differential equations.

3. *Shortcomings and limitations of MFGs*: MFGs are also associated with some shortcomings and challenges as well. Designing proper utility functions is a challenging task. The utility function needs to be designed so that it reflects the objective of the system application while at the same time satisfy the requirements of the game model. In addition, since there is no general technique to solve the mean field equations, it is challenging to obtaining the mean field equilibrium. Also, in MFG models, it is assumed that state-interchange among players does not affect the game's outcome. However, such an assumption may be unrealistic in many situations. Sometimes, this assumption can be relaxed through a careful design of the players' utility function so that it only depends on the player's state and the mean field. However, this is not an easy task because the existence and uniqueness of the mean field equilibrium are largely affected by the design of utility functions.

3.4 MFGs for D2D Communications in 5G

In 5G ultra-dense network, terminal devices will be supported by the Internet of Things (IoT) stemming from massive machine-to-machine (M2M), device-to-device (D2D), and vehicle-to-device (V2D) communication. Due to the proximity between devices and frequency reuse, other technical challenges such as intra-tier and inter-tier interferences are introduced. Moreover, because of the energy constraints of the devices, such as sensors, special technologies for energy harvesting are needed.

Because MFG can also incorporate nodes interference and other temporal dynamics, it poses as an appropriate approach for resource management in the 5G ultra-dense networks [42]. In this section, the applications of MFG in 5G ultra-dense D2D networks are surveyed to illustrate the usefulness of the game in the network. In addition, an example is given of how MFG can be applied to manage interference in future networks.

3.4.1 Applications of MFGs in 5G Ultra-dense D2D Networks

Among others, the interference emerges as an old acquaintance with new significance. The interference conditions and the role of aggressor and victim depend to a large extent on the density and the scenario. Because of the heterogeneous nature of 5G UDN, interference is a significant challenge. Figure 3.4 illustrates the interference problem in ultra-dense D2D communication, where devices communicating with each other or with radio access receive and also cause interference to and from the nearby communicating devices. In [50], different kinds of interference management techniques were surveyed from time, frequency, space domain, from the user, network, and a combination of them. Moreover, as pointed out, suitable interference mitigation technology for specific scenarios should largely depend on the interference distribution, which is why the dominant interference ratio (DIR), defined as the ratio between the DI and the rest of the perceived interference, is used.

As a suitable mitigation technique, an MFG framework for energy and interference-aware problem in ultra-dense D2D networks is formulated in [51]. The framework is formulated as a cost minimization problem by considering the effects of both the interference caused by the generic D2D transmitter to others and the interference from all others caused to the generic D2D receiver. To obtain MFG, the distributive iterative

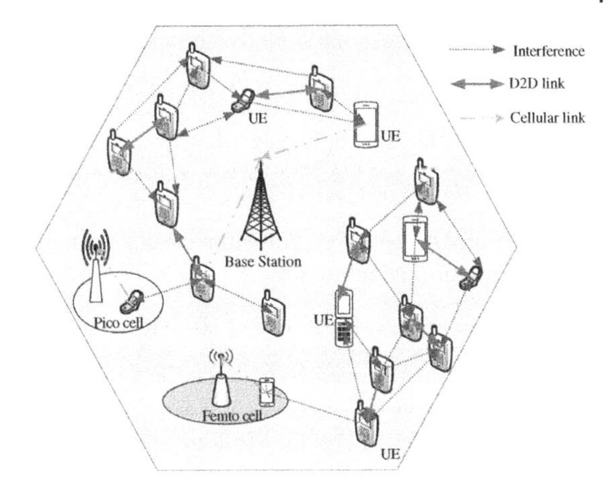

Figure 3.4 Interference in a 5G ultra-dense D2D communication network.

algorithm is derived. The proposed distributed power control policy has proven to optimize both the spectrum and energy efficiency of the network.

3.4.2 An Example of MFGs for Interference Management in UDN

The description of an MFG theoretic framework for ultra-dense D2D networks with the interference mean-field approximation is described as follows. Consider a differential game defined as

$$G_s = \{\Upsilon, \{P_i\}_{i\in N}, \{S_i\}_{i\in N}\{Q_i\}_{i\in N}\{c_i\}_{i\in N}\}, \tag{3.5}$$

where $\Upsilon = \{1...K...N\}$ represents the player set of massive communication pairs. They are rational policymakers, and the number of N is arbitrarily large and even goes to infinity (i.e. $N \to \infty$). $\{P_i\}_{i\in N}$ is a set of possible actions, e.g. transmit powers. Each player determines the action $P_i(t) \in \{P_i\}$ at any time $t \in [0, T]$ to minimize the cost function $\{c_i\}_{i\in N}$. $\{S_i\}_{i\in N}$ is the state space of player i as the interference introduced by other links. (Here the state as the interference from player i to other links can be considered as well.) Control policy $\{Q_i\}_{i\in N}$ is denoted by $\{Q_i(t)\}$, with $t \in [0, T]$. Control policy is designed to minimize the average cost over the time interval T with one-dimensional states.

1. *State dynamics of MFG*: The aggregate interference to player i introduced by other links is defined as follows:

$$\mu_i(t) = \sum_{j=1, j\neq 1}^{N} P_j(t) g_{j,i} \tag{3.6}$$

where $g_{j,i}$ is the channel gain from interference link $j \to i$. To simplify the notation, the equation above is represented as

$$\mu_i(t) = \varepsilon_i(t) P_j(t), \tag{3.7}$$

where

$$\varepsilon_i(t) = \sum_{j=1, j\neq 1}^{N} g_{j,i}. \tag{3.8}$$

Then the interference state dynamics is defined as

$$d\mu_i(t) = \varepsilon_i(t)dP_j(t) + P_j(t)\partial_t\varepsilon_i(t). \tag{3.9}$$

Therefore the state space for player i is defined as

$$s_i(t) = \{\mu_i(t)\}, i = \Upsilon. \tag{3.10}$$

2. *Cost function of MFG*: For interference management in UDN communications, the cost function is defined by

$$c_i(t) = (\gamma_i(t) - \gamma^{th}(t))^2 + \lambda p_i(t). \tag{3.11}$$

   where λ is introduced to balance the units of the achieved SINR difference and the consumed power. It is easy to prove that the cost function is convex with respect to $p_i(t)$. Moreover, $\gamma_i(t)$ is the SINR experienced by player i and γ^{th} is the SINR threshold.
3. *Function of MFG*: The general optimal control problem $Q_i^*(t)$ ranging from $t = 0$ to $t = T$ can be stated as follows:

$$Q_i^*(t) = \underset{P_i(t)}{\text{argmin}}\, E\left[\int_0^T c_i(t)dt + c_i(T)\right] \tag{3.12}$$

   where $c_i(T)$ is the cost at time T. Therefore, the value function is defined as follows:

$$u_i(t, s_i(t) = \underset{P_i(t)}{\text{argmin}}\, E\left[\int_t^T c_i(t)dt + u_i(T, s_i(T))\right] \tag{3.13}$$

   where $u_i(T, s_i(T))$ is a value at the final state $s_i(T)$ at time T.

To formulate the mean field game and solve the mean field equilibrium (MFE), the mean field equation is first defined as given in Eq. (3.1). The mean field approximation method is used to approach the aggregate interference of a device from other devices. The corresponding HJB and FPK differential equations are derived and can be solved using the finite difference method and Lagrange relaxation method to obtain the mean field equilibrium.

Figure 3.5 is an example of the simulation results obtained that describe mean field distribution and power policy at the state after convergence with varying time [51]. The results show that the randomness of the distribution of mean fields and power are introduced by the random interference space. Also the proposed scheme can achieve power equilibrium, which is the final power control policy at the final energy and time interval states after convergence.

3.5 MFGs for Radio Access Network in 5G

The 5G networks will be characterized with ultra-dense, heterogeneous, and dynamic radio access networks. Macrocells, small cells, femtocells, picocells, and unmanned aerial vehicle-mounted base stations (UAV-BSs) are among the radio access that will result in a multi-tier architecture of 5G networks. In the UDNs, typical problems include delay, interference, power control, backhaul resource allocation, energy, cost, handover, spectrum reuse, and limited infrastructure resources. Figure 3.6 illustrates the interference problem in UDN, where small cells (i.e. pico- and femtocells) and

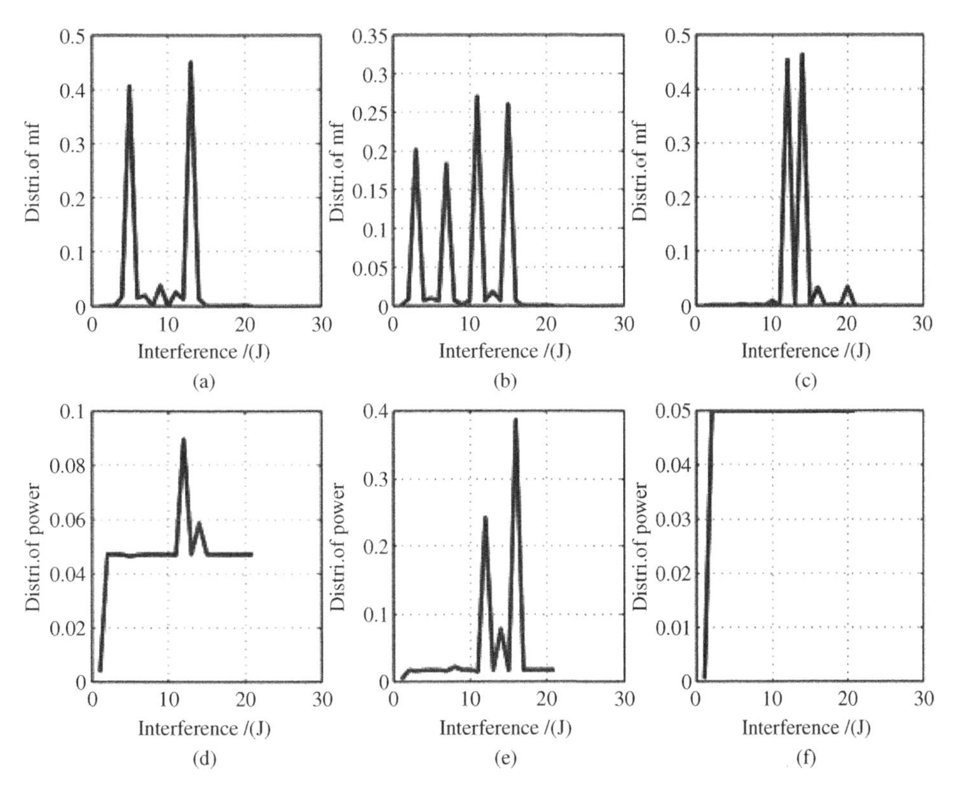

Figure 3.5 Cross-section of mean field distributions and power policy at the states after convergence but varying with time [51].

devices are densely deployed. The small cells coexist with the macro base station via shared spectrum access. Therefore, severe co-tier and cross-tier interference exist and thus impact the overall system performance. Due to the complexity, the utilization of RANs resources in UDNs becomes a significant challenge.

3.5.1 Application of MFGs for Radio Access Network in 5G

To solve the RAN resources management problem, MFG has been used to optimize caching, interference, and energy in RAN.

1. *Caching management*: Caching can improve the network performance and reduce the network traffic by minimizing congestion and workload. MFG has demonstrated to be a potential method for optimizing network performance by providing an advance caching mechanism.

   Using MFGs, authors in [52] studied the distributed caching problem in ultra-dense small cell networks. The base station in this game defines a caching policy that reduces the load on the capacity-limited backhaul links while taking into consideration the storage state of the other base stations to decide on the amount of content to cache. In [53], cellular edge caching control mechanisms, which maximize local caching gain while minimizing the replicated content caching for a massive number of users and small base stations (SBSs), is designed. The algorithm,

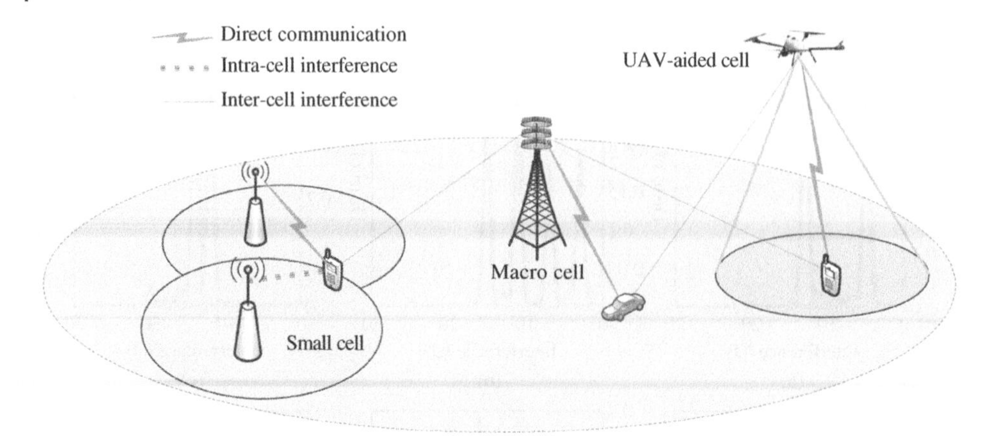

Figure 3.6 Interference in 5G ultra-dense networks.

which is independent of the number of base stations and users while including users' interference, storage constraints, and spatio-temporal demand dynamics, is developed using MFG and stochastic geometry. The proposed ultra-dense edge caching network (UDCN) control algorithm performed better compared to the popularity-based one in terms of average cost and replicated content.

2. *Interference management*: Interference can severely affect the network transmission capacity and performance. Therefore reducing interference can minimize conflict and retransmission of signals, thus reducing the waste of energy, improving the net throughput, and extending network lifetime. The exponential increase in demands of spectral efficiency, data rate, degrees of freedom, and network capacity due to the massive growth of end devices intensify the interference problem. Proper interference management schemes can be developed by utilizing MFG. Figure 3.6 shows the types of interference that UE can experience in 5G networks. UE can be affected by intra-cell interference where the unwanted signal from a nearby cell is received by the communicating UE. UE can also be affected by signal interference from adjacent cells, known as inter-cell interference.

   The inter-cell interference management issue in HetNets is formulated as two nested problems in [54], where the mean-field theory is exploited to help decouple a complex, large-scale optimization problem into a family of localized optimization problems. Thus, each small cell base station can implement its policy by using only its local information and some macroscopic information. The authors in [55] use a mean-field game theoretic approach for the interference problem in hyper-dense HetNets. The performance of two different cost functions for the mean field game formulation is analyzed. Both of these cost functions are designed using stochastic geometry analysis in such a way that the cost functions are valid for the MFG setting. A finite difference algorithm is then developed based on the Lax-Friedrichs scheme and Lagrange relaxation to solve the corresponding MFG. The authors in [56] formulated MFG control policy between cellular users to solve the Nash equilibrium strategy that minimizes a linear combination of the transmitted power and the quality of service (QoS). The localized and non-localized interference optimization problems are formulated to minimize the signal interference level and therefore improve QoS.

3. *Energy management*: In UDNs, modeling the complex interference among a massive number of interacting elements that are simultaneously transmitting signals is difficult. Therefore power control optimization becomes a challenge. However, because MFGs allow the inherent mathematical complexity of dynamic stochastic games, they can be an appropriate approach to UDN energy management.

   For instance, in [57], joint power control and user scheduling are proposed using MFG for optimizing energy efficiency (EE) in ultra-dense small cell networks in terms of bits per unit of energy. The resource management problem is formulated as a dynamic stochastic game between small cell base stations and is solved using the drift plus penalty approach in the framework of Lyapunov optimization. The MFG equilibrium was analyzed with the aid of low-complexity tractable partial differential equations and an equilibrium control policy per SBS, which maximizes the network utility and achieves high gain while ensuring that user quality-of-service is obtained. Using the concepts of MFG and stochastic geometry, authors in [58] modeled a power control scheme in a downlink ultra-dense cellular network where base stations (BSs) are uniformly distributed over a two-dimensional infinite Euclidean plane. In this model, users are distributed independently of BSs, and the BSs with no user within the coverage stays dormant so as to serve energy. The proposed MFG schemes achieve higher energy efficiency of more than 1.5 times compared to a fixed power baseline. In [59], a more tractable MFG is used to study the energy efficiency performance of optimal proactive scheduling strategies in UDN. The proposed method obtained optimal transmission powers that allowed every user request to be served during its required service time while minimizing the total energy consumed.

3.5.2 Energy Harvesting

Energy harvesting is one of the promising solutions for mitigating energy problems for the emerging Internet of Things (IoT)-driven ultra-dense small cell networks [60]. The small cell base stations (SBSs) can harvest energy from environmental sources whereas the macrocell base station (MBS) uses a conventional power supply [61]. When the number of SBSs tends to infinity (i.e. a highly dense network), the centralized scheme becomes infeasible, and therefore the mean field stochastic game is used to obtain a distributed power control policy for each SBS. The amount of energy available at the transmitter will vary according to the amount of the transmitted power and the energy arrival during each transmission interval.

3.5.3 An Example of MFGs for Radio Access Network in 5G

This example considers a scenario where small cells, denoted by SeNBs, in the ultra-dense small cell network provide services to end users, represented as SUEs. For N SeNBs (i.e. players in a game) that fully share a spectrum, the mutual intra-tier interference and power control problem can be formulated using MFG as follows.

1. *State dynamics*: Assuming SeNBi, $i \in \{1.....N\}$ is the generic player i chosen randomly from the N players. The perceived aggregate intra-tier interference of the

generic player i from other players $j \in N, j \neq i$, is given as

$$\mu_i(t) = \sum_{j=1}^{n} p_j(t) g_{j,i}(t) - p_i(t) g_{i,i}(t) \tag{3.14}$$

where $p_j(t)$ is the transmit power of SeNB$_j$ at time t, $p_i(t)$ is the transmit power of any player i, and $g_{j,i}(t)$ and $g_{i,i}(t)$ are the channel gains and stochastic processes, respectively.

The state dynamics s_i of each of the players i are described by the following equations:

$$s_i = d\mu_i(t) = p_j(t) d\left(\sum_{j=1}^{n} g_{j,i}(t)\right) - p_i(t) dg_{i,i}(t) + \xi_i(t) dt \tag{3.15}$$

In this equation, $\xi_i(t)dt$ is the introduced disturbance in state dynamics with the unit variance of the stochastic process $\xi_i(t)$, where $\xi_i(t)$ is the unknown parameter denoting an unknown disturbance entering into the dynamics at time t.

2. *Cost function*: With the defined system state dynamics $s(t) = [s_i(t)]$, the cost function $c(t, s, p)$ is minimized by the optimal power control of $p(t) = [p_i(t)]$ for the generic player i. With aggregate interference $\mu_i(t)$, the signal-to-noise ratio (SINR) of the generic player i is defined as

$$\gamma_i(t) = \frac{p_i(t) g_{i,i}(t)}{\mu_i(t) + \sigma^2(t)} \tag{3.16}$$

where $\sigma^2(t)$ represents the background noise power at time i. Each player i wants to determine the optimal power $p_i(t)$ that minimizes the total cost c_i given as

$$c_i(t, \mu_i(t), p_i(t)) = [p_i(t) g_{i,i}(t) - \gamma^{th}(\mu_i(t) + \sigma^2(t))]^2. \tag{3.17}$$

This cost function is derived from the inequality $\gamma_i(t) \geq \gamma^{th}$, where γ^{th} is the introduced threshold identical to SINR. It is the minimum SINR requirement predefined to meet the basic communication requirements. The cost function $c_i(t, \mu_i(t), p_i(t))$ is concave with constant interference $\mu_i(t)$ at any time t with respect to the power $p_i(t)$. Implementation of the optimal power control over a period of $[0, T]$ should fully consider the dynamics of the perceived interference $\mu_i(t)$ at any time t. Equation (3.16) presents the dynamic of the perceived interference in which the disturbance $\xi_i(t)$ is introduced. Therefore, the cost function of generic player i over $[0, T]$ is given as follows:

$$L_i(p_{i,} \mu_i) = \int_0^T c_i(t, \mu_i(t), p_i(t)) dt + c_i(T), \tag{3.18}$$

where $c_i(T)$ is the cost at the final time T. To make sure that the cost is finite when the disturbance is under the worst case, the integral of $\xi_i^2(t)$ is constrained. $\xi_i(t)$ is constrained by taking the ratio between L_i and the summation of $\xi_i^2(t)$ as follows:

$$\frac{L_i(p_{i,} \mu_i)}{\xi_i^2 + c_i(0)} \leq \rho^2, \tag{3.19}$$

where $c_i(0)$ is the cost at $t = 0$ and ρ measures the robustness level. Assuming $c_i(0) = 0$, the cost function can be written as

$$J_i^p(p_{i,} \mu_i, \xi_i) = L_i(p_{i,} \mu_i) - \rho^2 \int_0^T \xi_i^2(t) dt. \tag{3.20}$$

3. *Optimal control problem*: With the state space $s_i(t)$, the optimal power control policy $Q_i^*(t)$ will be determined by the generic player i to minimize the cost function $J_i^p(p_{i,}\mu_i, \xi_i)$ during the time interval $[0, T]$:

$$Q^*(t) = \arg\min_{p_i(t)} \max_{\xi_i(t)} \mathrm{E}[J_i^p(p_{i,}\mu_i, \xi_i)]. \tag{3.21}$$

With the defined robust cost function $J_i^p(p_{i,}\mu_i, \xi_i)$ in Eq. (3.18) and state dynamics $d\mu_i(t)$ in Eq. (3.15) the stochastic game problem is represented as $\arg\min_{p_i(t)} \max_{\xi_i(t)} \mathrm{E}[J_i^p(p_{i,}\mu_i, \xi_i)]$. This is the value of the function represented as

$$u(t, s(t)) = \arg\min_{p_i(t)} \max_{\xi_i(t)} \mathrm{E}[J_i^p(p_{i,}\mu_i, \xi_i)], t \in [0, T]. \tag{3.22}$$

The value function in Eq. (3.22) meets and gives a solution to the HJB partial differential equation according to Bellman's optimality principle following the optimal control theory. The value also gives the minimum cost value for the given dynamic system.
With the defined states $s_i(t)$ in Eq. (3.16), the corresponding FPK equation which satisfies the mean field $m(t, s)$ is defined. The derived FPK and HJB equations are combined to represent the MFG.

Figure 3.7 is an illustration of the distribution of the mean field with respect to the time and interference state of a generic player where Femtocells were used as SeNBs. It can be seen that the tendency with respect to time is rising or decreasing between

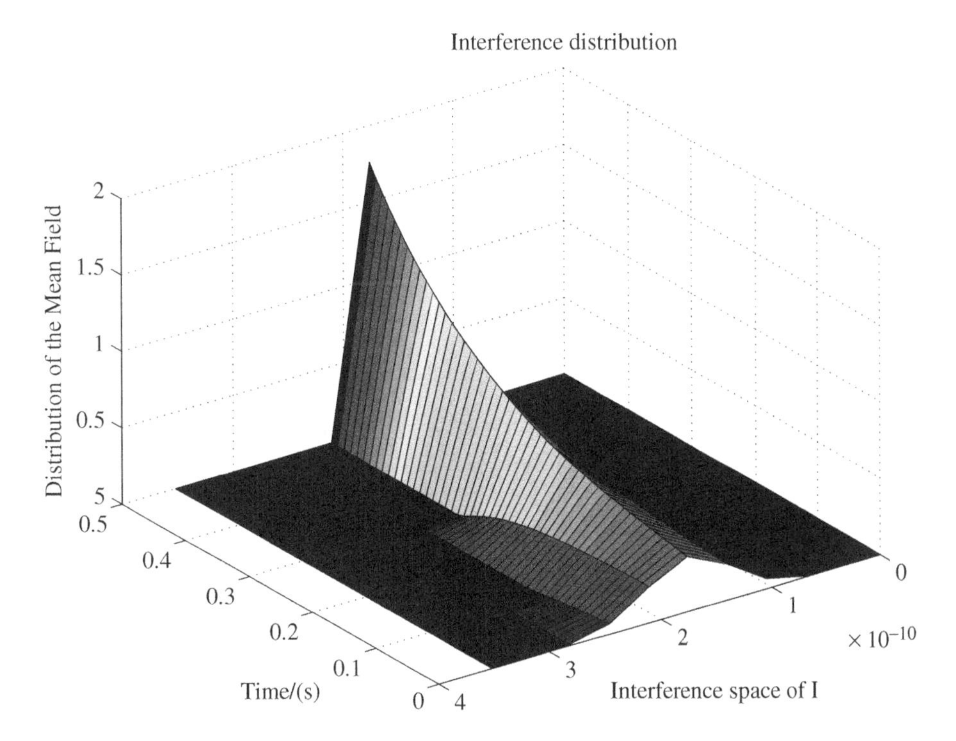

Figure 3.7 Distribution of the mean field at equilibrium [62].

different constant interference state values, which can be proved by using the iterative function of the mean field.

3.6 MFGs in 5G Edge Computing

The Internet of Things (IoT) is a promising paradigm, which integrates a large number of heterogeneous and pervasive objects with different connecting and computing capabilities, aimed at providing a view of the physical world through the network. The variety of devices that are capable of sending data about the real world to the Internet is ever increasing. As a result, the IoT can rely on a large range of objects, from dummy entities capable of providing just their positions (through attached tags) to objects capable of sensing the status of an environment, processing the data, and sending the results to the cloud, if believed meaningful. This is making the range of possible applications increase even more rapidly, so much so nowadays that our society can rely on powerful tools for introducing intelligence into living environments.

From a pervasive system perspective, the vision of the IoT suggests the integration between ubiquitous computing and wireless communications targeting reliable connectivity of things, i.e. computers, sensors, and everyday objects equipped with transceivers. From a backhaul bandwidth, network resource sharing, and optimization perspective, the cloud-based processing and radio access network virtualization provide a revolutionary approach toward balancing the degree of centralization of physical and virtual resources management. Figure 3.8 represents an example of 5G network architecture, where the cloud is extended to the network edge for cooperative networking and radio resource management for UDN to improve network capacity, edge content distribution, and data processing.

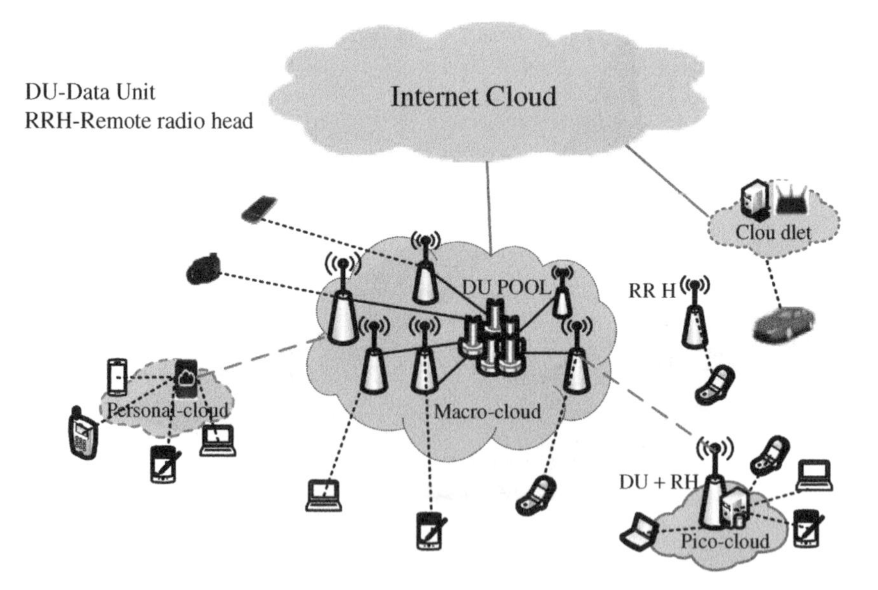

Figure 3.8 Example of edge cloud in 5G network architecture.

The concept of C-RAN is already being established as a potential mobile network architecture evolution. The concept can be expanded to include Radio Access and also network core functionalities, which is referred to as Network Visualization [40]. Network Visualization refers to the capability of pooling logical elements or underlying physical resources in a network and takes advantage of the Cloud Services and Software-Defined Networking technologies.

3.6.1 MFG Applications in Edge Cloud Communication

In the future 5G network, the cloud is expected to expand to the network edge [63]. In addition to the C-RAN, portable devices can become networking nodes instead of being just the convectional end-point that terminates the communication link. A personal cloud, which is the collection of electronic contents and services that are accessible from any device, will be formed. For example, in a personal cloud, a large number of wearable devices can form a local network cluster with the portable devices. As shown in Figure 3.8, personal clouds, macro-clouds, pico-clouds, C-RANs, and cloudlets will be densely deployed in the future network. Therefore, MFG can be utilized to address some of the challenges in such ultra-dense cloud-based networks.

To address the question of how to share limited capacity more efficiently, authors in [64] propose cloud-based networking. Using MFG concepts, the last level of cache sharing problems is considered in large-scale cloud networks. Assuming an infinite number of users wanting to access the cloud network, the closed-form expression of the optimal pricing that gives an efficient resource-sharing policy with the assumption of a weighted norm of action is provided. In the scheme, each player updates the learning strategy using a developed distributed iterative learning algorithm.

3.7 Conclusion

Ultra-dense networks have been promisingly recognized as the most effective way to enhance both the spectrum and the energy efficiency. However, many challenges arose due to the massive connections and billions of devices in the network. Among the difficulties is resource management. The mean field game has found extensive applications in modeling, analyzing, and designing ultra-dense networks. MFG has been applied to address resource management in ultra-dense D2D, small cells, and cloud-based networks.

Bibliography

1 Ericsson, "Ericsson Mobility Report: On the Pulse of the Networked Society," Nov. 2016.
2 "Cisco Visual Networking Index: Forecast and Methodology, 2016–2021," White Paper, 2017.
3 D. Lopez-Perez, M. Ding, H. Claussen, and A. H. Jafari, "Towards 1 Gbps/UE in cellular systems: understanding ultra-dense small cell deployments," *IEEE Communications Surveys & Tutorials*, vol. 17, no. 4, pp. 2078–2101, Fourth quarter 2015.

4 M. Kamel, W. Hamouda, and A. Youssef, "Ultra-dense and networks: a survey," *IEEE Communications Surveys & Tutorials*, vol. 18, no. 4, pp. 2522–2545, 2016.
5 Imt-2020 (5G) Promotion Group, "IMT Vision towards 2020 and beyond," Feb. 2014.
6 Nokia, "Ultra Dense Network (UDN)," White Paper, 2016.
7 S. Chen, F. Qin, B. Hu, X. Li, and Z., Chen, "User-centric ultra-dense networks for 5G: challenges, methodologies, and directions," *IEEE Wireless Communications*, vol. 23, no. 2, Apr. 2016.
8 J.G. Andrews, S. Singh, Q. Ye, X. Lin, and H.S. Dhillon, "An overview of load balancing in HetNets: old myths and open problems," *IEEE 21 Wireless Communications*, no. 2, pp. 18–25, Apr. 2014.
9 P. Semasinghe, K. Zhu, and E. Hossain, "Distributed resource allocation for self-organizing small cell networks: an evolutionary game approach," *Proc. Workshop Heterogeneous Small Cell Netw. (HetSNets) Conjunction IEEE Global Telecommun. Conf.*, 2013, pp. 702–707.
10 P.V. Klaine, M.A. Imran, O. Onireti, and R.D. Souza, "A survey of machine learning techniques applied to self-organizing cellular networks," *IEEE Communications Surveys and Tutuorials*, vol 19, no. 4, pp. 2392–2431, Jul. 2017.
11 A. Gupta and R.K. Jha, "A survey of 5G network: architecture and emerging technologies," *IEEE Access*, vol. 3, pp. 1206–1232, July 2015.
12 ITU-R M.2083, "IMT Vision-Framework and overall objectives of the future development of IMT for 2020 and beyond," Oct. 2015.
13 P. Agyapong, M. Iwamura, D. Staehle, W. Kiess, and A. Benjebbour, "Design considerations for a 5G network architecture," *IEEE Communications Magazine*, vol. 52, no. 11, pp. 65–75, Nov. 2014.
14 I.F. Akyildiz, S. Nie, S.C. Lin, and M. Chandrasekaran, "5g roadmap: 10 key enabling technologies," *Computer Networks*, vol.106, pp. 17–48, 2016.
15 E. Hossain and M. Hasan, "5G Cellular: Key Enabling Technologies and Research Challenges," *arXiv:1503.00674*, Feb. 2015.
16 S. Talwar, D. Choudhury, K. Dimou, E. Aryafar, B. Bangerter, and K. Stewart, "Enabling technologies and architectures for 5G wireless," *Proc. IEEE MTT-S Int. Microw. Symp.*, pp. 1–4, 2014.
17 A.G. Gotsis, S. Stefanatos, and A. Alexiou, "Ultra dense networks: the new wireless frontier for enabling 5G access," *EEE Vehicular Technology Magazine*, vol. 11, no. pp. 71–78, Jun. 2016.
18 N. Bhushan et al., "Network densification: the dominant theme for wireless evolution into 5G," *IEEE Commun. Mag.*, vol.52, no. 2, pp. 82–89, Feb. 2014.
19 X. Ge, S. Tu, G. Mao, C.X. Wang, and T. Han, "5G ultra-dense cellular networks," *IEEE Wireless Communications*, vol. 23, no. 1, pp. 72–79, Mar. 2016.
20 U.N. Kar and D.K. Sanyal, "An overview of device-to-device communication in cellular networks," *ICT Express*, 2017.
21 O. Bello and S. Zeadally, "Intelligent device-to-device communication in the Internet of Things," *IEEE Systems Journal*, vol. 10, no. 3, pp. 1172–1182, Sept. 2016.
22 S. Wang, X. Zhang, Y. Zhang, L. Wang, J. Yang, and W. Wang, "A survey on mobile edge networks: convergence of computing, caching and communications," *IEEE Access*, vol.5, pp. 6757–6779, Feb. 2017.
23 R.I. Ansari et al., "5G D2D networks: techniques, challenges, and future prospects," *IEEE Systems Journal*, vol. PP, no. 99, pp. 1–15, Dec 2017.

24 C. Hoymann et al., "Relaying operation in 3GPP LTE: challenges and solutions," *IEEE Commun. Magazine*, vol. 50, no. 2, pp. 156–162 Feb. 2012.
25 M.N. Tehrani, M. Uysal, and H. Yanikomeroglu, "Device-to-device communication in 5G cellular networks: challenges, solutions and future directions," *IEEE Communications Magazine*, pp. 86–92, May 2014.
26 J. Qiao, X.S. Shen, J.W. Mark, Q. Shen, Y. He, and L. Lei, "Enabling device-to-device communications in millimeter-wave 5G cellular networks," *IEEE Communications Magazine*, vol.53, no. 1, pp. 209–215, Jan. 2015.
27 J. Lianghai, B. Han, M. Liu, and H.D. Schotten, "Applying device-to-device communication to enhance IoT services," *IEEE Communications Standards Magazine*, vol. 1, no. 2, pp. 85–91, Jul. 2017.
28 S. Krajnovic, "5G and radio access network architecture," in *International Conference on Telecommunications (ConTEL)*, Zagreb, Croatia, Jun. 2017.
29 P. Marsch, et al., "5G radio access network architecture: design guidelines and key considerations," *IEEE Communications Magazine*, vol. 54, no. 11, pp. 24–32, Nov. 2016.
30 X. Gelabert, P. Legg, and C. Qvarfordt, "Small cell densification requirements in high capacity future cellular networks," in *IEEE International Conference on Communication Workshop (ICCW)*, Jun. 2013.
31 N.H.M. Adnan, I.M. Rafiqul, and A.H.M.Z. Alam, "Massive MIMO for Fifth Generation (5G): opportunities and challenges," in *International Conference on Computer and Communication Engineering (ICCCE)*, July 2016, Kuala Lumpur, Malaysia.
32 M. Giordani, M. Mezzavilla, and M. Zorzi, "Initial access in 5G mmWave cellular networks," *IEEE Communications Magazine*, vol. 54, no. 11, pp. 40–47, Nov. 2016.
33 Y. Ai, M. Peng, and K. Zhang, "Edge cloud computing technologies for Internet of Things: a primer," *Digital Communications and Networks*, Jul. 2017
34 F.Z. Yousaf, M. Bredel, S. Schaller, and F. Schneider, "NFV and SDN-Key technology enablers for 5G networks," *IEEE Journal on Selected Areas in Communications*, vol. 35, no. 11, pp. 2468–2478, Nov. 2017.
35 X. Wang, et al., "Virtualized Cloud Radio Access Network for 5G Transport," *IEEE Communications Magazine*, vol. 55, no. 9, pp. 202–209, Sept. 2017.
36 T.X. Tran, P. Pandey, A. Hajisami, and D. Pompili, "Collaborative multi-bitrate video caching and processing in Mobile-Edge Computing networks," in *13th Annual Conference on Wireless On-demand Network Systems and Services (WONS)*, pp. 165–172, 2017.
37 D. Pompili, A. Hajisami, and H. Viswanathan, "Dynamic provisioning and allocation in Cloud Radio Access Networks (C-RANs)," *Ad Hoc Networks*, vol.30, pp. 128–143, Mar. 2015.
38 N. Al-Falahy and O.Y. Alani, "Technologies for 5G Networks: challenges and opportunities," *IT Professional*, vol.19, no. 1, pp. 12–20, Feb. 2017.
39 J. Pedreno-Manresa, P.S. Khodashenas, M.S. Siddiqui, and P. Pavon-Marino, "On the need of joint bandwidth and NFV resource orchestration: a realistic 5G Access Network use case," *IEEE Communications Letters*, vol. 22, no. 1, pp. 145–148, Jan. 2018.
40 N. Zhang, N. Cheng, A.T. Gamage, K. Zhang, J.W. Mark, and X. Shen, "Cloud assisted HetNets toward 5G wireless networks," *IEEE Communications Magazine*, vol. 53, no. 6, pp. 59–65, Jun. 2015

41 O. Sallent, J. Perez-Romero, R. Ferrus, and R. Agusti, "On radio access network slicing from a radio resource management perspective," *IEEE Wireless Communications*, vol. 24, no. 5, pp. 166–174, Oct. 2017.
42 C. Yang, J. Li, Min Sheng, A. Anpalagan, and J. Xiao. "Mean field game-theoretic framework for interference and energy-aware control in 5G ultra-dense networks," *IEEE Wireless Communications Magazine*, vol. PP, no. 99, pp. 1–8, Sept. 2017.
43 P. Semasinghe, S. Maghsudi, and E. Hossain, "Game theoretic mechanisms for resource management in massive wireless IoT systems," *IEEE Communications Magazine*," vol. 55, no. 2, pp. 121–127, Feb. 2017.
44 J.-M. Lasry and P.L. Lions, "Mean field games," *Japanese Journal of Mathematics*, vol.2, pp. 229–260, 2007.
45 J.M. Lasry and P.L. Lions, "Mean field games. 1. The stated case," *Comptes Rendus Mathematique*, vol. 343 no. 9, pp. 619–625, Nov. 2006.
46 M. Huang, R.P. Malhamé, P.E. Caines, et al., *"Communications in Information and Systems*," vol. 6, p. 221, 2006.
47 O. Gueant, J.M. Lasry, and P.L. Lions, "Mean field games and applications," in *Paris-Princeton Lectures on Mathematical Finance 2010. Lecture Notes in Mathematics*, vol. 2003, pp. 205–266, Berlin, Heidelberg: Springer, 2011.
48 D. Gomes, L. Lafleche, and L. Nurbekyan, "A mean-field game economic growth model," *American Control Conference (ACC)*, Jul. 2016.
49 B. Djehiche, A.T. Siwe, and H. Tembine, "Mean-Field-Type Games in Engineering," *arXiv:1605.03281*, Nov. 2017.
50 B. Soret, K. Pedersen, N.T.K. Jrgensen, and F.-L. Vctor, "Interference coordination for dense wireless networks," *IEEE Communications Magazine*, vol.53, no. 1, pp. 102–109, 2015.
51 C. Yang, J. Li, P. Semasinghe, E. Hossain, S.M. Perlaza, and Z. Han, "Distributed interference and energy-aware power control for ultra-dense D2D networks: a mean field game," *IEEE Transactions on Wireless Communications*, vol. 16, no. 2, Feb. 2017.
52 K. Hamidouche, W. Saad, M. Debbah, and H.V. Poo, "Mean-field games for distributed caching in ultra-dense small cell networks," *American Control Conference (ACC)*, pp. 4699–4704, Jul. 2016.
53 H. Kim, J. Park, M. Bennis, S. Kim, and M. Debbah, "Ultra-dense edge caching under spatio-temporal demand and network dynamics," *IEEE International Conference on Communications (ICC)*, May 2017.
54 A.Y. Al-Zahrani, F.R. Yu, and M. Huang, "A joint cross-layer and co-layer interference management scheme in hyper-dense heterogeneous networks using mean-field game theory," *IEEE Transactions on Vehicular Technology*, vol. 65, no. 3, pp. 1522–1535, Mar. 2015.
55 P. Semasinghe and E. Hossain, "Downlink power control in self-organizing dense-small cells underlaying macrocells: a mean field game," *IEEE Transactions on Mobile Computing*, vol. 15, no. 2, pp. 350–363, Feb. 2015.
56 M. Aziz, and P.E. Caines, "A mean field game computational methodology for decentralized cellular network optimization," *IEEE Transactions on Control Systems Technology*, vol. 25, no. 2, pp. 563–576, Mar. 2017.

57 S. Samarakoon, M. Bennis, W. Saad, M. Debbah, and M. Latva-aho, "Ultra dense small cell networks: turning density into energy efficiency," *IEEE Journals of Selected Areas in Communication*, vol. 34, no. 5, May 2016.

58 J. Park, S. Kim, M. Bennis, and M. Debbah, "Spatio-temporal network dynamics framework for energy-efficient ultra-dense cellular networks," *IEEE Global Communications Conference (GLOBECOM)*, Dec. 2016.

59 M. de Mari, E.C. Strinati, M. Debbah, and T.Q.S. Quek, "Joint stochastic geometry and mean field game optimization for energy-efficient proactive scheduling in ultra dense networks," *IEEE Transactions on Cognitive Communications and Networking*, vol. 3, no. 4, pp. 766–781, Dec. 2017.

60 S. Maghsudi and E. Hossain, "Distributed Cell Association for Energy Harvesting IoT Devices in Dense Small Cell Networks: A Mean-Field Multi-Armed Bandit Approach," *arXiv:1605.00057v1*, Apr. 2016.

61 T.K. Thuc, E. Hossain, and H. Tabassum, "Downlink power control in two-tier cellular networks with energy-harvesting small cells as stochastic games," *IEEE Transactions on Communications*, 2015.

62 C. Yang, H. Dai, Y. Zhang, J. Li, and Z. Han, "Distributed interference-aware power control in ultra-dense small cell networks: a robust mean field game," *IEEE Access*, vol. 6, pp. 12608–12619, Jan. 2018.

63 Q. Li, H. Niu, A. Papathanassiou, and G. Wu, "Edge Cloud and underlay networks: empowering 5G cell-less wireless architecture," in *Proceedings of European Wireless 2014; 20th European Wireless Conference*, May 2014, Barcelona, Spain.

64 A.F. Hanif, H. Tembine, M. Assaad, and D. Zeghlache, "Mean-field games for resource sharing in Cloud-based networks," *IEEE/ACM Transactions on Networking*, 2015.

Part II

Ultra-dense Networks with Emerging 5G Technologies

4

Inband Full-duplex Self-backhauling in Ultra-dense Networks

Dani Korpi[1], Taneli Riihonen[2] and Mikko Valkama[2]

[1] *Nokia Bell Labs, Espoo, Finland*
[2] *Laboratory of Electronics and Communications Engineering, Tampere University of Technology, Finland*

4.1 Introduction

Traditionally, the backhauling of data in cellular networks has been handled by connecting the base station (BS) or access node (AN) to a core network via a physical cable. This ensures high data rates for the backhaul link, but requires the installation of cables, which entails a high cost. This is an especially significant issue in the ultra-dense networks, where the number of ANs is too high for a physical backhaul link to be commercially feasible. To this end, the prospect of wireless backhaul connections has been brought up. It would mean that no cables were required, making the corresponding cellular networks more easily scalable to higher densities.

In the current systems, the basic principle behind the wireless backhaul links has been to ensure line-of-sight (LOS) and to use a center frequency different to that utilized in the actual access link between the AN and the user equipments (UEs). Although some of the benefits of wireless connectivity can still be obtained with such an approach, careful planning and additional spectral resources are required. Together, these aspects reduce the cost-efficiency of the wireless backhauling solution, hindering the commercial feasibility of utilizing such an approach in very densely deployed networks.

To this end, one of the paradigms of the upcoming 5G standard is to integrate the access and backhaul links [1]. Therefore, the same spectral resources and radio access technology could be used for serving the UEs as well as for backhauling the data. This also facilitates non-line-of-sight (NLOS) backhaul connections, while not requiring any additional frequencies. Such self-backhauling radio access systems are an integral part of implementing commercially feasible ultra-dense networks as they significantly reduce the cost of the backhaul links.

In this chapter, we investigate how to utilize the recently developed inband full-duplex (IBFD) technology to further improve the efficiency of wireless self-backhauling. In particular, we consider a scenario where one macro BS serves several densely deployed ANs, each of which serves an individual UE. The ANs use a wireless link to backhaul the data from the UEs to the BS and vice versa. Three different self-backhauling strategies are evaluated under Quality-of-Service (QoS) requirements for the UEs by determining the minimal transmit powers with which the QoS requirements can be fulfilled. Two of these

Ultra-dense Networks for 5G and Beyond: Modelling, Analysis, and Applications,
First Edition. Edited by Trung Q. Duong, Xiaoli Chu and Himal A. Suraweera.

strategies rely on the IBFD technology, while one of them uses time-division duplexing (TDD) to communicate in a traditional half-duplex (HD) manner. In addition, these AN-based solutions are compared to a HD reference scheme where the BS communicates directly with the UEs.

4.2 Self-backhauling in Existing Literature

Several recent works have considered wireless inband self-backhauling as a possible option for decreasing the cost of the future ultra-dense cellular networks of the future [1–11]. As mentioned, the cost savings are incurred by enabling the AN to backhaul all the data with a macro BS without requiring any wired data link. While some works propose performing the wireless backhauling on a separate frequency band at the mmWave frequencies [10, 11], backhauling the data on the same frequency channel as the downlink (DL) and uplink (UL) transmissions will significantly reduce the overall cost of the radio system. What is more, combining this concept with the IBFD technology will result in various alternative solutions for performing the wireless backhauling, some of which could further improve the spectral efficiency of such a network. This makes IBFD-based self-backhauling an intriguing concept for the future 5G systems, as it facilitates higher data rates while also reducing the associated overall costs.

Thus far, most works have considered a relay-type AN that is directly forwarding the signals transmitted by the UL UEs to the BS, or vice versa [2, 3, 5–8, 12–15]. The reason for the popularity of this type of scheme is likely due to the fact that such a relay-type AN is more or less directly compatible with the existing networks, as it would essentially just extend the range of the macro BS. This, on the other hand, will obviously result in increased data rates and better coverage.

In particular, in [12], the power control of such a relay-type AN is investigated and the performance of both HD and IBFD operation modes is then compared. The obtained results indicate that the IBFD AN is capable of obtaining higher throughputs than the corresponding HD system, although a certain amount of self-interference (SI) suppression is obviously required. A similar analysis is performed in [2], where the BS is assumed to have a massive antenna array. There, the optimal power allocation for the BS and the AN is solved iteratively. The work in [7], on the other hand, investigates different beamforming solutions for a BS with massive antenna arrays, although no IBFD operation is assumed in any of the nodes therein.

Moreover, the effect of radio resource management (RRM) on the performance of the relay-type AN is investigated in [6]. There, the RRM tools are used to balance the SI with the other sources of interference, and the resulting solution is shown to outperform the HD benchmark scheme. In [13], the spectral efficiency of a similar system is maximized by solving the optimal power allocation for both IBFD and HD AN. While the power allocation is solved in closed form for the HD case, only an algorithm is proposed for optimizing the transmit powers of the IBFD scenario. Also the IBFD solution is shown to outperform the corresponding HD case. The work in [4], on the other hand, maximizes the user rates for ANs utilizing frequency-division duplexing (FDD) by optimizing the bandwidth allocation of the backhaul link. A somewhat similar analysis is carried out in [5], where the spectrum allocation is optimized for both half-duplex and full-duplex ANs. The results obtained therein indicate that, with sufficient SI cancellation, the full-duplex ANs outperform their half-duplex counterparts.

The DL coverage of a relay-type self-backhauling AN is then analyzed in [3] and [8]. The findings in [8] indicate that, while the throughput of the network with IBFD-capable ANs is almost doubled in comparison to the HD systems, the increased interference levels result in a somewhat smaller coverage. The results obtained in [3] suggest, on the other hand, that on a network level it may be better to have also some ANs that perform the self-backhauling on a different frequency band. This somewhat reduces the interference between the different backhaul links and the DL UEs. Finally, in [15], the throughput and outage probability of a relay-type IBFD AN is analyzed under an antenna selection scheme, where individual transmit (TX) and receive (RX) antennas are chosen at the AN based on a given criterion. Again, the IBFD AN is shown to usually outperform the corresponding HD AN, although this is not the case under all channel conditions.

In this chapter, we investigate different self-backhauling strategies for ultra-dense networks. Namely, as discussed earlier, we consider four different network architectures, three of which utilize ANs to relay the traffic between the UEs and the BS. Moreover, two of the strategies rely on the IBFD technology: one utilizes IBFD-capable ANs while the other assumes an IBFD-capable BS. Of these two, the former corresponds to the relay-type scenario mostly considered in the earlier literature. Therefore, this chapter will comprehensively evaluate the suitability of such a backhauling strategy in the context of ultra-dense networks by comparing it to various alternative solutions.

4.3 Self-backhauling Strategies

Herein, we describe the three different strategies for serving the UEs with an intermediate AN that is backhauling the data with a macro BS. For reference, we consider also a basic scheme where the BS is directly serving the UEs in a conventional macrocell fashion. Moreover, in two of the backhauling strategies, either the BS or the ANs must be IBFD capable, while one of the schemes has no such requirements. Table 4.1 provides a high-level description of each self-backhauling strategy, while the sections below give further details.

In the analysis presented in this chapter, the BS is assumed to have massive antenna arrays, alongside with perfect channel state information (CSI). The assumption of perfect CSI is obviously optimistic, but it allows the derivation of analytical data rate expressions that provide information about the ultimate performance limits of the considered system. Namely, this assumption means that, apart from SI, none of the signals received or transmitted by the BS interfere with each other, which represents a best-case scenario. Nevertheless, the effect of residual SI is still considered, as no full knowledge of the SI coupling channel is assumed. Furthermore, the ANs and the UEs will generate significant mutual interference as each of them has only a single antenna.

In this chapter, we assume that the wireless system employs TDD to separate the transmissions when operating in half-duplex mode. However, the analysis would be identical if FDD was utilized. Therefore, the results presented herein apply also to FDD systems as long as the division of frequencies between the two communication directions can be adjusted.

Table 4.1 Summary of the considered self-backhauling strategies, where it is shown whether the node transmits (TX), receives (RX), or is idle (–) within the time slot in question.

Strategy	Time slot 1					Time slot 2				
Half-duplex BS without ANs	**BS**	**DL ANs**	**UL ANs**	**DL UEs**	**UL UEs**	**BS**	**DL ANs**	**UL ANs**	**DL UEs**	**UL UEs**
	TX	n/a	n/a	RX	–	RX	n/a	n/a	–	TX
Half-duplex BS with half-duplex ANs	**BS**	**DL ANs**	**UL ANs**	**DL UEs**	**UL UEs**	**BS**	**DL ANs**	**UL ANs**	**DL UEs**	**UL UEs**
	TX	RX	RX	–	TX	RX	TX	TX	RX	–
Full-duplex BS with half-duplex ANs	**BS**	**DL ANs**	**UL ANs**	**DL UEs**	**UL UEs**	**BS**	**DL ANs**	**UL ANs**	**DL UEs**	**UL UEs**
	TX+RX	RX	TX	–	–	–	TX	RX	RX	TX
Half-duplex BS with full-duplex ANs	**BS**	**DL ANs**	**UL ANs**	**DL UEs**	**UL UEs**	**BS**	**DL ANs**	**UL ANs**	**DL UEs**	**UL UEs**
	TX	TX+RX	–	RX	–	RX	–	TX+RX	–	TX

4.3.1 Half-duplex Base Station without Access Nodes

In general, the considered scenario involves a macro BS that has massive antenna arrays at its disposal and is exchanging data with UEs both in UL and DL. Therefore, an obvious reference solution is the case where a half-duplex-capable BS serves these UEs directly itself, without any intermediate nodes. This strategy is depicted in Figure 4.1, where both the DL and UL time slots are shown. All of the alternative solutions presented in this section will be compared to this basic reference scheme.

4.3.2 Half-duplex Base Station with Half-duplex Access Nodes

In all of the forthcoming alternative strategies, the basic idea is to introduce so-called ANs into the network to act as intermediate nodes between the UEs and the BS. Moreover, deploying the ANs densely will ensure that each UE is close to at least one AN, resulting in reduced path loss. Figure 4.2 illustrates such a backhauling strategy, where all the nodes are legacy half-duplex radio devices. This strategy is particularly suitable for very lost-cost deployments, where only half-duplex devices are available.

In principle, using the ANs to relay the traffic of the access links results in reduced TX powers for all parties, especially so as the dense deployment of ANs ensures smaller path losses. However, the challenge of this type of scheme are the various sources of interference. Namely, since all the nodes use the same frequency band, they will also interfere with each other when transmitting and receiving simultaneously. The BS avoids this as it has large antenna arrays that facilitate accurate beamforming, but the ANs and the UEs have no such benefit. Therefore, referring to Figure 4.2, in the first time slot the UL UEs interfere with the reception of the DL ANs, while in the second time slot the DL ANs interfere with each other and the DL UEs. In order to manage these various interference sources while also fulfilling the QoS requirements, careful transmit power allocation is needed. Such power allocation is especially crucial in ultra-dense networks where the significance of interference is much larger. This is investigated in more detail in Section 4.4.

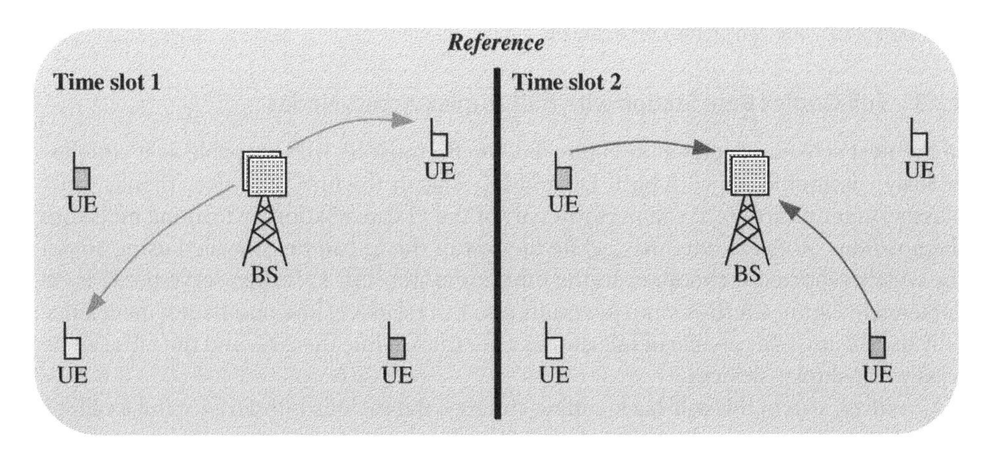

Figure 4.1 The reference scenario, where the BS serves the UEs directly using TDD to separate UL and DL.

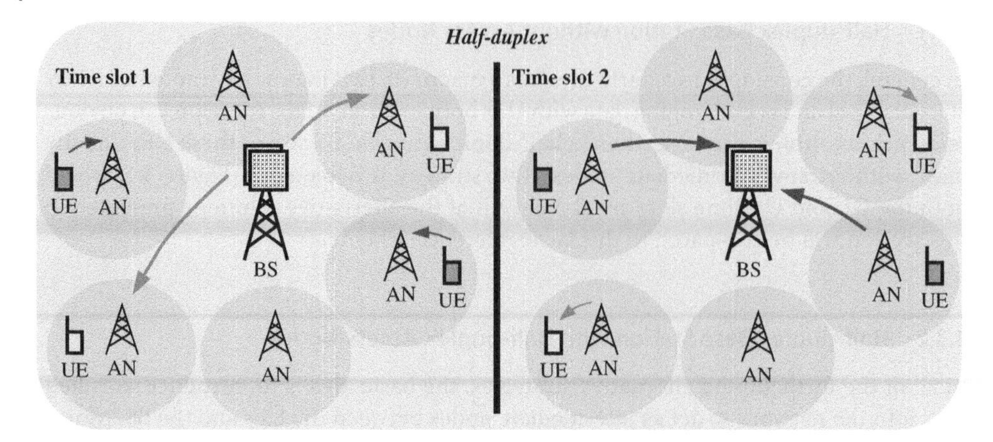

Figure 4.2 The half-duplex self-backhauling solution, where the densely deployed ANs handle the traffic between the BS and the UEs such that no IBFD capabilities are required.

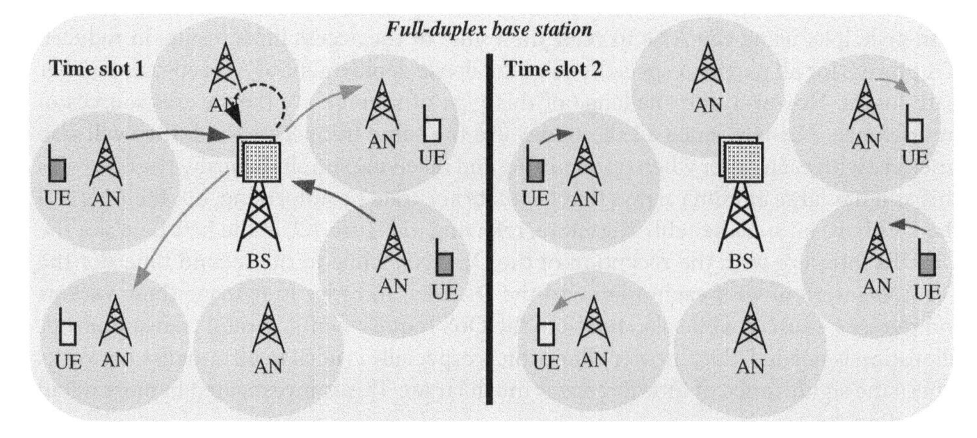

Figure 4.3 A full-duplex self-backhauling solution, where the ANs communicate first with an IBFD-capable BS, after which they serve the UEs.

4.3.3 Full-Duplex Base Station with Half-Duplex Access Nodes

In the next scheme, depicted in Figure 4.3, the BS must be IBFD capable as it simultaneously communicates with both DL and UL ANs in the first time slot. Thanks to its massive antenna arrays, it can perform part of the SI cancellation by forming nulls into the positions of its RX antennas, while the rest of the SI can be cancelled using any of the widely reported techniques. In the other time slot, the ANs then serve the UEs, as depicted in Figure 4.3. This strategy results also in a relatively low-cost deployment since only the BS must be capable of full-duplex operation, while the ANs and the UEs can be legacy half-duplex devices.

Moreover, also in this self-backhauling strategy, the various interference links call for careful transmit power allocation, especially when considering ultra-dense networks. In particular, in the first time slot, the UL ANs interfere with the reception of the DL ANs, while in the second time slot the UL UEs and the DL ANs produce interference at

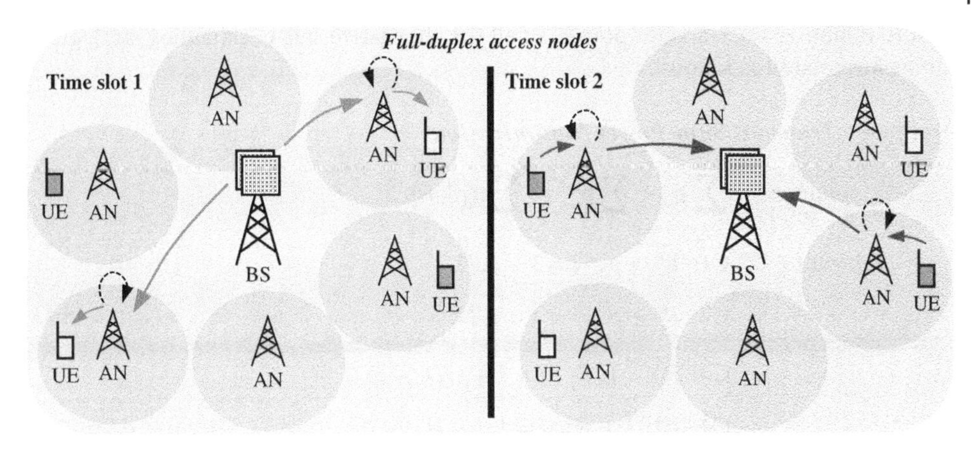

Figure 4.4 A full-duplex self-backhauling solution, where IBFD-capable ANs relay the traffic between the UEs and the BS.

the DL UEs and the UL ANs. However, as opposed to the purely half-duplex strategy, now the residual SI at the BS must also be taken into account, in addition to the other interference terms.

4.3.4 Half-duplex Base Station with Full-duplex Access Nodes

The fourth considered backhauling solution is the one depicted in Figure 4.4, where the ANs are essentially IBFD relays forwarding the data between the BS and the UEs. Therefore, in this case, the ANs must be capable of IBFD operation. The time slots are divided between DL and UL such that in the first time slot the AN forwards the signals from the BS to the DL UEs, while in the second time slot it forwards the UL signals to the BS. Utilizing this strategy will result in an increased cost of deployment since now each AN must be equipped with the necessary SI cancellation capability. However, as will be shown in Section 4.5.2, the benefit of this strategy is the decreased transmit power consumption, which is likely to outweigh the additional costs incurred by the IBFD-capable ANs.

Now, in addition to the residual SI at the AN, the ANs and UEs again produce interference to each other, as some nodes must receive data while others are transmitting it on the same frequency band. Since this analysis considers a very dense deployment of the ANs, it is crucial to manage such interference within the network. The transmit power allocation scheme discussed next in Section 4.4 takes these interference links into consideration and aims at minimizing their harmful effects.

4.4 Transmit Power Optimization under QoS Requirements

Herein, the heterogeneous network is analyzed in terms of transmit power minimization under QoS requirements, defined as a minimum data rate for each UE. This facilitates the comparison of the transmit power efficiencies of the alternative solutions under the same circumstances and requirements. The generic optimization problem, which can be used to minimize the transmit powers of all the alternative self-backhauling strategies, can be formulated as follows.

Problem ***Transmit Sum-Power Minimization***

$$\begin{aligned}
\underset{\mathbf{p}_\mathrm{u},\, \mathbf{p}_\mathrm{u}^\mathrm{AN},\, \mathbf{p}_\mathrm{d}^\mathrm{AN},\, p_\mathrm{d}^\mathrm{BS},\, \eta}{\text{minimize}} \quad & \sum \mathbf{p}_\mathrm{u} + \sum \mathbf{p}_\mathrm{u}^\mathrm{AN} + \sum \mathbf{p}_\mathrm{d}^\mathrm{AN} + p_\mathrm{d}^\mathrm{BS} \\
\text{subject to} \quad & \text{C1: } R_i^\mathrm{d} \geq \rho_\mathrm{d},\ i = 1, \ldots, D, \\
& \text{C2: } R_j^\mathrm{u} \geq \rho_\mathrm{u},\ j = 1, \ldots, U, \\
& \text{C3: } R_i^\mathrm{d,AN} \geq R_i^\mathrm{d},\ i = 1, \ldots, D, \\
& \text{C4: } R_j^\mathrm{u,AN} \geq R_j^\mathrm{u},\ j = 1, \ldots, U,
\end{aligned} \tag{4.1}$$

where

- $\mathbf{p}_\mathrm{u}$ is a vector containing the transmit powers of the UL UEs;
- $\mathbf{p}_\mathrm{u}^\mathrm{AN}$ is a vector containing the transmit powers of the UL ANs;
- $\mathbf{p}_\mathrm{d}^\mathrm{AN}$ is a vector containing the transmit powers of the DL ANs;
- p_d^BS is the DL transmit power of the BS;
- R_i^d is the DL data rate of the ith DL UE in bps/Hz;
- R_j^u is the UL data rate of the jth UL UE in bps/Hz;
- ρ_d is the DL data rate requirement of an individual UE in bps/Hz;
- ρ_u is the UL data rate requirement of an individual UE in bps/Hz;
- $R_i^\mathrm{d,AN}$ is the data rate of the backhaul link between the BS and the ith DL AN in bps/Hz;
- $R_j^\mathrm{u,AN}$ is the data rate of the backhaul link between the BS and the jth UL AN in bps/Hz;
- D is the total amount of DL UEs;
- U is the total amount of UL UEs;
- η is the duplexing parameter that determines the relative lengths of the two time slots.

The constraints C1 and C2 ensure the QoS of the UEs, while the constraints C3 and C4 ensure sufficient backhauling capability in the ANs. Due to the ultra-dense deployment of the ANs, this type of optimization procedure is crucial in facilitating any data transfer within the network since the interference will be intolerably powerful, unless dealt with properly. Therefore, determining the optimal transmit power allocation, and consequently minimizing the effect of interference, ensures that the QoS requirements can be fulfilled.

The objective function can be obtained in a rather straightforward manner by determining the relationship between the achieved data rates, the related transmit powers, and the duplexing parameter. The alternative self-backhauling strategies differ mostly in the structure of the interference terms, as each of them divides the transmissions between the two time slots in a different way. Moreover, the residual SI only affects the strategies where full-duplex operation is utilized. As a result, the exact form of the optimization problem is different for each self-backhauling strategy. Although similar optimization problems have been solved in closed form for somewhat more simpler systems [16], in this work the optimization will be performed numerically.

In principle, the optimization procedure involves first determining the minimum transmit powers for a given duplexing parameter η. This results in a linear system of equations, from which the solution of this subproblem can be obtained. Then, the obtained optimal transmit powers for any given duplexing parameter are used to form another objective function with respect to η, the solving of which gives the global optimum solution of the complete optimization problem.

It should be noted that this formulation of the optimization problem minimizes the overall transmit power consumption within the whole cell, without considering the transmit powers of the individual nodes. Therefore, if the objective is to minimize the transmit powers of, say, the UEs, the optimization problem in Eq. (4.1) can be modified by increasing the cost of the UE transmit powers in the objective function. Alternately, to provide another example, if the transmit power of the BS is of no interest, it can be entirely removed from the objective function. Nevertheless, in this chapter, such variations of the optimization procedure are omitted for brevity.

In some cases, however, the required QoS requirements cannot be fulfilled with any finite transmit powers. In the above optimization problem, these cases manifest themselves as negative optimal transmit powers, which are obviously physically impossible. Such scenarios are referred to as *infeasible* network geometries, since then the required minimum data rates cannot be achieved due to the strength of the various interference links.

4.5 Performance Analysis

4.5.1 Simulation Setup

Next, the proposed system is evaluated with the help of Monte Carlo simulations, considering the different self-backhauling strategies. To concentrate on a simple and straightforward scenario for this initial analysis, in the simulations the ANs and UEs are randomly and uniformly positioned into a circular cell of given size, at the center of which is the BS. Each UE is then allocated an AN based on the closest distance. In this work, it is assumed that an individual AN only serves one UE, as it is unlikely to have two UEs in the same cell due to the high density of the ANs. However, if it happens that two UEs share the same cell, then one of the UEs is allocated to the next closest AN, the preference being given to the DL UEs. For simplicity, it is assumed that there are no external interference sources, e.g. from adjacent cells. Extending this analysis to consider different node distributions and interference from other cells is an important future research direction in the area of ultra-dense networks.

One example realization of the network architecture is presented in Figure 4.5, where the macrocell area has been divided into smaller cells based on which AN is the closest. Furthermore, the radius of the macrocell is represented by the large circle. The DL and UL UE positions are also illustrated, together with the allocation of the ANs. Note that each UE gets served both in the DL and in the UL by having them alternate between the two modes at regular intervals. The pathlosses for the different links are calculated based on the corresponding distances and the adopted pathloss model. By calculating

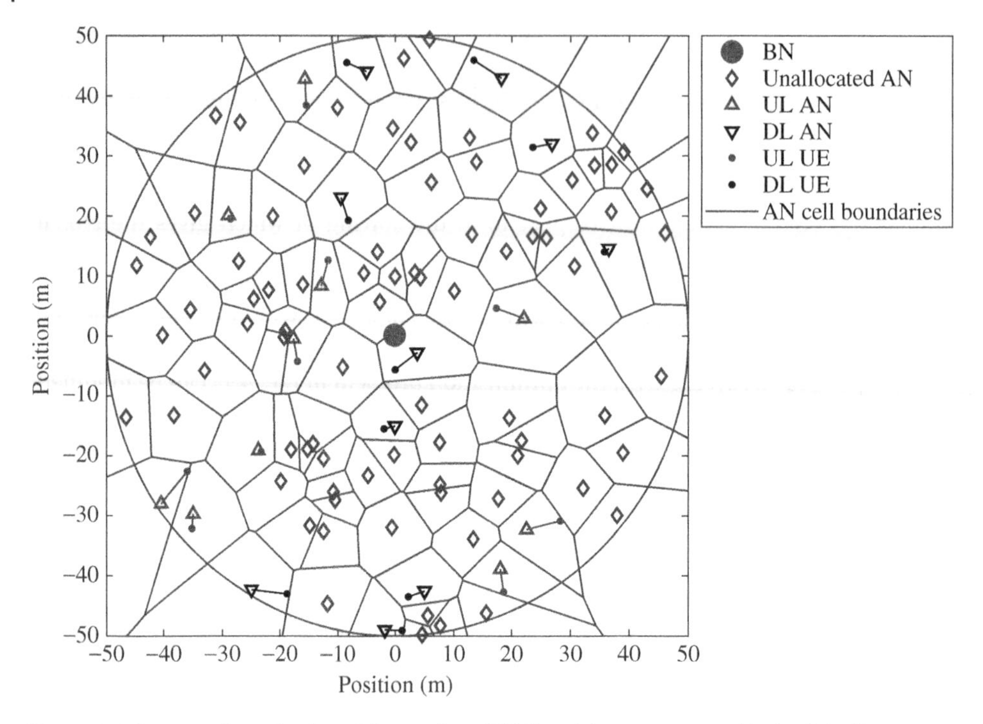

Figure 4.5 An example realization of the random AN/UE positions, together with the AN cell boundaries.

the optimal transmit powers over various random network realizations, the cumulative distribution functions (CDFs) of the corresponding quantities can then be obtained. The CDF describes the probability with which the transmit power is lower than the abscissa.

Table 4.2 lists all the default system parameters, which are used in the simulations unless otherwise mentioned. Moreover, the pathloss model is taken from [17], where a measurement-based model for a center frequency of 3.5 GHz is reported, considering both LOS and NLOS conditions. Denoting the distance in meters by d_{m}, the former pathloss model is defined as

$$L_{\mathrm{LOS}} = 42.93 + 20\log_{10}(d_{\mathrm{m}}),$$

while the latter is

$$L_{\mathrm{NLOS}} = 33.5 + 40\log_{10}(d_{\mathrm{m}}).$$

In the simulations, the LOS model is applied to the links between the AN and the BS and between the AN and the allocated UE, while the NLOS model is used for all the other links. The forthcoming CDFs are obtained by generating 10^4 random UE locations for which the optimal transmit powers are calculated. Furthermore, to ensure a fair comparison between the different schemes, the transmit powers of the different schemes are weighted by the proportion of time spent in the corresponding time slot, as this more realistically illustrates their overall transmit power usage.

Table 4.2 The essential default system parameters. Many of the parameter values are also varied in the evaluations.

Parameter	Value
Number of BS TX/RX antennas	200/100
Number of DL and UL UEs	10
Number of ANs	100
Receiver noise floor	−90 dBm
Amount of SI cancellation in the AN	−100 dB
Amount of SI cancellation in the BS	−100 dB
Per-UE DL/UL rate requirement	2/0.5 bps/Hz
Cell radius	50 m
Number of Monte Carlo simulation runs	10^4

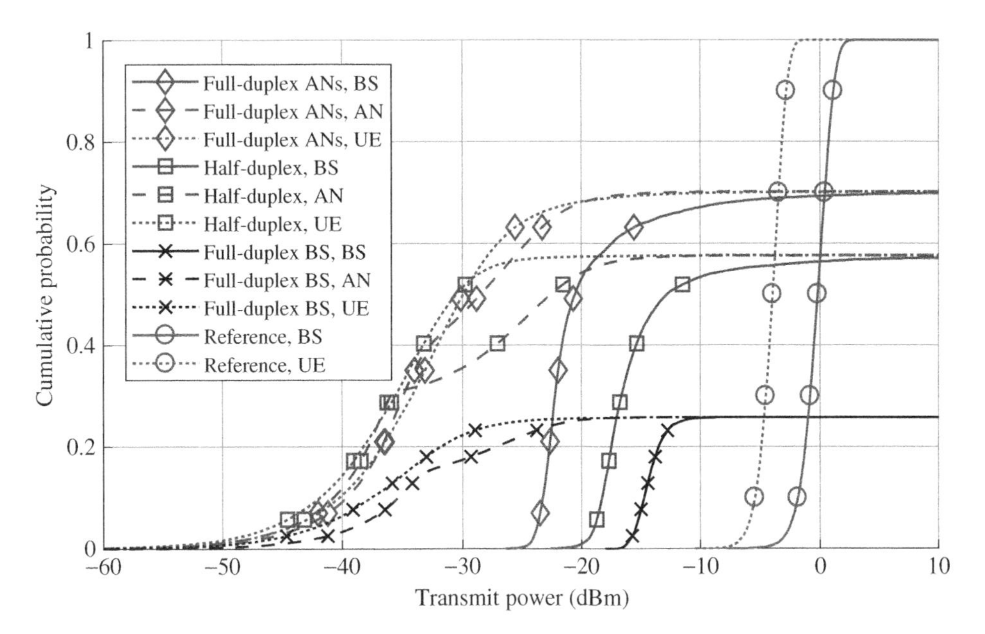

Figure 4.6 CDFs of the individual parties' transmit powers with the different self-backhauling strategies.

4.5.2 Numerical Results

Considering then the realized transmit powers of the different self-backhauling strategies, Figure 4.6 shows the transmit powers of the individual communicating parties using the default system parameters. In terms of the UE transmit power consumption, the half-duplex solution is very closely matched with the one with full-duplex ANs. However, having full-duplex capable ANs significantly reduces the transmit power requirements of the BS, thereby rendering it the preferable option in terms of the total power usage. The strategy with a full-duplex BS is inferior to both of these solutions, as it cannot obtain the QoS requirements for most network geometries. This is evident from the CDFs saturating to a value below 1, indicating that roughly 75% of the random network geometries are such that the QoS requirements cannot be fulfilled with any finite transmit powers. In fact, the solutions with half-duplex devices and full-duplex ANs also suffer from this same phenomenon, the former obtaining the minimum data rates for nearly 60% of the network geometries while the latter obtains them for 70% of the geometries. As opposed to this, the reference scheme without any ANs can always obtain the QoS requirements, although it uses significantly higher transmit powers to achieve this compared to the solutions utilizing the ANs. Therefore, as long as the infeasible network geometries can be identified and avoided, the solution with full-duplex ANs is clearly the most favorable option.

To analyze the effect of the UE density, Figure 4.7 illustrates the self-backhauling solutions from a different perspective, showing the CDFs of their total transmit power usage for different amounts of UEs. In terms of the total transmit power, the solution with full-duplex ANs is clearly the preferable option, as it outperforms the other solutions both in obtaining a lower transmit power and in fulfilling the QoS requirements. It should be noted, however, that with 24 UEs in the cell in total, even this solution cannot obtain the required data rates for more than 55% of the network geometries. Similar to the observations made in Figure 4.6, the reference scheme without any ANs can always fulfill the data rate requirements, albeit with a considerable increase in the total transmit power.

Let us then investigate the SI cancellation requirements of the ANs in more detail. To this end, Figure 4.8 illustrates how the total transmit power usage of the solution with full-duplex ANs is affected by the amount of SI cancellation in the ANs. For reference,

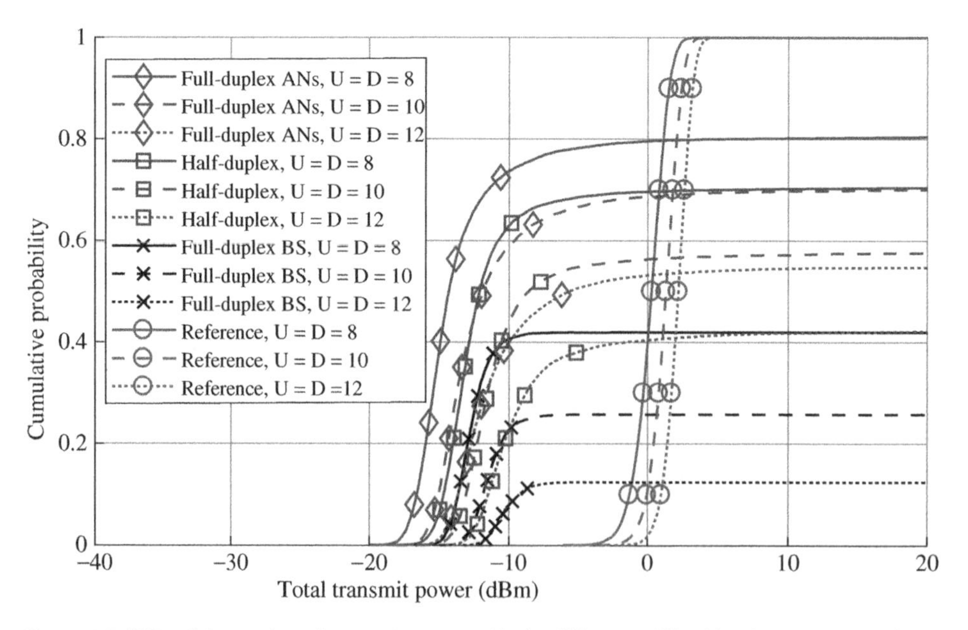

Figure 4.7 CDFs of the total used transmit power with the different self-backhauling strategies shown for different numbers of UEs.

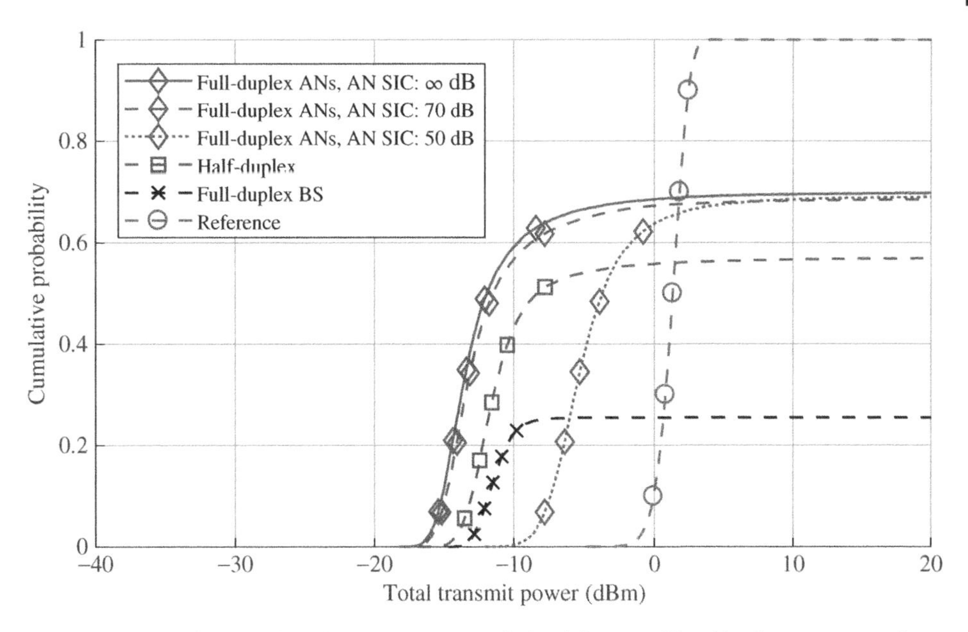

Figure 4.8 CDFs of the total used transmit power with the different self-backhauling strategies, shown for different amounts of AN SI cancellation.

the total transmit powers of the other solutions are also shown, using the default system parameters. The main observation from Figure 4.8 is that a very small amount of SI cancellation suffices for the full-duplex ANs. Namely, while 50 dB of cancellation in the AN is not enough to outperform the half-duplex solution, there is no significant difference in the transmit power consumption between the cases of 70 dB of SI cancellation and perfect SI cancellation. The primary reason for this is the interference produced by the other nodes, which is already dominating the SI with 70 dB of cancellation. Moreover, another important reason for the low SI cancellation requirement is the fact that the transmit powers are minimized, which means that the SI power is lower to begin with. This is an encouraging finding, as such a low amount of SI suppression can likely be obtained even when using a shared TX/RX antenna and performing only digital SI cancellation [18, 19]. Such full-duplex ANs can therefore be implemented without a significant increase in the cost over half-duplex ANs.

Considering then the QoS requirements, Figure 4.9 shows the CDFs of the total transmit power consumption for different UE data rate requirements. As can be expected, lower data rate requirements translate to lower transmit power consumption, the system with full-duplex ANs providing again the highest performance. With the lowest considered rate requirement of $\rho_d = 1$ and $\rho_u = 0.25$, the different AN-based solutions are quite closely matched, but as the rate requirements are increased, the gap between the networks with full-duplex ANs and those with half-duplex ANs becomes wider. Therefore, utilizing full-duplex ANs for such a self-backhauling radio access system is especially beneficial with higher data rate requirements. Nevertheless, it should also be noted that the data rate requirements cannot be fulfilled under all network geometries. A possible solution would be to downgrade to the reference scheme under these infeasible scenarios. This would mean that higher transmit powers would be used when necessary

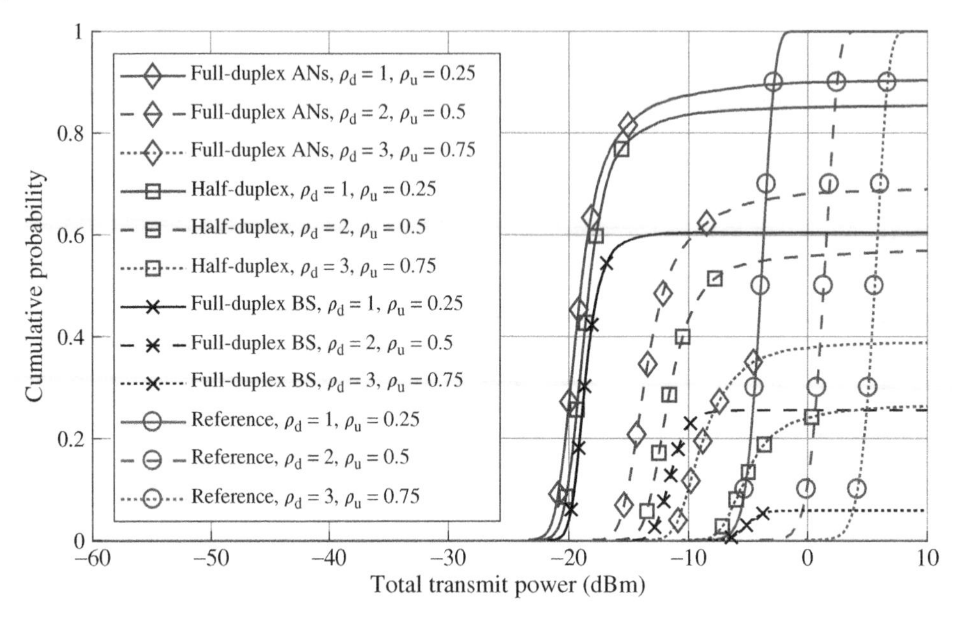

Figure 4.9 CDFs of the total used transmit power with the different self-backhauling strategies, shown for different data rate requirements.

to ensure that the QoS remains on the required level, while utilizing the full-duplex ANs whenever they can provide the required data rates with smaller transmit powers.

Next, Figure 4.10a illustrates the effect of the UE density by showing the probability of feasibility of the different self-backhauling solutions with respect to the number of UEs in the cell.[1] The medians of the corresponding sum transmit powers are shown in Figure 4.10b. First, it can be observed from Figure 4.10a that the number of supported UEs is higher the larger the cell is. This indicates that the limiting factor of the considered system is the interference between the different communicating parties, whose effect is decreased when the cell radius is increased. As for the individual self-backhauling strategies, the network with full-duplex ANs is again the best solution, supporting always the highest number of UEs for a given probability of feasibility. For instance, with a cell radius of 50 m, full-duplex ANs can provide the required data rates for 10 UL and 10 DL UEs with a probability of 70%, while the half-duplex ANs can support the same number of UEs with a probability of 55%.

Investigating the medians of the sum transmit powers in Figure 4.10b, the network with full-duplex ANs can be observed to be also the most power efficient solution under most circumstances. The only exceptions are the scenarios where the number of UEs approaches the boundary after which more than 50% of the network geometries are infeasible. Beyond this point, the median sum transmit power tends to infinity and the corresponding curves in Figure 4.10b become undefined. As opposed to this, the reference scheme is never infeasible, but it requires considerably higher transmit powers. Altogether, utilizing full-duplex ANs to provide the radio access seems therefore the

1 Note that the reference scheme is always feasible and is therefore excluded from Figures 4.10a and 4.11a.

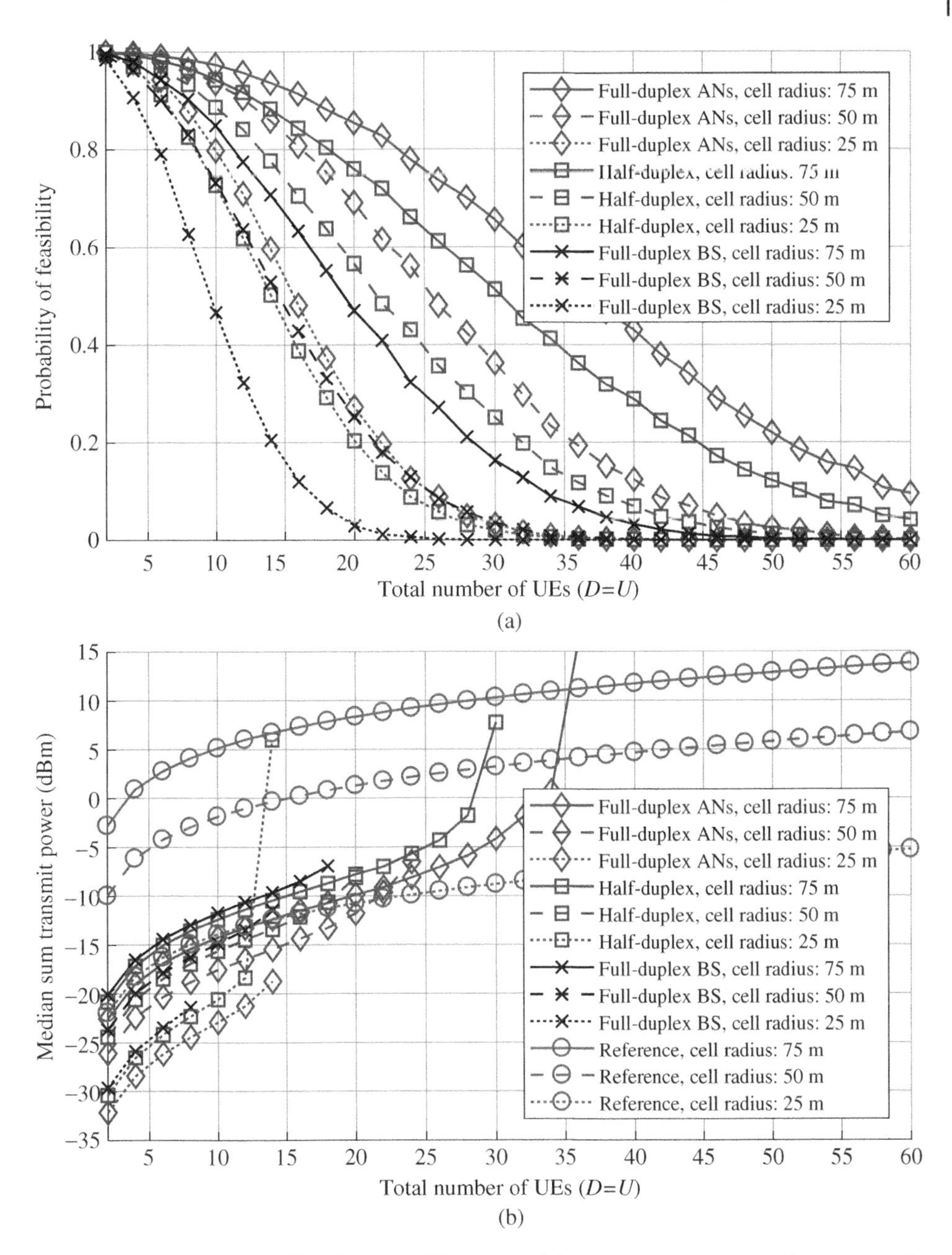

Figure 4.10 (a) Probability of feasibility and (b) median of the sum transmit power with respect to the number of UEs, shown for different cell radii.

most preferable option in terms of transmit power consumption and number of supported UEs, as long as steps are taken to ensure the feasibility of the network geometry.

Finally, Figure 4.11a shows the probability of feasibility with respect to the number of ANs, while Figure 4.11b shows the medians of the corresponding sum transmit powers. First, Figure 4.11a indicates that the probability of feasibility is not dramatically affected

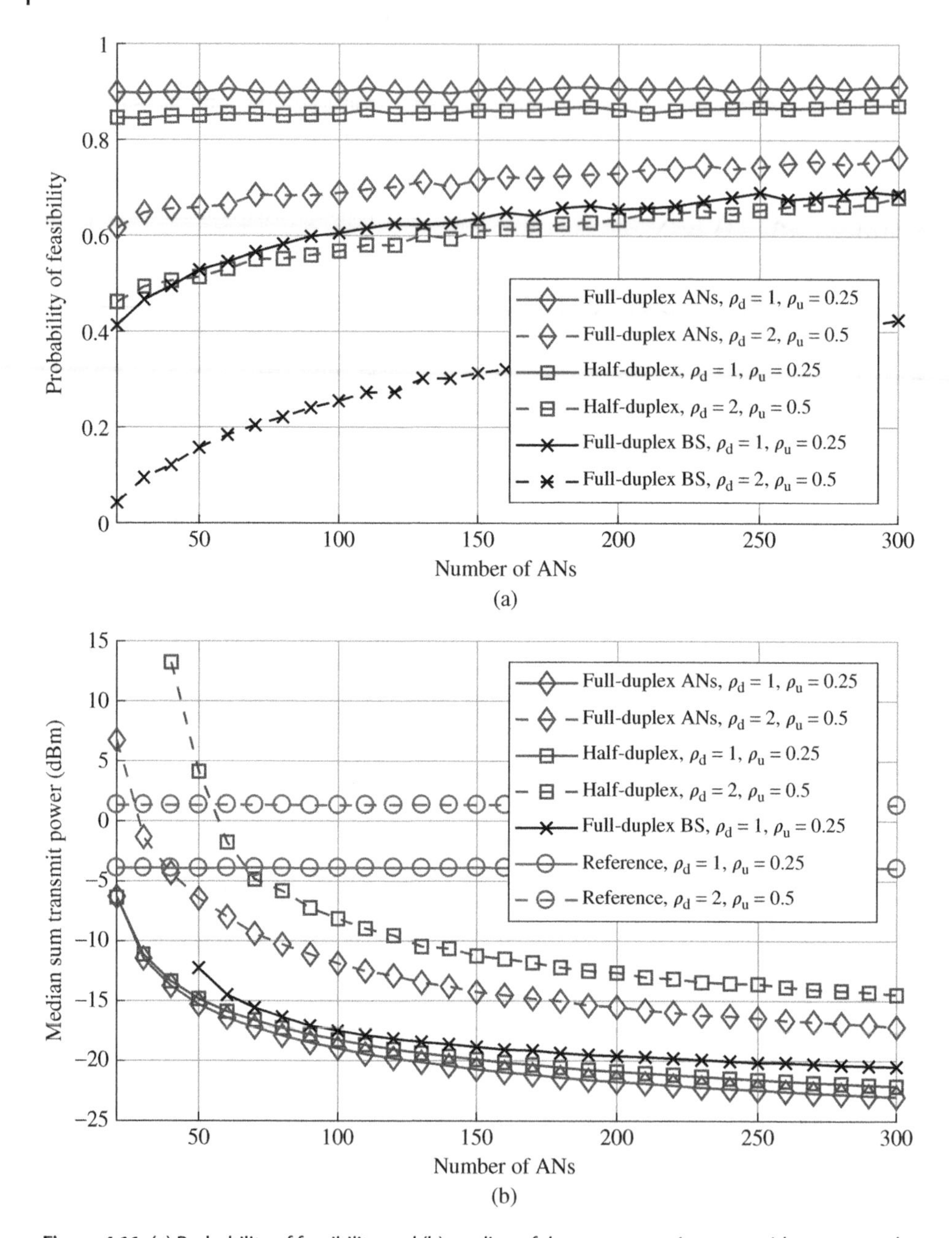

Figure 4.11 (a) Probability of feasibility and (b) median of the sum transmit power with respect to the number of ANs, shown for different data rate requirements.

by the density of the ANs, although more ANs do translate to a slightly higher probability of feasibility. However, with the considered system parameters, it seems that 100 ANs is enough to obtain a sufficiently high probability of feasibility with most self-backhauling strategies. Furthermore, again, the case with full-duplex ANs obtains the highest probability of feasibility among the considered self-backhauling strategies, especially with the higher data rate requirements.

As for overall transmit power consumption, Figure 4.11b indicates that a larger number of ANs always translates to a lower median sum transmit power. This is an intuitive result as more densely deployed ANs mean that the distance between a randomly positioned UE and the closest AN is smaller, resulting in reduced path losses. The lowest transmit powers are obtained by utilizing full-duplex ANs, while the reference scheme without any ANs requires the highest transmit powers under most circumstances. However, if the number of ANs is very small, the transmit power consumption of the AN-based strategies might be higher than that of the reference scheme. In fact, Figure 4.11b shows that, in many cases, the median transmit powers of the networks with half-duplex ANs tend to infinity if the number of ANs is too small. The most extreme example of this is the network with a full-duplex BS, as its median transmit power tends to infinity regardless of the number of ANs when the data rate requirements are $\rho_\mathrm{d} = 2$ and $\rho_\mathrm{u} = 0.5$. Therefore, the corresponding curve is not visible in Figure 4.11b.

To conclude, the most important observations based on the numerical results are the following.

- Utilizing ANs to relay the traffic between the UEs and the BS reduces the transmit power consumption under most circumstances compared to a case where the BS serves the UEs directly. However, this requires careful transmit power allocation and favorable network geometry.
- Full-duplex ANs can obtain the QoS with lower overall transmit power consumption than half-duplex ANs, requiring only 70 dB of SI cancellation in doing so.
- Having a full-duplex capable BS is not helpful under the considered scenarios. Therefore, the BS and the UEs should rely on legacy half-duplex processing.
- The AN density should be significantly higher than the UE density to ensure that the interference links remain sufficiently weak in comparison to the data links.

4.6 Summary

In this chapter, we investigated different self-backhauling strategies for cellular networks with densely deployed ANs that relay data between the UEs and a macro BS. Three different self-backhauling strategies were considered: one where all the nodes are half-duplex-capable, one where only the BS is full-duplex-capable, and one where all the ANs are full-duplex-capable. The optimal transmit power allocations under minimum QoS requirements were then determined for all these strategies as well as for a reference scheme without any ANs. The results indicate that utilizing full-duplex ANs is the most transmit power efficient solution, and it outperforms also the case where the BS communicates directly with the UEs. However, the drawback of utilizing the ANs as intermediary nodes is the fact that the QoS requirements cannot be obtained at all under some network geometries, due to the interference links. Nevertheless, as long as the system is designed such that the network geometry remains favorable, the findings of this chapter clearly demonstrate the benefits of deploying the ANs densely to minimize the transmit power consumption of ultra-dense wireless networks.

As for the potential future research directions, the analysis presented in this chapter could be extended to consider a multi-cell scenario with several BSs. In addition, the effect of alternative node distributions and pathloss conditions should also be investigated to ensure that the results obtained herein can be generalized. Moreover, to address the problem of infeasible network geometries, a hybrid solution utilizing the optimal self-backhauling strategy for the prevailing node locations could be developed and evaluated.

Bibliography

1 A.B. Ericsson, "5G radio access," White Paper, 2015 [Online]. Available at: https://www.ericsson.com/assets/local/publications/white-papers/wp-5g.pdf.

2 L. Chen, F.R. Yu, H. Ji, V.-C.M. Leung, X. Li, and B. Rong, "A full-duplex self-backhaul scheme for small cell networks with massive MIMO," in *Proc. IEEE International Conference on Communications (ICC)*, May 2016.

3 H. Tabassum, A.H. Sakr, and E. Hossain, "Analysis of massive MIMO-enabled downlink wireless backhauling for full-duplex small cells," *IEEE Transactions on Communications*, vol. 64, no. 6, pp. 2354–2369, Jun. 2016.

4 N. Wang, E. Hossain, and V.K. Bhargava, "Joint downlink cell association and bandwidth allocation for wireless backhauling in two-tier hetnets with large-scale antenna arrays," *IEEE Transactions on Wireless Communications*, vol. 15, no. 5, pp. 3251–3268, May 2016.

5 U. Siddique, H. Tabassum, and E. Hossain, "Downlink spectrum allocation for in-band and out-band wireless backhauling of full-duplex small cells," *IEEE Transactions on Communications*, vol. 65, no. 8, pp. 3538–3554, Aug. 2017.

6 R.-A. Pitaval, O. Tirkkonen, R. Wichman, K. Pajukoski, E. Lähetkangas, and E. Tiirola, "Full-duplex self-backhauling for small-cell 5G networks," *IEEE Wireless Communications*, vol. 22, no. 5, pp. 83–89, Oct. 2015.

7 Z. Zhang, X. Wang, K. Long, A. Vasilakos, and L. Hanzo, "Large-scale MIMO-based wireless backhaul in 5G networks," *IEEE Wireless Communications*, vol. 22, no. 5, pp. 58–66, Oct. 2015.

8 A. Sharma, R.K. Ganti, and J.K. Milleth, "Joint backhaul-access analysis of full duplex self-backhauling heterogeneous networks," *IEEE Transactions on Wireless Communications*, vol. 16, no. 3, pp. 1727–1740, Mar. 2017.

9 Huawei Technologies Co. Ltd., "5G: A technology vision," White Paper, 2013 [Online]. Available at: http://www.huawei.com/ilink/en/download/HW_314849.

10 P.H. Huang and K. Psounis, "Efficient mmWave wireless backhauling for dense small-cell deployments," in *Proc. 13th Annual Conference on Wireless On-demand Network Systems and Services (WONS)*, Feb. 2017, pp. 88–95.

11 D.T. Phan-Huy, P. Ratajczak, R. D'Errico, J. Järveläinen, D. Kong, K. Haneda, B. Bulut, A. Karttunen, M. Beach, E. Mellios, M. Castaneda, M. Hunukumbure,

and T. Svensson, "Massive multiple input massive multiple output for 5G wireless backhauling," in *Proc. IEEE Globecom Workshops (GC Wkshps)*, Dec. 2017, pp. 1–6.
12 I. Harjula, R. Wichman, K. Pajukoski, E. Lähetkangas, E. Tiirola, and O. Tirkkonen, "Full duplex relaying for local area," in *Proc. 24th Annual IEEE International Symposium on Personal, Indoor, and Mobile Radio Communications (PIMRC)*, Sep. 2013, pp. 2684–2688.
13 X. Huang, K. Yang, F. Wu, and S. Leng, "Power control for full-duplex relay-enhanced cellular networks with QoS guarantees," *IEEE Access*, vol. 5, pp. 4859–4869, Mar. 2017.
14 S. Hong, J. Brand, J. Choi, M. Jain, J. Mehlman, S. Katti, and P. Levis, "Applications of self-interference cancellation in 5G and beyond," *IEEE Communications Magazine*, vol. 52, no. 2, pp. 114–121, Feb. 2014.
15 G. Chen, Y. Gong, P. Xiao, and R. Tafazolli, "Dual antenna selection in self-backhauling multiple small cell networks," *IEEE Communications Letters*, vol. 20, no. 8, pp. 1611–1614, Aug. 2016.
16 D. Korpi, "Full-duplex wireless: self-interference modeling, digital cancellation, and system studies," PhD dissertation, Tampere University of Technology, Dec. 2017.
17 I. Rodriguez, H.C. Nguyen, N.T.K. Jørgensen, T.B. Sørensen, J. Elling, M.B. Gentsch, and P. Mogensen, "Path loss validation for urban micro cell scenarios at 3.5 GHz compared to 1.9 GHz," in *Proc. IEEE Global Communications Conference (GLOBECOM)*, Dec. 2013, pp. 3942–3947.
18 D. Korpi, J. Tamminen, M. Turunen, T. Huusari, Y.-S. Choi, L. Anttila, S. Talwar, and M. Valkama, "Full-duplex mobile device: pushing the limits," *IEEE Communications Magazine*, vol. 54, no. 9, pp. 80–87, Sep. 2016.
19 D. Korpi, M. Heino, C. Icheln, K. Haneda, and M. Valkama, "Compact inband full-duplex relays with beyond 100 dB self-interference suppression: enabling techniques and field measurements," *IEEE Transactions on Antennas and Propagation*, vol. 65, pp. 960–965, Feb. 2017.

5

The Role of Massive MIMO and Small Cells in Ultra-dense Networks

Qi Zhang[1], *Howard H. Yang*[2] *and Tony Q. S. Quek*[2]

[1] *The Jiangsu Key Laboratory of Wireless Communications, Nanjing University of Posts and Telecommunications, China*
[2] *Information Systems Technology and Design Pillar, Singapore University of Technology and Design, Singapore*

5.1 Introduction

To deal with the explosive increase of mobile data traffic, the fifth generation (5G) communications system has come at the forefront of wireless communications theoretical research [1]. Two main approaches in 5G are massive antennas and dense deployments of access points, which lead to the massive multiple-input multiple-output (MIMO) and small cell techniques [2].

Massive MIMO employs hundreds of antenna elements at the base station (BS) to serve tens of users simultaneously at the same time-frequency resource block (RB) [3–5]. The large size of transmit antenna array not only significantly increases the capacity with excessive spatial dimensions [6–9] but also averages out the effect of fast channel fading and provides extremely sharp beamforming concentrated into small areas [3, 4]. Aside from these, the huge degrees-of-freedom offered by massive MIMO can also reduce the transmit power [10].

On the other track, a small cell improves the system capacity by densely deploying low-power access points into traditional high-power macrocells [2, 11]. In this fashion, the distance between the transmitter and receiver can be significantly reduced, which results in remarkably enhanced rate gains. As small cells do not always have direct links to the macro BS, they can be intelligently deployed in accordance with the traffic demand without much cost on the fiber usage and real estate. In 5G, it is being planned to ensure seamless coverage, and a larger number of small cells will be densely deployed for cellular networks to form an ultra-dense network [12].

As both massive MIMO and small cells have attractive attributes in capacity enhancement, it is natural to wonder which one performs better under different scenarios. Initial comparisons between massive MIMO and small cells in terms of spectral and energy efficiency have been addressed in [13] to [15]. However, [13] and [14] model and analyze the massive MIMO and small cell network separately. In 5G, the ultra-dense network built with small cells is designed to offload traffic from the macro BS by overlying the existing infrastructure. Hence, an interactive multi-tier framework is necessary. In [15], a two-tier architecture including a massive-antenna macro BS and small cell

Ultra-dense Networks for 5G and Beyond: Modelling, Analysis, and Applications,
First Edition. Edited by Trung Q. Duong, Xiaoli Chu and Himal A. Suraweera.

access points is explored, but it only considers the single-cell scenario and ignores the randomness of the BS locations. Since the flexible and intelligent deployment is one of the most important advantages of small cells, it is necessary to involve the BS distributions into the analysis. Therefore, a reasonable framework to compare massive MIMO and small cells should be a heterogeneous cellular network (HCN) containing different types of randomly located BSs with multiple antennas. Recently, many works have utilized the stochastic geometry tool in [16] to investigate the HCN with Poisson point process (PPP) distributed BSs [17–23]. A general multi-tier framework is proposed in [17], the coverage probability with flexible biased cell association is analyzed in [18], and the multi-antenna transmission has been presented in [19] and [20]. However, none of these works can be directly applied to the comparison of massive MIMO and small cells, since they do not take into account channel estimation and the further effect of pilot contamination, which is the main limiting factor of massive MIMO. Furthermore, they adopt the single-slope pathloss model with only non-line-of-sight (NLOS) transmissions, which is not fit for dense small cell networks since the shorter propagation distance results in more line-of-sight (LOS) transmissions [24] and the dual-slope pathloss affects the benefit of BS densification in a very negative way [25].

In this chapter, we propose a downlink HCN where the user location as well as the deployment of multi-antenna macro and small BSs are modeled as independent PPPs. Users are flexibly associated with the strongest BS, and each BS simultaneously serves multiple users associated with it on the same time-frequency RB. Further, the signal propagation experiences a pathloss model differentiating LOS and NLOS transmissions, and the channel state information at BSs is acquired through uplink training. Our main contributions are summarized as:

- We propose a general framework for the analysis of downlink HCN that consists of PPP distributed macro and small BSs with multiple antennas. Our analysis captures the essential keypoints of both massive MIMO and small cells, including LOS/NLOS transmission, BS deployment density, imperfect channel estimation, and random network topology.
- Based on our analytical results, we compare the system performance between densifying small cells and expanding macro BS antennas. It is found that adding small cells into the network can improve the downlink rate much faster than expanding antenna arrays at the macro BS. However, when the small cell density exceeds a critical threshold, the densification will stop benefiting and further impair the network capacity. In contrast, the downlink rate always increases with growing antenna size and saturates to an upper bound caused by pilot contamination. This upper bound is also larger than the peak rate obtained from deployment of small cells.
- We provide the optimal small cell density that maximizes the average downlink rate, which can be used as a guideline for practical small cell deployment. Moreover, the optimal bias factor is also provided for reference of the practical massive MIMO configuration.

The remainder of this chapter is organized as follows. Section 5.2 introduces the HCN framework, the LOS/NLOS transmission model, and the user association policy. In Section 5.3, we derive a tight approximation of the average downlink rate accounting for channel estimation with uplink training. In Section 5.4, we provide numerical results to validate the analytical results and further study the performance of the HCN. Finally, Section 5.5 summarizes the main results of this chapter.

Notation Throughout the chapter, vectors are expressed in lowercase boldface letters while matrices are denoted by uppercase boldface letters. We use $\mathbf{X}^H$ to denote the conjugate-transpose of $\mathbf{X}$ and use $[\mathbf{X}]_{ij}$ to denote the (i, j)th entry of $\mathbf{X}$. Finally, $\mathbb{1}(e)$ is the indicator function for logic e, $\mathbb{E}\{\cdot\}$ is the expectation operator, and $\|\cdot\|$ is the Euclidean norm.

5.2 System Model

5.2.1 Network Topology

We consider the downlink of a two-tier heterogeneous cellular network, where high-power macro BSs (MBSs) are overlaid with successively denser and lower-power small BSs (SBSs), as illustrated in Figure 5.1. We assume that the MBSs and SBSs are deployed on a plane according to independent PPPs Φ_{m} and Φ_{s} with spatial densities λ_{m} and λ_{s}, respectively. All MBSs and SBSs are equipped with M_{m} and M_{s} antennas, respectively, whereas each MBS transmits with power P_{m} and each SBS transmits with power to be P_{s}. In the light of high spectral utilization, we consider a co-channel deployment of small cells with the macro cell tier; i.e. MBSs and SBSs share the same frequency band for transmission. We model the mobile users as another independent PPP Φ_{u} with spatial density λ_{u}, where each user is equipped with a single antenna. The users associate with their targeted MBSs or SBSs according to the association policy described in Section 5.2.3. In practice, due to the finite number of antennas, an MBS cannot serve more users than its available number of antennas in one time-frequency

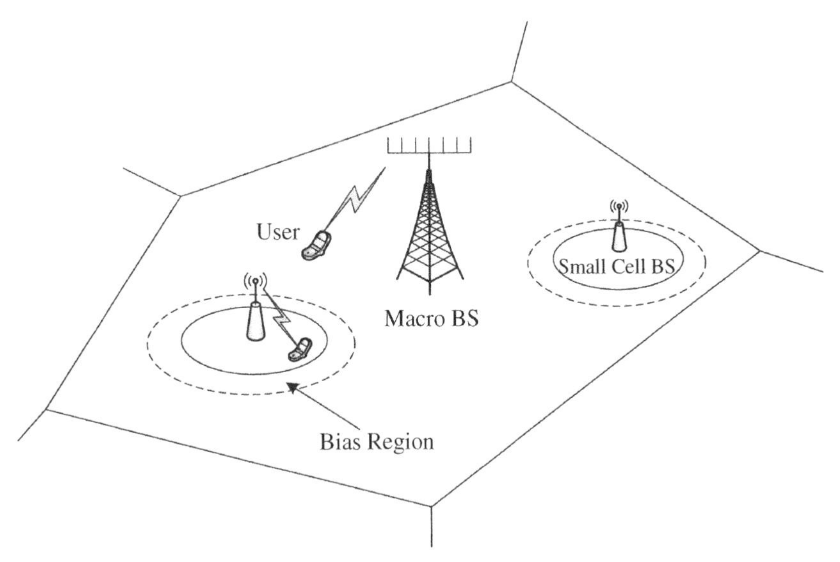

Figure 5.1 A two-tier heterogeneous network utilizing a mix of macro and small cell BSs. Reprinted with permission from [28].

RB. Therefore, we limit the maximum number of users to be simultaneously served by one MBS in each RB as $N \leq M_{\rm m}$. Moreover, to avoid severe pilot contamination, we consider only MBSs to be responsible for pilot allocation. Hence, for each RB, an MBS randomly selects N users in its coverage, including MBS users as well as SBS users, to assign orthogonal pilot sequences. As such, N should also be less than the available pilot length. Additionally, we assume the user density is sufficiently large, i.e. $\lambda_{\rm u} \gg \lambda_{\rm m}$, such that every MBS has at least N users in its coverage to be served.[1]

5.2.2 Propagation Environment

We model the channels between any pair of antennas as independent and identically distributed (i.i.d.) and quasi-static, i.e. the channel is constant during a sufficiently long coherence block and varies independently from block to block. Moreover, we assume that each channel is narrowband and affected by two attenuation components, namely small-scale Rayleigh fading and large-scale pathloss. Regarding the practical LOS and NLOS transmissions, we model the pathloss with two parts, i.e. the LOS and NLOS path. More formally, the pathloss φ_{inl} between U_{nl} and BS i can be written as

$$\varphi_{inl} = \begin{cases} \varphi_{inl}^{\rm L} = L^{\rm L} r_{inl}^{-\alpha^{\rm L}}, & \text{if LOS}, \\ \varphi_{inl}^{\rm NL} = L^{\rm NL} r_{inl}^{-\alpha^{\rm NL}}, & \text{if NLOS}, \end{cases} \tag{5.1}$$

where r_{inl} is the distance between user U_{nl} and BS i, $L^{\rm L}$ and $L^{\rm NL}$ denote the pathlosses evaluated at a reference distance 1 for LOS and NLOS, respectively, and $\alpha^{\rm L}$ and $\alpha^{\rm NL}$ are the LOS and NLOS pathloss exponents, respectively. As the probability of a wireless link being LOS or NLOS is mainly affected by the distance between the transmitter and receiver, we model such probability as a homogeneous event in the following analysis [29]. Therefore, the channel coefficient between U_{nl} and the mth antenna of MBS i can be formulated as

$$g_{minl}^{\rm m} = \begin{cases} h_{minl}^{\rm m} \sqrt{\varphi_{inl}^{\rm m,L}}, & \text{with probability } p_{\rm m}^{\rm L}(r_{inl}^{\rm m}), \\ h_{minl}^{\rm m} \sqrt{\varphi_{inl}^{\rm m,NL}}, & \text{with probability } 1 - p_{\rm m}^{\rm L}(r_{inl}^{\rm m}), \end{cases} \tag{5.2}$$

and the channel coefficient between U_{nl} and the mth antenna of SBS j is

$$g_{mjnl}^{\rm s} = \begin{cases} h_{mjnl}^{\rm s} \sqrt{\varphi_{jnl}^{\rm s,L}}, & \text{with probability } p_{\rm s}^{\rm L}(r_{jnl}^{\rm s}), \\ h_{mjnl}^{\rm s} \sqrt{\varphi_{jnl}^{\rm s,NL}}, & \text{with probability } 1 - p_{\rm s}^{\rm L}(r_{jnl}^{\rm s}), \end{cases} \tag{5.3}$$

where $h_{minl}^{\rm m}$ and $h_{mjnl}^{\rm s}$ denote the small-scale fading with $h_{minl}^{\rm m}, h_{mjnl}^{\rm s} \sim \mathcal{CN}(0,1)$ and $p_{\rm m}^{\rm L}(\cdot)$ and $p_{\rm s}^{\rm L}(\cdot)$ are the LOS probabilities functions with the MBS and SBS, respectively. Note that the LOS probabilities can be different for macro and small cells due to the assorted propagation environment as well as various antenna heights.

---

1 We make this assumption for the sake of analytical tractability. Note that removing this assumption does not change the main outcomes of this chapter since the probability of having less numbers of users than N is very small for large $\lambda_{\rm u}$ [20, 26, 27].

5.2.3 User Association Policy

For load balancing, we adopt cell range expansion for user associations in this network. Specifically, all the SBSs employ a bias factor B for the control of cell range expansion, and users associate to the BS that provides the largest average biased received power. Note that with the existence of an LOS path, it is possible that a user is associated to a far away LOS BS instead of a nearby NLOS BS. Due to the stationary property of PPP, we can evaluate the performance of a typical user located at the origin, denoted as U_{00}, thanks to Slivyark's theorem. Therefore, the average biased-received power of the typical user from the MBS i is

$$\mathcal{P}_{\mathrm{m}}(r_{i00}^{\mathrm{m}}) = \begin{cases} P_{\mathrm{m}} L^{\mathrm{L}}(r_{i00}^{\mathrm{m}})^{-\alpha^{\mathrm{L}}}, & \text{with probability } p_{\mathrm{m}}^{\mathrm{L}}(r_{i00}^{\mathrm{m}}), \\ P_{\mathrm{m}} L^{\mathrm{NL}}(r_{i00}^{\mathrm{m}})^{-\alpha^{\mathrm{NL}}}, & \text{with probability } 1 - p_{\mathrm{m}}^{\mathrm{L}}(r_{i00}^{\mathrm{m}}), \end{cases} \tag{5.4}$$

and the average biased-received power of the typical user from the SBS j is

$$\mathcal{P}_{\mathrm{s}}(r_{j00}^{\mathrm{s}}) = \begin{cases} P_{\mathrm{s}} L^{\mathrm{L}}(r_{j00}^{\mathrm{s}})^{-\alpha^{\mathrm{L}}} B, & \text{with probability } p_{\mathrm{s}}^{\mathrm{L}}(r_{j00}^{\mathrm{s}}), \\ P_{\mathrm{s}} L^{\mathrm{NL}}(r_{j00}^{\mathrm{s}})^{-\alpha^{\mathrm{NL}}} B, & \text{with probability } 1 - p_{\mathrm{s}}^{\mathrm{L}}(r_{j00}^{\mathrm{s}}), \end{cases} \tag{5.5}$$

where B is the small-cell bias factor, which can be set greater or smaller than one to extend or shrink the coverage.

5.3 Average Downlink Rate

In this section, we analyze the average downlink rate of the HCN with LOS/NLOS transmissions. Particularly, we utilize tools from stochastic geometry to derive the user association probabilities, the distribution of serving distance, and finally the tight approximation of the average downlink rate. For a better readability, most proofs have been relegated to the Appendix.

5.3.1 Association Probabilities

With LOS/NLOS transmission, the coverage area of BSs no longer forms weighted Voronoi cells because a user can associate to a far away BS with LOS path instead of a nearby BS with an NLOS path. To analyze this more challenging user association, we start with a decomposition of the PPPs Φ_{m} and Φ_{s}. To be precise, if an MBS has an LOS path to the typical user located at the origin, we classify it as in the LOS MBS set $\Phi_{\mathrm{m}}^{\mathrm{L}}$; otherwise, we put it into the set of NLOS MBS $\Phi_{\mathrm{m}}^{\mathrm{NL}}$. Since this operation is performed independently for each point in Φ_{m}, from the Thinning Theorem [[30], Theorem 2.36] we know that $\Phi_{\mathrm{m}}^{\mathrm{L}}$ and $\Phi_{\mathrm{m}}^{\mathrm{NL}}$ are two independent inhomogeneous PPPs with densities $\lambda_{\mathrm{m}} p_{\mathrm{m}}^{\mathrm{L}}(r)$ and $\lambda_{\mathrm{m}}(1 - p_{\mathrm{m}}^{\mathrm{L}}(r))$, respectively, where r denotes the distance from an MBS to the typical user. Similarly, we can also decompose Φ_{s} into the sets of LOS SBS $\Phi_{\mathrm{s}}^{\mathrm{L}}$ and NLOS SBS $\Phi_{\mathrm{s}}^{\mathrm{NL}}$, which are two inhomogeneous PPPs with densities $\lambda_{\mathrm{s}} p_{\mathrm{s}}^{\mathrm{L}}(r)$ and $\lambda_{\mathrm{s}}(1 - p_{\mathrm{s}}^{\mathrm{L}}(r))$, respectively.

We define the event that the typical user is associated with the MBS in LOS and NLOS, and associated with the SBS in LOS and NLOS, as $E_{\mathrm{m}}^{\mathrm{L}}$, $E_{\mathrm{m}}^{\mathrm{NL}}$, $E_{\mathrm{s}}^{\mathrm{L}}$, and $E_{\mathrm{s}}^{\mathrm{NL}}$, respectively. The probabilities of these events are given as follows.

Theorem 5.1 The probabilities that the typical user is associated with an MBS in the LOS and NLOS path are given by

$$\mathcal{A}_{\mathrm{m}}^{\mathrm{L}} \triangleq \mathbb{P}[E_{\mathrm{m}}^{\mathrm{L}}] = 2\pi\lambda_{\mathrm{m}} \int_0^{\infty} rp_{\mathrm{m}}^{\mathrm{L}}(r)\zeta_1\left(r, k_1 r^{\alpha^{\mathrm{L}}/\alpha^{\mathrm{NL}}}\right)\zeta_2\left(k_2 r, k_1 k_3 r^{\alpha^{\mathrm{L}}/\alpha^{\mathrm{NL}}}\right) dr, \tag{5.6}$$

$$\mathcal{A}_{\mathrm{m}}^{\mathrm{NL}} \triangleq \mathbb{P}[E_{\mathrm{m}}^{\mathrm{NL}}] = 2\pi\lambda_{\mathrm{m}} \int_0^{\infty} r(1-p_{\mathrm{m}}^{\mathrm{L}}(r))\zeta_1\left(k_4 r^{\alpha^{\mathrm{NL}}/\alpha^{\mathrm{L}}}, r\right)\zeta_2\left(k_2 k_4 r^{\alpha^{\mathrm{NL}}/\alpha^{\mathrm{L}}}, k_3 r\right) dr, \tag{5.7}$$

and the probabilities that the typical user is associated with the SBS in the LOS and NLOS path are given by

$$\mathcal{A}_{\mathrm{s}}^{\mathrm{L}} \triangleq \mathbb{P}[E_{\mathrm{s}}^{\mathrm{L}}] = 2\pi\lambda_{\mathrm{s}} \int_0^{\infty} rp_{\mathrm{s}}^{\mathrm{L}}(r)\zeta_1\left(\frac{r}{k_2}, \frac{k_1}{k_3} r^{\alpha^{\mathrm{L}}/\alpha^{\mathrm{NL}}}\right)\zeta_2(r, k_1 r^{\alpha^{\mathrm{L}}/\alpha^{\mathrm{NL}}}) dr, \tag{5.8}$$

$$\mathcal{A}_{\mathrm{s}}^{\mathrm{NL}} \triangleq \mathbb{P}[E_{\mathrm{s}}^{\mathrm{NL}}] = 2\pi\lambda_{\mathrm{s}} \int_0^{\infty} r(1-p_{\mathrm{s}}^{\mathrm{L}}(r))\zeta_1\left(\frac{k_4}{k_2} r^{\alpha^{\mathrm{NL}}/\alpha^{\mathrm{L}}}, \frac{r}{k_3}\right)\zeta_2\left(k_4 r^{\alpha^{\mathrm{NL}}/\alpha^{\mathrm{L}}}, r\right) dr, \tag{5.9}$$

where

$$\zeta_1(x_1, x_2) \triangleq \exp\left(-\int_0^{x_1} \lambda_{\mathrm{m}} p_{\mathrm{m}}^{\mathrm{L}}(u) 2\pi u du - \int_0^{x_2} \lambda_{\mathrm{m}}(1-p_{\mathrm{m}}^{\mathrm{L}}(u)) 2\pi u du\right), \tag{5.10}$$

$$\zeta_2(x_1, x_2) \triangleq \exp\left(-\int_0^{x_1} \lambda_{\mathrm{s}} p_{\mathrm{s}}^{\mathrm{L}}(u) 2\pi u du - \int_0^{x_2} \lambda_{\mathrm{s}}(1-p_{\mathrm{s}}^{\mathrm{L}}(u)) 2\pi u du\right), \tag{5.11}$$

with $k_1 \triangleq \left(\frac{L^{\mathrm{NL}}}{L^{\mathrm{L}}}\right)^{1/\alpha^{\mathrm{NL}}}$, $k_2 \triangleq \left(\frac{BP_{\mathrm{s}}}{P_{\mathrm{m}}}\right)^{1/\alpha^{\mathrm{L}}}$, $k_3 \triangleq \left(\frac{BP_{\mathrm{s}}}{P_{\mathrm{m}}}\right)^{1/\alpha^{\mathrm{NL}}}$, and $k_4 \triangleq \left(\frac{L^{\mathrm{L}}}{L^{\mathrm{NL}}}\right)^{1/\alpha^{\mathrm{L}}}$.

Proof: See Appendix A.1.

The accuracy of Theorem 5.1 will be validated later in Figure 5.2 in Section 5.4. From Theorem 5.1, it is easy to obtain the probability that the typical user is associated with an MBS as $\mathcal{A}_{\mathrm{m}} = \mathcal{A}_{\mathrm{m}}^{\mathrm{L}} + \mathcal{A}_{\mathrm{m}}^{\mathrm{NL}}$, and with an SBS as $\mathcal{A}_{\mathrm{s}} = \mathcal{A}_{\mathrm{s}}^{\mathrm{L}} + \mathcal{A}_{\mathrm{s}}^{\mathrm{NL}}$. Based on the above results, we can further derive the average number of users associated with each BS as follows.

Corollary 5.1 The average number of users associated with an MBS is $\mathcal{N}_{\mathrm{m}} = \mathcal{A}_{\mathrm{m}} N$, and with an SBS is $\mathcal{N}_{\mathrm{s}} = \mathcal{A}_{\mathrm{s}} \lambda_{\mathrm{m}} N/\lambda_{\mathrm{s}}$.

Proof: See Appendix A.2.

Remark 5.1 Larger λ_{s} results in a higher probability for a user to be associated with an SBS and reduces the probability to be associated with an MBS, i.e. $\mathcal{A}_{\mathrm{s}}$ grows and $\mathcal{A}_{\mathrm{m}}$ decreases. Hence, as λ_{s} increases, the decline of $\mathcal{N}_{\mathrm{m}}$ is obvious from Corollary 5.1, and after applying Eqs. (5.8) and (5.9) in $\mathcal{N}_{\mathrm{s}}$, we know that the average number of users associated with an SBS reduces as λ_{s} grows.

When the typical user is associated with the MBS in an LOS or NLOS path, and the SBS in an LOS or NLOS path, the distance between the typical user and its serving BS are

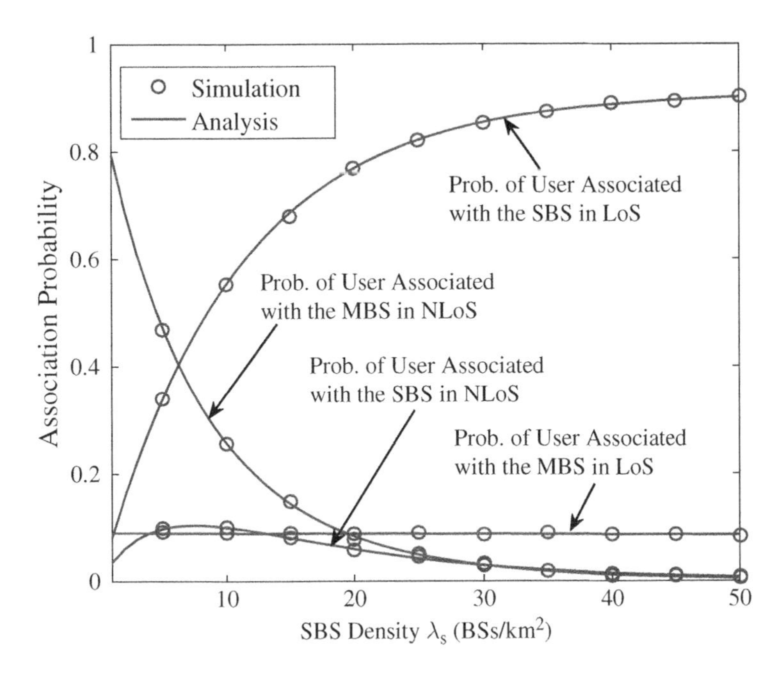

Figure 5.2 User association probability versus SBS density λ_s, where $B = 1$.

denoted as R_m^L, R_m^{NL}, R_s^L, and R_m^{NL}, respectively. The next theorem provides the probability density function (PDF) for each of these distances.

Theorem 5.2 The PDF of R_m^L, R_m^{NL}, R_s^L, and R_s^{NL} are given as follows:

$$f_{R_m^L}(r) = \frac{2\pi\lambda_m}{A_m^L} r p_m^L(r) \zeta_1\left(r, k_1 r^{\alpha^L/\alpha^{NL}}\right) \zeta_2\left(k_2 r, k_1 k_3 r^{\alpha^L/\alpha^{NL}}\right), \tag{5.12}$$

$$f_{R_m^{NL}}(r) = \frac{2\pi\lambda_m}{A_m^{NL}} r\left(1 - p_m^L(r)\right) \zeta_1\left(k_4 r^{\alpha^{NL}/\alpha^L}, r\right) \zeta_2\left(k_2 k_4 r^{\alpha^{NL}/\alpha^L}, k_3 r\right), \tag{5.13}$$

$$f_{R_s^L}(r) = \frac{2\pi\lambda_s}{A_s^L} r p_s^L(r) \zeta_1\left(r/k_2, k_1/k_3 r^{\alpha^L/\alpha^{NL}}\right) \zeta_2\left(r, k_1 r^{\alpha^L/\alpha^{NL}}\right), \tag{5.14}$$

$$f_{R_s^{NL}}(r) = \frac{2\pi\lambda_s}{A_s^{NL}} r\left(1 - p_s^L(r)\right) \zeta_1\left(k_4/k_2 r^{\alpha^{NL}/\alpha^L}, r/k_3\right) \zeta_2\left(k_4 r^{\alpha^{NL}/\alpha^L}, r\right). \tag{5.15}$$

Proof: See Appendix A.3.

The above results will be applied in the derivation of the average downlink rate. Next, we investigate the channel estimation procedure from uplink training.

5.3.2 Uplink Training

During a dedicated uplink training phase, users in each macrocell simultaneously transmit mutually orthogonal pilot sequences that allow the BSs to estimate channels of users associated with them. Due to the limited pilot length, we further assume that the same set of orthogonal pilot sequences is reused in every macrocell. In particular, the MBS assigns orthogonal pilots of length τ symbols for the N users in its cell ($\tau \geq N$) and notifies each SBS the pilot sequence of users associated with it.

The pilot sequence used by U_{nl} is a $\tau \times 1$ vector $\sqrt{\tau}\boldsymbol{\epsilon}_{nl}$, which satisfies $\boldsymbol{\epsilon}_{nl}^{H}\boldsymbol{\epsilon}_{cl} = \delta[n-c]$, with $\delta[\cdot]$ being the Kronecker delta function. Furthermore, we assume that for any $i \neq l$, $\boldsymbol{\epsilon}_{ni} = \boldsymbol{\epsilon}_{nl}$. By transmitting these pilot signals over τ symbols in the uplink, the collective $M_{\mathrm{m}} \times \tau$ received pilot signal at MBS i is

$$\mathbf{Y}_i^{\mathrm{m}} = \sqrt{\tau p_p} \sum_{l\in\Phi_{\mathrm{m}}} \sum_{n=1}^{N} \mathbf{g}_{inl}^{\mathrm{m}} \boldsymbol{\epsilon}_{nl} + \mathbf{N}_i^{\mathrm{m}}, \tag{5.16}$$

where p_p is the pilot power, $\mathbf{g}_{inl}^{\mathrm{m}} = [g_{1inl}^{\mathrm{m}}, \dots, g_{M_{\mathrm{m}}inl}^{\mathrm{m}}]^T$, and $\mathbf{N}_i^{\mathrm{m}}$ is the additive white Gaussian noise (AWGN) matrix with i.i.d. zero-mean elements and variance σ^2. With the minimum mean-square-error (MMSE) estimation method, the estimate of $\hat{\mathbf{g}}_{ini}^{\mathrm{m}}$ is given by

$$\hat{\mathbf{g}}_{ini}^{\mathrm{m}} = \eta_{ini}^{\mathrm{m}} \frac{1}{\sqrt{\tau p_p}} \mathbf{Y}_i^{\mathrm{m}} \boldsymbol{\epsilon}_{ni}^{H}, \tag{5.17}$$

where $\eta_{ini}^{\mathrm{m}} \triangleq \varphi_{ini}^{\mathrm{m}} / \left(\sum_{l\in\Phi_{\mathrm{m}}} \varphi_{inl}^{\mathrm{m}} + \sigma^2/\tau p_p \right)$. Similarly, the MMSE estimate of $\mathbf{g}_{jni}^{\mathrm{s}} = [g_{1jni}^{\mathrm{s}}, \dots, g_{M_{\mathrm{s}}jni}^{\mathrm{s}}]^T$, i.e. the channel vector from U_{ni} to SBS j, is given by

$$\hat{\mathbf{g}}_{jni}^{\mathrm{s}} = \eta_{jni}^{\mathrm{s}} \frac{1}{\sqrt{\tau p_p}} \mathbf{Y}_j^{\mathrm{s}} \boldsymbol{\epsilon}_{ni}^{H}, \tag{5.18}$$

where $\eta_{jni}^{\mathrm{s}} \triangleq \varphi_{jni}^{\mathrm{s}} / \left(\sum_{l\in\Phi_{\mathrm{m}}} \varphi_{jnl}^{\mathrm{s}} + \sigma^2/\tau p_p \right)$.

Note that due to pilot reuse, the estimated channel vector is polluted by channels from users in other cells who share the same pilot, thus causing the pilot contamination. As mentioned in [6], pilot contamination is a main limiting factor for the performance of massive MIMO. This impact with LOS/NLOS transmissions will be further explored in Section 5.4.

5.3.3 Downlink Data Transmission

Let $\mathcal{U}_i^{\mathrm{m}}$ and $\mathcal{U}_j^{\mathrm{s}}$ be the collection of users associated with the MBS i and the SBS j, respectively, and $|\mathcal{U}_i^{\mathrm{m}}|$ and $|\mathcal{U}_j^{\mathrm{s}}|$ denote the corresponding cardinalities. Each BS utilizes the estimated channel obtained from uplink training to establish the downlink precoder. The received signal at the typical user is

$$s_0 = \sqrt{P_{\mathrm{m}}} \sum_{l\in\Phi_{\mathrm{m}}} \sum_{U_{nl}\in\mathcal{U}_l^{\mathrm{m}}} (\mathbf{g}_{l00}^{\mathrm{m}})^H \mathbf{f}_{lnl}^{\mathrm{m}} x_{lnl}^{\mathrm{m}} + \sqrt{P_{\mathrm{s}}} \sum_{j\in\Phi_{\mathrm{s}}} \sum_{l\in\Phi_{\mathrm{m}}} \sum_{U_{nl}\in\mathcal{U}_j^{\mathrm{s}}} (\mathbf{g}_{j00}^{\mathrm{s}})^H \mathbf{f}_{jnl}^{\mathrm{s}} x_{jnl}^{\mathrm{s}} + n_0, \tag{5.19}$$

where $\mathbf{f}_{lnl}^{\mathrm{m}}$ is the $M_{\mathrm{m}} \times 1$ precoding vector of MBS l to U_{nl}, and $\mathbf{f}_{jnl}^{\mathrm{s}}$ is the $M_{\mathrm{s}} \times 1$ precoding vector of SBS j to U_{nl}, x_{lnl}^{m} and x_{jnl}^{m} are the signals intended for U_{nl} from MBS l and the SBS j, respectively, and n_0 is the AWGN.

Let $\tilde{\mathbf{g}}_{ini}^{\mathrm{m}} = \hat{\mathbf{g}}_{ini}^{\mathrm{m}} - \mathbf{g}_{ini}^{\mathrm{m}}$ and $\tilde{\mathbf{g}}_{jni}^{\mathrm{s}} = \hat{\mathbf{g}}_{jni}^{\mathrm{s}} - \mathbf{g}_{jni}^{\mathrm{s}}$ denote the channel estimation error. If the typical user is associated with MBS 0, the downlink SINR of the typical user is $\mathtt{SINR}_{\mathrm{m}}$

$= P_{\mathrm{m}}|(\hat{\mathbf{g}}^{\mathrm{m}}_{000})^H \mathbf{f}^{\mathrm{m}}_{000}|^2/(\mathtt{I}_{\mathrm{m}} + \sigma^2)$, where

$$\begin{aligned}\mathtt{I}_{\mathrm{m}} \triangleq\ & P_{\mathrm{m}} \sum_{U_{n0}\in\mathcal{U}_0^{\mathrm{m}}\backslash U_{00}} |(\hat{\mathbf{g}}^{\mathrm{m}}_{000})^H \mathbf{f}^{\mathrm{m}}_{0n0}|^2 + P_{\mathrm{m}} \sum_{U_{n0}\in\mathcal{U}_0^{\mathrm{m}}} |(\tilde{\mathbf{g}}^{\mathrm{m}}_{000})^H \mathbf{f}^{\mathrm{m}}_{0n0}|^2 \\ & + P_{\mathrm{m}} \sum_{l\in\Phi_{\mathrm{m}}\backslash 0} \sum_{U_{nl}\in\mathcal{U}_l^{\mathrm{m}}} |(\mathbf{g}^{\mathrm{m}}_{l00})^H \mathbf{f}^{\mathrm{m}}_{lnl}|^2 + P_{\mathrm{s}} \sum_{j\in\Phi_{\mathrm{s}}} \sum_{l\in\Phi_{\mathrm{m}}} \sum_{U_{nl}\in\mathcal{U}_j^{\mathrm{s}}} |(\mathbf{g}^{\mathrm{s}}_{j00})^H \mathbf{f}^{\mathrm{s}}_{jnl}|^2. \end{aligned} \quad (5.20)$$

If the typical user is associated with the SBS, denoted as q_0, the downlink SINR of the typical use is $\mathtt{SINR}_{\mathrm{s}} = P_{\mathrm{s}}|(\hat{\mathbf{g}}^{\mathrm{s}}_{q_000})^H \mathbf{f}^{\mathrm{s}}_{q_000}|^2/(\mathtt{I}_{\mathrm{s}} + \sigma^2)$, where

$$\begin{aligned}\mathtt{I}_{\mathrm{s}} \triangleq\ & P_{\mathrm{s}} \sum_{l\in\Phi_{\mathrm{m}}} \sum_{U_{nl}\in\mathcal{U}_{q_0}^{\mathrm{s}}\backslash U_{00}} |(\hat{\mathbf{g}}^{\mathrm{s}}_{q_000})^H \mathbf{f}^{\mathrm{s}}_{q_0nl}|^2 + P_{\mathrm{s}} \sum_{l\in\Phi_{\mathrm{m}}} \sum_{U_{nl}\in\mathcal{U}_{q_0}^{\mathrm{s}}} |(\tilde{\mathbf{g}}^{\mathrm{s}}_{q_000})^H \mathbf{f}^{\mathrm{s}}_{q_0nl}|^2 \\ & + P_{\mathrm{m}} \sum_{l\in\Phi_{\mathrm{m}}} \sum_{U_{nl}\in\mathcal{U}_l^{\mathrm{m}}} |(\mathbf{g}^{\mathrm{m}}_{l00})^H \mathbf{f}^{\mathrm{m}}_{lnl}|^2 + P_{\mathrm{s}} \sum_{j\in\Phi_{\mathrm{s}}\backslash q_0} \sum_{l\in\Phi_{\mathrm{m}}} \sum_{U_{nl}\in\mathcal{U}_j^{\mathrm{s}}} |(\mathbf{g}^{\mathrm{s}}_{j00})^H \mathbf{f}^{\mathrm{s}}_{jnl}|^2. \end{aligned} \quad (5.21)$$

5.3.4 Approximation of Average Downlink Rate

Based on the SINR of the typical user described above, the average downlink rate of the typical user can be expressed as

$$\begin{aligned}\mathcal{R} =\ & \mathbb{E}\{\log_2(1 + \mathtt{SINR}^{\mathrm{L}}_{\mathrm{m}})\}\mathcal{A}^{\mathrm{L}}_{\mathrm{m}} + \mathbb{E}\{\log_2(1 + \mathtt{SINR}^{\mathrm{NL}}_{\mathrm{m}})\}\mathcal{A}^{\mathrm{NL}}_{\mathrm{m}} \\ & + \mathbb{E}\{\log_2(1 + \mathtt{SINR}^{\mathrm{L}}_{\mathrm{s}})\}\mathcal{A}^{\mathrm{L}}_{\mathrm{s}} + \mathbb{E}\{\log_2(1 + \mathtt{SINR}^{\mathrm{NL}}_{\mathrm{s}})\}\mathcal{A}^{\mathrm{NL}}_{\mathrm{s}}, \end{aligned} \quad (5.22)$$

where $\mathtt{SINR}^{\mathrm{L}}_{\mathrm{m}}$, $\mathtt{SINR}^{\mathrm{NL}}_{\mathrm{m}}$, $\mathtt{SINR}^{\mathrm{L}}_{\mathrm{s}}$, and $\mathtt{SINR}^{\mathrm{NL}}_{\mathrm{s}}$ denote the received SINR of the typical user when it is associated with the MBS in the LOS path and NLOS path, and with the SBS in the LOS path and NLOS path, respectively. To facilitate the rate derivation, we make the following assumptions.

Assumption 5.1 We approximate the coverage region of the MBS $\mathcal{M}$ as a ball centered at $\mathcal{M}$ with radius $C_{\mathrm{v}} = 1/\sqrt{\pi\lambda_{\mathrm{m}}}$ [31].

Let Θ_n be the point process formed by locations of the user n in each macrocell. Note that Θ_n is a perturbation of the process Φ_{m} and thus not a PPP. Obtaining the exact correlation between Φ_{m} and Θ_n requires complicated mathematical derivations and is highly intractable. Therefore, we use the similar method in [32] to model the interfering users in Θ_n, denoted by Θ'_n, as an inhomogeneous PPP. Motivated by the fact that the probability that a user scheduled by the MBS $\mathcal{M}$ in an LOS path is $\zeta_1(r, k_1 r^{\alpha^{\mathrm{L}}/\alpha^{\mathrm{NL}}})$, and in an NLOS path is $\zeta_1(k_4 r^{\alpha^{\mathrm{NL}}/\alpha^{\mathrm{L}}}, r)$, where r is the distance of this user to $\mathcal{M}$, we make the following assumption.

Assumption 5.2 The point process Θ'_n can be approximated as an inhomogeneous PPP with density being $\lambda_{\Theta'_n}(r) = \lambda_{\mathrm{m}}\left[1 - \zeta_1\left(r, k_1 r^{\frac{\alpha^{\mathrm{L}}}{\alpha^{\mathrm{NL}}}}\right) - \zeta_1\left(k_4 r^{\frac{\alpha^{\mathrm{NL}}}{\alpha^{\mathrm{L}}}}, r\right)\right]$. Moreover, for $n_1 \neq n_2$, Θ'_{n_1} and Θ'_{n_2} are independent.

Based on all the analysis and assumptions mentioned above, a tight approximation of the average downlink rate is given in the following theorem.

Theorem 5.3 The average downlink rate of typical user is approximated as

$$\begin{aligned}
\mathcal{R} &\approx \tilde{\mathcal{R}} \\
&= \mathcal{A}_{\mathrm{m}}^{\mathrm{L}} \int_0^{\infty} \int_0^{\infty} \frac{e^{-z}}{z \ln 2} \Xi(r, k_1 r^{\alpha^{\mathrm{L}}/\alpha^{\mathrm{NL}}}, k_2 r, k_1 k_3 r^{\alpha^{\mathrm{L}}/\alpha^{\mathrm{NL}}}) \Psi\left(P_{\mathrm{m}}, \mathcal{K}_{\mathrm{m}}, M_{\mathrm{m}}, \chi_1, \frac{L^{\mathrm{L}}}{r^{\alpha^{\mathrm{L}}}}\right) f_{R_{\mathrm{m}}^{\mathrm{L}}}(r) dz dr \\
&+ \mathcal{A}_{\mathrm{m}}^{\mathrm{NL}} \int_0^{\infty} \int_0^{\infty} \frac{e^{-z}}{z \ln 2} \Xi(k_4 r^{\alpha^{\mathrm{NL}}/\alpha^{\mathrm{L}}}, r, k_2 k_4 r^{\alpha^{\mathrm{NL}}/\alpha^{\mathrm{L}}}, k_3 r) \Psi\left(P_{\mathrm{m}}, \mathcal{K}_{\mathrm{m}}, M_{\mathrm{m}}, \chi_1, \frac{L^{\mathrm{NL}}}{r^{\alpha^{\mathrm{NL}}}}\right) f_{R_{\mathrm{m}}^{\mathrm{NL}}}(r) dz dr \\
&+ \mathcal{A}_{\mathrm{s}}^{\mathrm{L}} \int_0^{\infty} \int_0^{\infty} \frac{e^{-z}}{z \ln 2} \Xi\left(\frac{r}{k_2}, \frac{k_1}{k_3} r^{\alpha^{\mathrm{L}}/\alpha^{\mathrm{NL}}}, r, k_1 r^{\alpha^{\mathrm{L}}/\alpha^{\mathrm{NL}}}\right) \Psi\left(P_{\mathrm{s}}, \mathcal{K}_{\mathrm{s}}, M_{\mathrm{s}}, \chi_2, \frac{L^{\mathrm{L}}}{r^{\alpha^{\mathrm{L}}}}\right) f_{R_{\mathrm{s}}^{\mathrm{L}}}(r) dz dr \\
&+ \mathcal{A}_{\mathrm{s}}^{\mathrm{NL}} \int_0^{\infty} \int_0^{\infty} \frac{e^{-z}}{z \ln 2} \Xi\left(\frac{k_4}{k_2} r^{\alpha^{\mathrm{NL}}/\alpha^{\mathrm{L}}}, \frac{r}{k_3}, k_4 r^{\alpha^{\mathrm{NL}}/\alpha^{\mathrm{L}}}, r\right) \Psi\left(P_{\mathrm{s}}, \mathcal{K}_{\mathrm{s}}, M_{\mathrm{s}}, \chi_2, \frac{L^{\mathrm{NL}}}{r^{\alpha^{\mathrm{NL}}}}\right) f_{R_{\mathrm{s}}^{\mathrm{NL}}}(r) dz dr,
\end{aligned} \tag{5.23}$$

where

$$\begin{aligned}
&\Xi(x_1, x_2, x_3, x_4) \triangleq \\
&\exp\left(-2\pi\lambda_{\mathrm{m}}\left[\int_{x_1}^{\infty} (1 - e^{-zP_{\mathrm{m}}L^{\mathrm{L}}/\rho_1 u^{\alpha^{\mathrm{L}}}}) p_{\mathrm{m}}^{\mathrm{L}}(u) u du + \int_{x_2}^{\infty} (1 - e^{-zP_{\mathrm{m}}L^{\mathrm{NL}}/\rho_1 u^{\alpha^{\mathrm{NL}}}})(1 - p_{\mathrm{m}}^{\mathrm{L}}(u)) u du\right]\right) \\
&\times \exp\left(-2\pi\lambda_{\mathrm{s}}\left[\int_{x_3}^{\infty} (1 - e^{-zP_{\mathrm{s}}L^{\mathrm{L}}/\rho_1 u^{\alpha^{\mathrm{L}}}}) p_{\mathrm{s}}^{\mathrm{L}}(u) u du + \int_{x_4}^{\infty} (1 - e^{-zP_{\mathrm{s}}L^{\mathrm{NL}}/\rho_1 u^{\alpha^{\mathrm{NL}}}})(1 - p_{\mathrm{s}}^{\mathrm{L}}(u)) u du\right]\right),
\end{aligned} \tag{5.24}$$

and

$$\begin{aligned}
\Psi(x_1, x_2, x_3, x_4, x_5) \triangleq \exp&\left(-\frac{z x_1}{\rho_1}\left[x_5 - \frac{x_5^2}{x_2(x_5 + x_4 + \sigma^2/\tau p_p)}\right]\right) \\
- \exp&\left(-\frac{z x_1}{\rho_1}\left[x_5 + \frac{x_3 x_5^2}{x_2(x_5 + x_4 + \sigma^2/\tau p_p)}\right]\right),
\end{aligned} \tag{5.25}$$

while $\chi_1 \triangleq 2\pi \int_{C_{\mathrm{v}}}^{\infty} u \lambda_{\Theta_0'}(u) [L^{\mathrm{L}} u^{-\alpha^{\mathrm{L}}} p_{\mathrm{m}}^{\mathrm{L}}(u) + L^{\mathrm{NL}} u^{-\alpha^{\mathrm{NL}}} (1 - p_{\mathrm{m}}^{\mathrm{L}}(u))] du$, and χ_2 is obtained by replacing $p_{\mathrm{m}}^{\mathrm{L}}$ in χ_1 with $p_{\mathrm{s}}^{\mathrm{L}}$, $\mathcal{K}_{\mathrm{m}} \triangleq \mathcal{A}_{\mathrm{m}}(\lambda_{\mathrm{m}} N - 1)/\lambda_{\mathrm{m}} + 1$, $\mathcal{K}_{\mathrm{s}} \triangleq \mathcal{A}_{\mathrm{s}}(\lambda_{\mathrm{m}} N - 1)/\lambda_{\mathrm{s}} + 1$, $\rho_1 \triangleq \tilde{\mu}_1 + \sigma^2$, where

$$\begin{aligned}
\tilde{\mu}_1 \triangleq &\frac{\mathcal{A}_{\mathrm{m}}^{\mathrm{L}} P_{\mathrm{m}} M_{\mathrm{m}} \xi_1}{\mathcal{K}_{\mathrm{m}}\left(\nu_1 + \chi_1 + \frac{\sigma^2}{\tau p_p}\right)} + \frac{\mathcal{A}_{\mathrm{m}}^{\mathrm{NL}} P_{\mathrm{m}} M_{\mathrm{m}} \xi_1}{\mathcal{K}_{\mathrm{m}}\left(\nu_2 + \chi_1 + \frac{\sigma^2}{\tau p_p}\right)} \\
&+ \frac{\mathcal{A}_{\mathrm{s}}^{\mathrm{L}} P_{\mathrm{s}} M_{\mathrm{s}} \xi_2}{\mathcal{K}_{\mathrm{s}}\left(\nu_3 + \chi_2 + \frac{\sigma^2}{\tau p_p}\right)} + \frac{\mathcal{A}_{\mathrm{s}}^{\mathrm{NL}} P_{\mathrm{s}} M_{\mathrm{s}} \xi_2}{\mathcal{K}_{\mathrm{s}}\left(\nu_4 + \chi_2 + \frac{\sigma^2}{\tau p_p}\right)},
\end{aligned} \tag{5.26}$$

with $\xi_1 \triangleq 2\pi\lambda_{\mathrm{m}} \int_{C_{\mathrm{v}}}^{\infty} u[(L^{\mathrm{L}})^2 u^{-2\alpha^{\mathrm{L}}} p_{\mathrm{m}}^{\mathrm{L}}(u) + (L^{\mathrm{NL}})^2 u^{-2\alpha^{\mathrm{NL}}} (1 - p_{\mathrm{m}}^{\mathrm{L}}(u))] du$, ξ_2 is obtained by replacing $p_{\mathrm{m}}^{\mathrm{L}}$ in ξ_1 with $p_{\mathrm{s}}^{\mathrm{L}}$, $\nu_1 \triangleq \int_0^{\infty} L^{\mathrm{L}} u^{-\alpha^{\mathrm{L}}} f_{R_{\mathrm{m}}^{\mathrm{L}}}(u) du$, $\nu_2 \triangleq \int_0^{\infty} L^{\mathrm{NL}} u^{-\alpha^{\mathrm{NL}}} f_{R_{\mathrm{m}}^{\mathrm{NL}}}(u) du$, $\nu_3 \triangleq \int_0^{\infty} L^{\mathrm{L}} u^{-\alpha^{\mathrm{L}}} f_{R_{\mathrm{s}}^{\mathrm{L}}}(u) du$, and $\nu_4 \triangleq \int_0^{\infty} L^{\mathrm{NL}} u^{-\alpha^{\mathrm{NL}}} f_{R_{\mathrm{s}}^{\mathrm{NL}}}(u) du$.

Proof: See Appendix A.4.

Equation (5.23) quantifies how all the key features of an HCN, i.e. LOS/NLOS transmissions, interference, and deployment strategy, affect the average downlink rate. From Theorem 5.3, we can analyze the effects of λ_{s} and M_{m} on the average downlink rate in the following propositions. Note that increasing λ_{s} means deploying more small cells and increasing M_{m} means expanding the antenna array at the MBS.

Proposition 5.1 For massive MIMO systems that do not have small cells, the downlink rate increases with M_{m} and saturates to a constant when $M_{\mathrm{m}} \to \infty$.

Proof: See Appendix A.5.

Note that by successively adding small cells into the macrocell, more users will be associated with SBSs, so $\mathcal{A}_{\mathrm{s}}^{\mathrm{L}}$ and $\mathcal{A}_{\mathrm{s}}^{\mathrm{NL}}$ grow and $\mathcal{A}_{\mathrm{m}}^{\mathrm{L}}$ and $\mathcal{A}_{\mathrm{m}}^{\mathrm{NL}}$ reduce, which will decay the impact of M_{m} on improving the downlink rate.

Proposition 5.2 The downlink rate increases with λ_{s} up to a certain value, after which it decreases and eventually approaches zero as $\lambda_{\mathrm{s}} \to \infty$.

Proof: See Appendix A.6.

Proposition 5.2 reveals that a larger λ_{s} can improve the downlink rate, but when λ_{s} exceeds a critical threshold, it will instead impair the rate performance, which is different from the monotonic behavior of expanding BS antennas in Proposition 5.1.

A more precise comparison between massive MIMO and small cells will be given in Section 5.4. Moreover, the validation of our analysis as well as several numerical results based on Eq. (5.23) will also be shown in Section 5.4 to give more practical insights into the design of HCN.

5.4 Numerical Results

In this section, we validate the accuracy of our analysis through simulations and evaluate the performance of the HCN via numerical results. In particular, we compare the performance of the downlink rate achieved by massive MIMO and small cells. Then, the optimal network configuration parameters are provided as a useful guidance for practical implementations. Finally, we explore the antenna allocation between centralized MBSs and distributed SBSs.

From a practical perspective, we use the linear LOS probability function adopted in 3GPP [33] for both the MBS and SBS pathlosses, which is

$$p_{\mathrm{m}}^{\mathrm{L}}(r) = p_{\mathrm{s}}^{\mathrm{L}}(r) = \begin{cases} 1 - r/d_{\mathrm{L}}, & 0 < r \le d_{\mathrm{L}}, \\ 0, & r > d_{\mathrm{L}}. \end{cases} \tag{5.27}$$

According to [33] and [34], parameters used in our simulation are set as follows: $d_{\mathrm{L}} = 0.3$ km, $L^{\mathrm{L}} = 10^{-10.38}$, $L^{\mathrm{NL}} = 10^{-14.54}$, $\alpha^{\mathrm{L}} = 2.09$, $\alpha^{\mathrm{NL}} = 3.75$, $P_{\mathrm{m}} = 53$ dBm, $P_{\mathrm{s}} = 33$ dBm, and $\sigma^2 = -104$ dBm. Moreover, $p_p = 24$ dBm and $\lambda_{\mathrm{m}} = 1$ BSs/km^2.

5.4.1 Validation of Analytical Results

In Figure 5.2, the simulated user association probability is compared with our analytical results in Theorem 5.1. Clearly, we can see good agreement between the simulated and analytical values, which justifies the accuracy of our analysis. When λ_{s} is small, users are likely to be associated with the MBS due to its large transmit power and especially in NLOS because of the long transmission distance caused by the sparse deployment of MBSs. As λ_{s} increases, more users will be associated with SBSs; thus the probability of a user associated with an MBS reduces, where, however, only the probability in NLOS declines remarkably and the probability in LOS barely changes. This is because the users associated with MBSs in LOS are located closely to the MBSs such that the addition of low-power SBSs cannot change their associations very much. Moreover, the probability of a user associated with an SBS obviously increases with λ_s. However, we can also observe that the probability of a user associated with an SBS in NLOS decreases after some density point. This is because, for small λ_s, connections between users and SBSs can be in NLOS, while as λ_s grows large, NLOS transmissions fade away. When $\lambda_s \to \infty$, no NLOS connections exist, and both the probabilities of users associated with the MBS and SBS saturate to constant values.

In Figure 5.3, the simulated average downlink rate is compared with the analytical approximation in Eq. (5.23) under different values of M_m and λ_s. We can see that the analytical approximation and simulation results match and follow the same trend fairly well, thus verifying Theorem 5.3.

Due to the tightness between the simulations and analysis, we will use the latter for our following investigations. Note that the number of BS antennas should be larger than the number of users it serves. Hence, M_m is set to be larger than N and M_s is set according to the average number of users associated with each SBS, as given in Corollary 5.1. From Remark 5.1, we know that $\mathcal{N}_s$ reduces as λ_s grows. Hence, the maximum $\mathcal{N}_s$ is acquired with the minimum λ_s, which is equal to λ_m. From Figure 5.2, we find that when $\lambda_s = 1$ BSs/km^2, $\mathcal{A}_s^L + \mathcal{A}_s^{NL} = 0.11$, and thus the minimum $\mathcal{N}_s = 0.11\lambda_m N$.

5.4.2 Comparison between Massive MIMO and Small Cells

In this subsection, we aim to compare the performance of massive MIMO and small cells. In particular, Figure 5.4 reveals the effect of increasing λ_s with fixed M_m, while

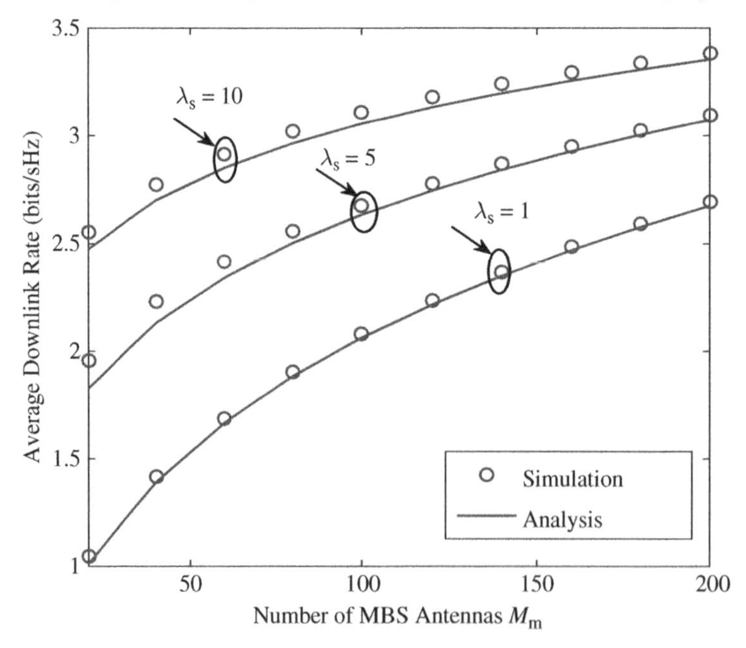

Figure 5.3 Average downlink rate versus number of MBS antennas M_m, where $N = 10$, $M_s = 5$, and $B = 1$.

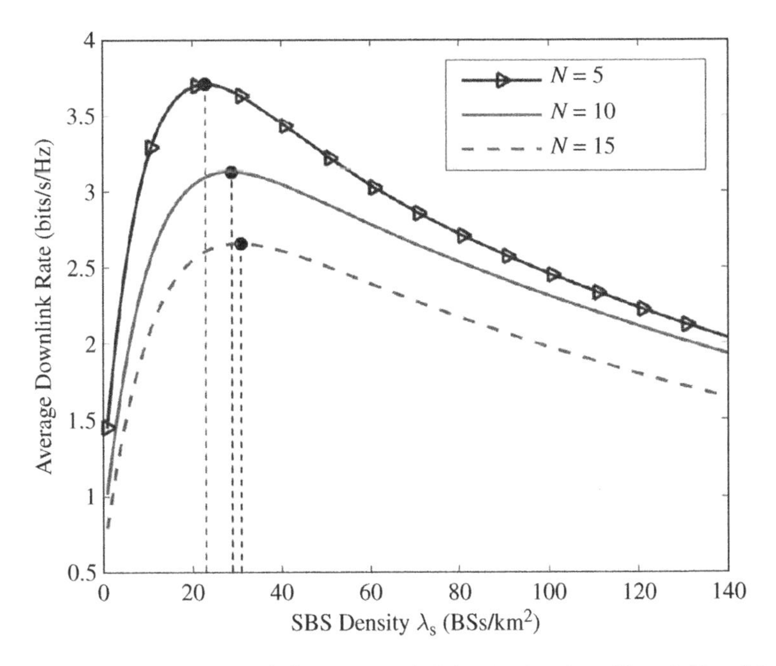

Figure 5.4 Average downlink rate versus SBS density λ_s, where $M_s = 5$, $M_m = 20$, and $B = 1$.

Figure 5.5 shows the effect of increasing M_m with fixed λ_s. For fairness, we set the same starting configuration for these two cases, where the MBS antenna array is $M_m = 20$ and the SBS density is $\lambda_s = 1$ BSs/km^2.

Figure 5.4 shows that the downlink rate burgeons with the increment of λ_s until reaching a critical threshold, after which the expectancy for further improvement breaks into a slow decline. This is because when λ_s is small, the network can benefit greatly from the small cell densification due to the reduced distance between transceivers. However, when λ_s becomes large, more and more interference paths switch from NLOS to LOS, resulting in an aggregated interference that significantly impairs the rate performance. This observation is consistent with the conclusion in [25] where the setup is with single antenna BSs. The critical SBS density threshold, i.e. the optimal λ_s that maximizes the downlink rate, is marked out by black dots. Note that as the scheduled user number N grows, the optimal λ_s also grows, since more small cells are needed in a more crowded environment. The optimal λ_s for different N is summarized in Figure 5.6.

Figure 5.5 shows that the downlink rate increases monotonically with respect to M_m, but it cannot grow without bounds and saturates to a constant value limited by pilot contamination. It can also be observed that the smaller N leads to a better rate performance, because we are considering the downlink rate per user, and less number of users gives less interferences.

By comparing Figures 5.4 and 5.5, we can see that adding small cells into a sparse network is more effective in boosting up the downlink rate than expanding antenna arrays at MBSs. Densification is remarkably beneficial when λ_s is low, where the rate can be enhanced almost linearly. For example, when $N = 10$, we observe that increasing λ_s by 10 can provide almost 100% rate gain. That means, on average, adding 10 small cells per macrocell can double the downlink rate. In contrast, for the massive MIMO system in

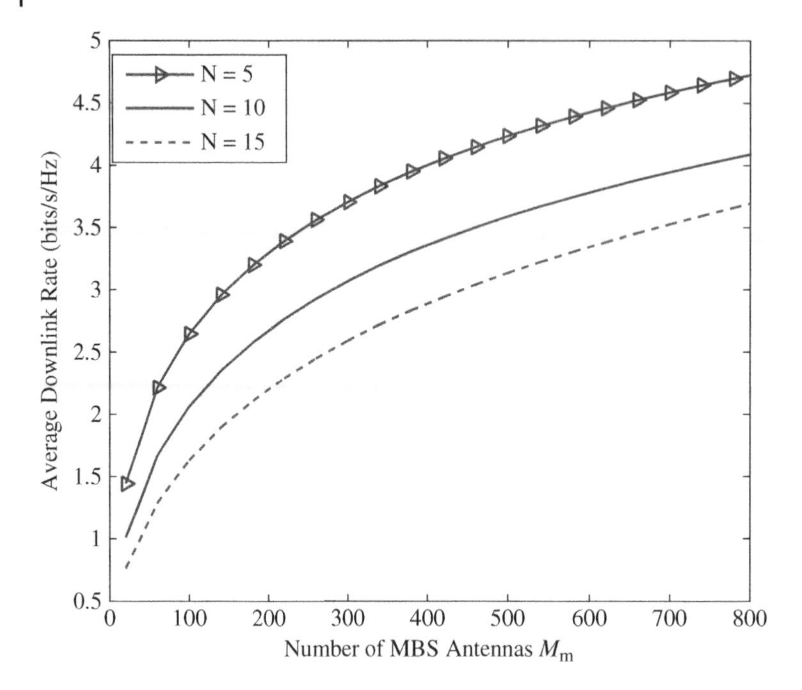

Figure 5.5 Average downlink rate versus the number of MBS antennas M_m, where $\lambda_s = 1$ BSs/km^2, $M_s = 5$, and $B = 1$.

Figure 5.5, we need to add more than 100 antennas to have the same rate gain. However, the peak rate obtained from small cell densification is lower than that from antenna array extension. Therefore, we conclude that deploying small cells can improve the downlink rate fast and effectively, but it will stop benefiting and further impair the network capacity when the SBS density exceeds a critical threshold. On the contrary, though the massive MIMO technique improves the system performance slowly, expanding the antenna size can always benefit the network capacity, and the maximum rate with large M_m is greater than that obtained from small cell deployment. In summary, if the rate demand is low, deploying small cells is preferred due to its rapid rate gain; however, if the rate requirement is high, the massive MIMO technique is more preferable due to the higher rate it provides.

5.4.3 Optimal Network Configuration

In this subsection, we are interested in obtaining the optimal SBS density λ_s that can maximize the average downlink rate as shown in Figure 5.4. Moreover, with massive antennas available at MBSs, the biasing policy should be adjusted to fully exploit the excessive degrees-of-freedom. Therefore, we also investigate the impact of a biasing factor in this subsection.

Figure 5.6 illustrates how the optimal λ_s varies with the number of users scheduled by the MBS, N, under different BS antenna numbers. We can see that the optimal λ_s increases monotonically with N, since the most effective way to offload traffic caused by crowded users is deploying more SBSs. We also find that the optimal λ_s increases with

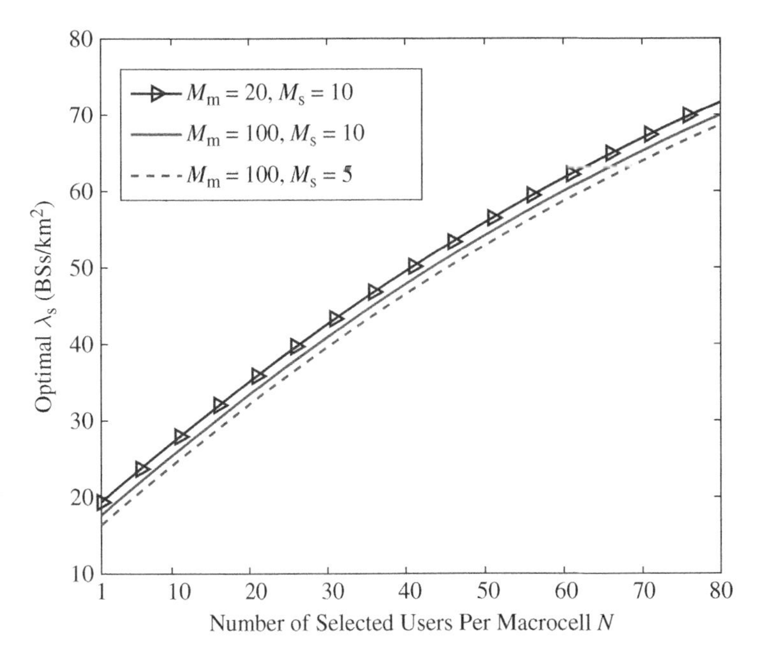

Figure 5.6 The optimal λ_s versus the number of selected users per macrocell N, where $B = 1$.

M_s and decreases with M_m, which coincides with a general intuition that a powerful MBS with a large antenna array can provide a sufficient data rate for users in its coverage and thus requires less SBS deployment, while SBSs with more antennas are desirable for rate enhancement and thus optimal λ_s increases.

Figure 5.7 presents the optimal bias factor B under different MBS antenna numbers M_m. Clearly, the optimal B decreases as M_m grows, since larger M_m can bring a more significant rate gain; thus a smaller B is desired to push more users associated with MBSs and benefit from the vast degrees-of-freedom. We also note that the optimal B decreases with increments of λ_s due to a similar argument. All these results can serve as guidance for practical network design.

5.5 Conclusion

In this chapter, we have developed a framework for downlink HCN that consists of randomly distributed MBSs and SBSs with multiple antennas, and the LOS and NLOS transmissions are differentiated. Using stochastic geometry, we have derived a tight approximation of downlink rates to compare the performance between densifying small cells and expanding BS antenna arrays. Interestingly, we have found that adding small cells into the network is more effective in boosting the downlink rate than expanding antenna arrays at MBS. However, when the small cell density exceeds a critical threshold value, the spatial densification stops benefiting and further impairs the network capacity. In contrast, an expanding BS antenna array can always improve the capacity until reaching an upper bound caused by pilot contamination, and this upper bound is larger than the peak rate obtained from the deployment of small cells. Hence, for low rate

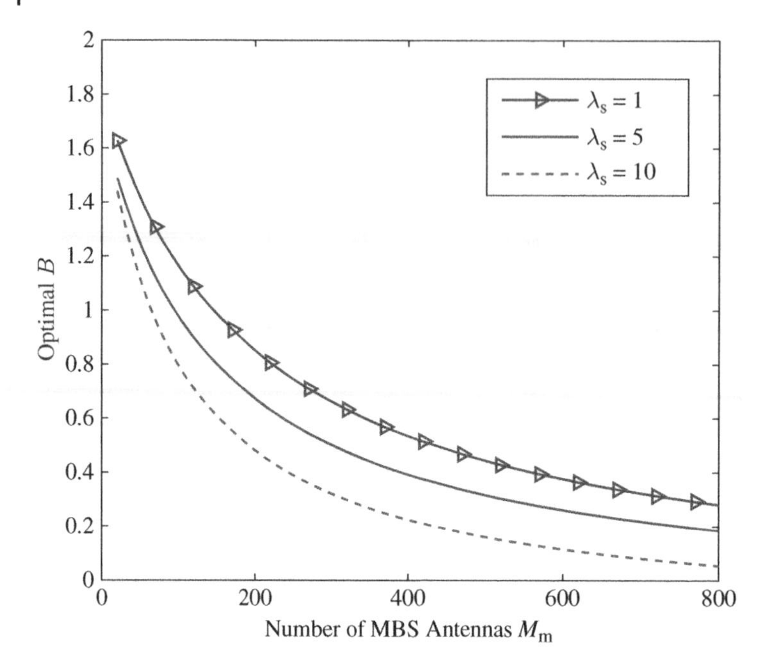

Figure 5.7 The optimal biased factor B versus the number of MBS antennas M_{m}, where $M_{\mathrm{s}} = 5$ and $N = 10$.

requirements, the small cell is preferred due to its sheer rate gain, but for higher rate requirements, the massive MIMO is preferred due to a higher achievable rate. In conclusion, this work has provided a further understanding on deploying small cells and massive MIMO in future HCNs.

Appendix

A.1 Proof of Theorem 5.1

The distance from the typical user to its nearest BS in each $\Phi_{\mathrm{m}}^{\mathrm{L}}$, $\Phi_{\mathrm{m}}^{\mathrm{NL}}$, $\Phi_{\mathrm{s}}^{\mathrm{L}}$, and $\Phi_{\mathrm{s}}^{\mathrm{NL}}$ are denoted as $S_{\mathrm{m}}^{\mathrm{L}}$, $S_{\mathrm{m}}^{\mathrm{NL}}$, $S_{\mathrm{s}}^{\mathrm{L}}$, and $S_{\mathrm{s}}^{\mathrm{NL}}$, respectively. In general, the typical user can receive four types of transmit powers, i.e. from the MBS through the LOS and NLOS path, and from SBS through the LOS and NLOS path, respectively. The typical user is associated with the MBS in a LOS path means that this received power is higher than the other three cases, which can be formulated from Eqs. (5.4) and (5.5) as follows:

$$\begin{aligned}\mathbb{P}[E_{\mathrm{m}}^{\mathrm{L}}] &= \mathbb{P}\left[\left\{\frac{P_{\mathrm{m}}L^{\mathrm{L}}}{(S_{\mathrm{m}}^{\mathrm{L}})^{\alpha^{\mathrm{L}}}} > \frac{P_{\mathrm{m}}L^{\mathrm{NL}}}{(S_{\mathrm{m}}^{\mathrm{NL}})^{\alpha^{\mathrm{NL}}}}\right\} \bigcap \left\{\frac{P_{\mathrm{m}}L^{\mathrm{L}}}{(S_{\mathrm{m}}^{\mathrm{L}})^{\alpha^{\mathrm{L}}}} > \frac{BP_{\mathrm{s}}L^{\mathrm{L}}}{(S_{\mathrm{s}}^{\mathrm{L}})^{\alpha^{\mathrm{L}}}}\right\}\right. \\ &\qquad \left. \bigcap \left\{\frac{P_{\mathrm{m}}L^{\mathrm{L}}}{(S_{\mathrm{m}}^{\mathrm{L}})^{\alpha^{\mathrm{L}}}} > \frac{BP_{\mathrm{s}}L^{\mathrm{NL}}}{(S_{\mathrm{s}}^{\mathrm{NL}})^{\alpha^{\mathrm{NL}}}}\right\}\right] \\ &= \int_0^{\infty} \mathbb{P}\left[S_{\mathrm{m}}^{\mathrm{NL}} > k_1 r^{\alpha^{\mathrm{L}}/\alpha^{\mathrm{NL}}} | S_{\mathrm{m}}^{\mathrm{L}} = r\right] \cdot \mathbb{P}\left[S_{\mathrm{s}}^{\mathrm{L}} > k_2 r | S_{\mathrm{m}}^{\mathrm{L}} = r\right] \\ &\qquad \mathbb{P}\left[S_{\mathrm{s}}^{\mathrm{NL}} > k_1 k_3 r^{\alpha^{\mathrm{L}}/\alpha^{\mathrm{NL}}} | S_{\mathrm{m}}^{\mathrm{L}} = r\right] \cdot f_{S_{\mathrm{m}}^{\mathrm{L}}}(r)dr. \end{aligned} \tag{5.28}$$

With results from a void probability, the first term in the integral of Eq. (5.28) can be calculated as

$$\mathbb{P}[S_{\mathrm{m}}^{\mathrm{NL}} > k_1 r^{\alpha^{\mathrm{L}}/\alpha^{\mathrm{NL}}} | S_{\mathrm{m}}^{\mathrm{L}} = r] = \exp\left(- \int_0^{k_1 r^{\alpha^{\mathrm{L}}/\alpha^{\mathrm{NL}}}} \lambda_{\mathrm{m}} (1 - p_{\mathrm{m}}^{\mathrm{L}}(u)) 2\pi u du \right), \tag{5.29}$$

and the left terms in the integral of Eq. (5.28) can also be acquired similarly. Moreover, the PDF of $S_{\mathrm{m}}^{\mathrm{L}}$ can be obtained as

$$f_{S_{\mathrm{m}}^{\mathrm{L}}}(r) = \frac{d(1 - \mathbb{P}[S_{\mathrm{m}}^{\mathrm{L}} > r])}{dr} = \exp\left(- \int_0^r \lambda_{\mathrm{m}} p_{\mathrm{m}}^{\mathrm{L}}(u) 2\pi u du \right) p_{\mathrm{m}}^{\mathrm{L}}(r) 2\pi r \lambda_{\mathrm{m}}. \tag{5.30}$$

Therefore, Eq. (5.6) follows from the substitution of Eqs. (5.29) and (5.30) into Eq. (5.28). By performing the same procedure, we can obtain $\mathbb{P}[E_{\mathrm{m}}^{\mathrm{NL}}]$, $\mathbb{P}[E_{\mathrm{s}}^{\mathrm{L}}]$, and $\mathbb{P}[E_{\mathrm{s}}^{\mathrm{NL}}]$.

A.2 Proof of Corollary 5.1

Since each MBS randomly selects N users into its cell, the density of scheduled users are $\lambda_{\mathrm{m}} N$. Let S be the area of the entire network. Then, the total number of scheduled users in the network is $\lambda_{\mathrm{m}} NS$, where, on average, $\mathcal{A}_{\mathrm{m}} \lambda_{\mathrm{m}} NS$ users are associated with the MBS and $\mathcal{A}_{\mathrm{s}} \lambda_{\mathrm{m}} NS$ users are associated with the SBS. Therefore, since the total number of MBSs in the network is $\lambda_{\mathrm{m}} S$ and the total number of SBSs is $\lambda_{\mathrm{s}} S$, we can get an average number of users associated with each MBS as $\mathcal{N}_{\mathrm{m}} = \mathcal{A}_{\mathrm{m}} \lambda_{\mathrm{m}} NS / \lambda_{\mathrm{m}} S = \mathcal{A}_{\mathrm{m}} N$ and the average number of users associated with each SBS as $\mathcal{N}_{\mathrm{s}} = \mathcal{A}_{\mathrm{s}} \lambda_{\mathrm{m}} NS / \lambda_{\mathrm{s}} S = \mathcal{A}_{\mathrm{s}} \lambda_{\mathrm{m}} N / \lambda_{\mathrm{s}}$.

A.3 Proof of Theorem 5.2

The distance from the typical user to its nearest BS in each $\Phi_{\mathrm{m}}^{\mathrm{L}}$, $\Phi_{\mathrm{m}}^{\mathrm{NL}}$, $\Phi_{\mathrm{s}}^{\mathrm{L}}$, and $\Phi_{\mathrm{s}}^{\mathrm{NL}}$ are denoted as $S_{\mathrm{m}}^{\mathrm{L}}$, $S_{\mathrm{m}}^{\mathrm{NL}}$, $S_{\mathrm{s}}^{\mathrm{L}}$, and $S_{\mathrm{s}}^{\mathrm{NL}}$, respectively. To derive the PDF of $R_{\mathrm{m}}^{\mathrm{L}}$, we first investigate its complementary cumulative distribution function as $\overline{F}_{R_{\mathrm{m}}^{\mathrm{L}}}(r) = \mathbb{P}[R_{\mathrm{m}}^{\mathrm{L}} > r]$. The event $R_{\mathrm{m}}^{\mathrm{L}} > r$ is equivalent to that of $S_{\mathrm{m}}^{\mathrm{L}} > r$ given that the typical user is associated with the MBS in an LOS path, i.e.

$$\mathbb{P}[R_{\mathrm{m}}^{\mathrm{L}} > r] = \mathbb{P}[S_{\mathrm{m}}^{\mathrm{L}} > r | E_{\mathrm{m}}^{\mathrm{L}}] = \mathbb{P}[S_{\mathrm{m}}^{\mathrm{L}} > r, E_{\mathrm{m}}^{\mathrm{L}}] / \mathbb{P}[E_{\mathrm{m}}^{\mathrm{L}}]. \tag{5.31}$$

A similar argument as in Eq. (5.28) leads us to the following calulation:

$$\mathbb{P}[S_{\mathrm{m}}^{\mathrm{L}} > r, E_{\mathrm{m}}^{\mathrm{L}}] = \int_r^{\infty} \mathbb{P}[S_{\mathrm{m}}^{\mathrm{NL}} > k_1 x^{\alpha^{\mathrm{L}}/\alpha^{\mathrm{NL}}} | S_{\mathrm{m}}^{\mathrm{L}} = x] \cdot \mathbb{P}[S_{\mathrm{s}}^{\mathrm{L}} > k_2 x | S_{\mathrm{m}}^{\mathrm{L}} = x]$$
$$\mathbb{P}[S_{\mathrm{s}}^{\mathrm{NL}} > k_1 k_3 x^{\alpha^{\mathrm{L}}/\alpha^{\mathrm{NL}}} | S_{\mathrm{m}}^{\mathrm{L}} = x] \cdot f_{S_{\mathrm{m}}^{\mathrm{L}}}(x) dx, \tag{5.32}$$

By using results in Eqs. (5.29) and (5.30), we have

$$\mathbb{P}[R_{\mathrm{m}}^{\mathrm{L}} > r] = \frac{2\pi \lambda_{\mathrm{m}}}{\mathcal{A}_{\mathrm{m}}^{\mathrm{L}}} \int_r^{\infty} x p_{\mathrm{m}}^{\mathrm{L}}(x) \zeta_1(x, k_1 x^{\alpha^{\mathrm{L}}/\alpha^{\mathrm{NL}}}) \zeta_2(k_2 x, k_1 k_3 x^{\alpha^{\mathrm{L}}/\alpha^{\mathrm{NL}}}) dx. \tag{5.33}$$

The PDF of $R_{\mathrm{m}}^{\mathrm{L}}$ then follows from taking the derivative of $1 - \overline{F}_{R_{\mathrm{m}}^{\mathrm{L}}}(r)$ with respect to r. The PDFs of $R_{\mathrm{m}}^{\mathrm{NL}}$, $R_{\mathrm{s}}^{\mathrm{L}}$, and $R_{\mathrm{s}}^{\mathrm{NL}}$ can be obtained from the same procedure, which are omitted due to space limitations.

A.4 Proof of Theorem 5.3

In this chapter, we consider the maximal-ratio-transmission (MRT) precoding[2] where the precoding vector $\mathbf{f}_{lnl}^{\mathrm{m}}$ is given by $\mathbf{f}_{lnl}^{\mathrm{m}} = \kappa_l^{\mathrm{m}} \hat{\mathbf{g}}_{lnl}^{\mathrm{m}} / \sqrt{\mathbb{E}_{\mathbf{h}}\{\|\hat{\mathbf{g}}_{lnl}^{\mathrm{m}}\|^2\}}$, where $\mathbb{E}_{\mathbf{h}}$ means the average over fast fading and κ_l^{m} is a power normalization factor that conforms to the constraint[3] $\mathbb{E}_{\mathbf{h}}\{\mathtt{tr}(\mathbf{F}_l^{\mathrm{m}}(\mathbf{F}_l^{\mathrm{m}})^H)\} = 1$, where $\mathbf{F}_l^{\mathrm{m}} = \{[\ldots, \mathbf{f}_{lnl}^{\mathrm{m}}, \ldots] \,|U_{nl} \in \mathcal{U}_l^{\mathrm{m}}\}$. Therefore, we find that $\kappa_l^{\mathrm{m}} = \sqrt{1/|\mathcal{U}_l^{\mathrm{m}}|}$. Similarly, the precoding vector $\mathbf{f}_{jnl}^{\mathrm{s}}$ for SBS is given as $\mathbf{f}_{jnl}^{\mathrm{s}} = \kappa_j^{\mathrm{s}} \hat{\mathbf{g}}_{jnl}^{\mathrm{s}} / \sqrt{\mathbb{E}_{\mathbf{h}}\{\|\hat{\mathbf{g}}_{jnl}^{\mathrm{s}}\|^2\}}$, where $\kappa_j^{\mathrm{s}} = \sqrt{1/|\mathcal{U}_j^{\mathrm{s}}|}$.

We take the derivation of $\mathtt{SINR}_{\mathrm{m}}^{\mathrm{L}}$, for example. With the approximation in [[8], Lemma 1], Assumption 5.1, and Campbell's theorem [36] as well as some algebraic operations, we can find that

$$\mathbb{E}\{\log_2(1 + \mathtt{SINR}_{\mathrm{m}}^{\mathrm{L}})\} \approx \mathbb{E}_{\varphi}\left\{\log_2\left(1 + \frac{\frac{P_{\mathrm{m}}(M_{\mathrm{m}}+1)(\varphi_{000}^{\mathrm{m}})^2}{\mathcal{K}_{\mathrm{m}}(\varphi_{000}^{\mathrm{m}}+\chi_1+\sigma^2/\tau p_p)}}{P_{\mathrm{m}}\sum\limits_{l\in\Phi_{\mathrm{m}}}\varphi_{l00}^{\mathrm{m}} + P_{\mathrm{s}}\sum\limits_{j\in\Phi_{\mathrm{s}}}\varphi_{j00}^{\mathrm{s}} - \frac{P_{\mathrm{m}}(\varphi_{000}^{\mathrm{m}})^2}{\mathcal{K}_{\mathrm{m}}(\varphi_{000}^{\mathrm{m}}+\chi_1+\sigma^2/\tau p_p)} + \rho_1}\right)\right\}, \tag{5.34}$$

where $\mathbb{E}_{\varphi}$ denotes the average over pathloss. Then, the rate of Eq. (5.34) can be further calculated with the continuous mapping theorem [37] and the probability generating functional of the PPP [38]. Following the same procedure, we can also derive the rates $\mathbb{E}\{\log_2(1 + \mathtt{SINR}_{\mathrm{m}}^{\mathrm{NL}})\}$, $\mathbb{E}\{\log_2(1 + \mathtt{SINR}_{\mathrm{s}}^{\mathrm{L}})\}$, and $\mathbb{E}\{\log_2(1 + \mathtt{SINR}_{\mathrm{s}}^{\mathrm{NL}})\}$. Then the desired result can be obtained from Eq. (5.22). A more detailed proof of Theorem 5.3 is given in Appendix D of [38].

A.5 Proof of Proposition 5.1

For massive MIMO systems that do not have small cells, Eq. (5.23) reduces to the rate with $\lambda_{\mathrm{s}} = 0$. From Eq. (5.34), we know that $\mathtt{SINR}_{\mathrm{m}}^{\mathrm{L}}$ is a monotonic function of M_{m}, while when $M_{\mathrm{m}} \gg 1$, $\mathtt{SINR}_{\mathrm{m}}^{\mathrm{L}}$ grows with the increment of M_{m}. Therefore, it can be deduced that $\mathtt{SINR}_{\mathrm{m}}^{\mathrm{L}}$ monotonically increases with M_{m} over the whole feasible region. After getting similar behavior for $\mathtt{SINR}_{\mathrm{m}}^{\mathrm{NL}}$, we know that the downlink rate with $\lambda_{\mathrm{s}} = 0$ is a monotonic increasing function of M_{m}. When $M_{\mathrm{m}} \to \infty$, it is easy to observe from Eq. (5.34) that the downlink rate will saturate to a constant.

A.6 Proof of Proposition 5.2

Since $\mathcal{A}_{\mathrm{m}}^{\mathrm{L}\prime}$, $\mathcal{A}_{\mathrm{m}}^{\mathrm{NL}\prime}$, ν_1', and ν_2' are all equal to 0 at $\lambda_{\mathrm{s}} = 0$, we know that ρ_1' and further that $\Xi'(\cdot)$ and $\Psi'(\cdot)$ are also equal to 0 at $\lambda_{\mathrm{s}} = 0$. Hence, with $\mathcal{A}_{\mathrm{s}}^{\mathrm{L}\prime}$ and $\mathcal{A}_{\mathrm{s}}^{\mathrm{NL}\prime} > 0$ at $\lambda_{\mathrm{s}} = 0$, we know that $\tilde{R}' > 0$ at $\lambda_{\mathrm{s}} = 0$, which denotes that the downlink rate is improved at the beginning of adding small cells.

---

2 Note that a more sophisticated precoding method like zero-forcing precoding can also be applied here. The maximum-ratio-transmission precoding is used in this work to provide more insights on our main targeted problem.

3 Here we consider an average power normalization over the fast fading [35], which can be an instantaneous constraint for each large-scale realization.

When $\lambda_{\mathrm{s}} \rightarrow \infty$, $\mathcal{A}_{\mathrm{m}}^{\mathrm{L}}$, $\mathcal{A}_{\mathrm{m}}^{\mathrm{NL}}$, and $\mathcal{A}_{\mathrm{s}}^{\mathrm{NL}}$ tend to 0 while $\mathcal{A}_{\mathrm{s}}^{\mathrm{L}}$ tends to 1. That is, the typical user is almost certainly associated with the SBS in LOS, and thus from Eqs. (5.22) and (5.34), the downlink rate becomes

$$\mathcal{R} \approx \tilde{\mathcal{R}} = \mathbb{E}_{\varphi}\left\{\log_2\left(1 + \frac{\frac{P_{\mathrm{s}}(M_{\mathrm{s}}+1)(\varphi_{q_0 00}^{\mathrm{s}})^2}{\mathcal{K}_{\mathrm{s}}(\varphi_{q_0 00}^{\mathrm{m}}+\chi_2+\sigma^2/\tau p_p)}}{P_{\mathrm{m}} \sum_{l\in\Phi_{\mathrm{m}}} \varphi_{l00}^{\mathrm{m}} + P_{\mathrm{s}} \sum_{j\in\Phi_{\mathrm{s}}} \varphi_{j00}^{\mathrm{s}} - \frac{P_{\mathrm{s}}(\varphi_{q_0 00}^{\mathrm{s}})^2}{\mathcal{K}_{\mathrm{s}}(\varphi_{q_0 00}^{\mathrm{s}}+\chi_2+\sigma^2/\tau p_p)} + \rho_1}\right)\right\}. \tag{5.35}$$

Moreover, as $\lambda_{\mathrm{s}} \rightarrow \infty$, the signals from SBSs become so strong that they absolutely dominate the propagation. Hence, the left terms, i.e. signals from MBSs and pilot-contaminated users as well as the noise, can be omitted. Then, with $K_{\mathrm{s}} \rightarrow 1$, Eq. (5.35) is equivalent to

$$\tilde{\mathcal{R}}^{\mathrm{E}} = \mathbb{E}_{\varphi}\left\{\log_2\left(1 + \frac{(M_{\mathrm{s}}+1)\varphi_{q_0 00}^{\mathrm{s}}}{\sum_{j\in\Phi_{\mathrm{s}}\backslash q_0} \varphi_{j00}^{\mathrm{s}}}\right)\right\}. \tag{5.36}$$

From [16], we know that when densifying SBSs, the distance to the closest SBS and to the interfering SBSs decrease at the same rate. In addition, we know the transmission from interfered SBSs will change from NLOS to LOS during the SBS densification, and thus the interference in Eq. (5.36), i.e. $\sum_{j\in\Phi_{\mathrm{s}}\backslash q_0} \varphi_{j00}^{\mathrm{s}}$, grows faster than the desired signal $\varphi_{q_0 00}^{\mathrm{s}}$. Hence, Eq. (5.36) tends to zero when $\lambda_{\mathrm{s}} \rightarrow \infty$, which denotes that overly densifying SBSs will impair the rate performance until zero.

Combining these two results, it can be concluded that the downlink rate first increases and then decreases as λ_{s} grows.

Bibliography

1 V. Jungnickel, K. Manolakis, W. Zirwas, B. Panzner, V. Braun, M. Lossow, M. Sternad, R. Apelfrojd, and T. Svensson, "The role of small cells, coordinated multipoint, and massive MIMO in 5G," *IEEE Commun. Mag.*, vol. 52, no. 5, pp. 44–51, May 2014.

2 I. Hwang, B. Song, and S.S. Soliman, "A holistic view on hyper-dense heterogeneous and small cell networks," *IEEE Commun. Mag.*, vol. 51, no. 6, pp. 20–27, June 2013.

3 F. Rusek, D. Persson, B.K. Lau, E.G. Larsson, T.L. Marzetta, O. Edfors, and F. Tufvesson, "Scaling up MIMO: opportunities and challenges with very large arrays," *IEEE Signal Process. Mag.*, vol. 30, no. 1, pp. 40–60, Jan. 2013.

4 E.G. Larsson, O. Edfors, F. Tufvesson, and T.L. Marzetta, "Massive MIMO for next generation wireless systems," *IEEE Commun. Mag.*, vol. 52, no. 2, pp. 186–195, Feb. 2014.

5 A. Pitarokoilis, S.K. Mohammed, and E.G. Larsson, "On the optimality of single-carrier transmission in large-scale antenna systems," *IEEE Wireless Commun. Lett.*, vol. 1, no. 4, pp. 276–279, Aug. 2012.

6 T.L. Marzetta, "Noncooperative cellular wireless with unlimited numbers of base station antennas," *IEEE Trans. Wireless Commun.*, vol. 9, no. 11, pp. 3590–3600, Nov. 2010.
7 J. Hoydis, S. ten Brink, and M. Debbah, "Massive MIMO in the UL/DL of cellular networks: How many antennas do we need?" *IEEE J. Sel. Areas Commun.*, vol. 31, no. 2, pp. 160–171, Feb. 2013.
8 Q. Zhang, S. Jin, K.-K. Wong, H. Zhu, and M. Matthaiou, "Power scaling of uplink massive MIMO systems with arbitrary-rank channel means," *IEEE J. Sel. Topics Signal Process.*, vol. 8, no. 5, pp. 966–981, Oct. 2014.
9 Q. Zhang, S. Jin, M. McKay, D. Morales-Jimenez, and H. Zhu, "Power allocation schemes for multicell massive MIMO systems," *IEEE Trans. Wireless Commun.*, vol. 14, no. 11, pp. 5941–5955, Nov. 2015.
10 H. Q. Ngo, E.G. Larsson, and T.L. Marzetta, "Energy and spectral efficiency of very large multiuser MIMO systems," *IEEE Trans. Commun.*, vol. 61, no. 4, pp. 1436–1449, Apr. 2013.
11 T.Q.S. Quek, G. de la Roche, İ. Güvenç, and M. Kountouris, *Small Cell Networks: Deployment, PHY Techniques, and Resource Allocation*. Cambridge University Press, 2013.
12 M. Kamel, W. Hamouda, and A. Youssef, "Ultra-dense networks: a survey," *IEEE Commun. Surveys Tuts.*, vol. 18, no. 4, pp. 2522–2545, Qtr. 2016.
13 H.D. Nguyen and S. Sun, "Massive MIMO versus small-cell systems: spectral and energy efficiency comparison," in *Proc. IEEE International Conference on Communications* (ICC 2016), Kuala Lumpur, Malaysia, May 2016, pp. 1–6.
14 W. Liu, S. Han, and C. Yang, "Energy efficiency comparison of massive MIMO and small cell network," in *Proc. IEEE GlobalSIP*, Atlanta, GA, USA, Dec. 2014, pp. 617–621.
15 E. Björnson, M. Kountouris, and M. Debbah, "Massive MIMO and small cells: improving energy efficiency by optimal soft-cell coordination," in *Proc. IEEE ICT*, Jeju Islang, Korea, Oct. 2013, pp. 1–5.
16 J.G. Andrews, F. Baccelli, and R.K. Ganti, "A tractable approach to coverage and rate in cellular networks," *IEEE Trans. Commun.*, vol. 59, no. 11, pp. 3122–3134, Nov. 2011.
17 H.S. Dhillon, R. K. Ganti, F. Baccelli, and J.G. Andrews, "Modeling and analysis of K-tier downlink heterogeneous cellular networks," *IEEE J. Sel. Areas Commun.*, vol. 30, no. 3, pp. 550–560, Apr. 2012.
18 H.-S. Jo, Y.J. Sang, P. Xia, and J.G. Andrews, "Heterogeneous cellular networks with flexible cell association: a comprehensive downlink SINR analysis," *IEEE Trans. Wireless Commun.*, vol. 11, no. 10, pp. 3484–3495, Oct. 2012.
19 A.K. Gupta, H.S. Dhillon, S. Vishwanath, and J.G. Andrews, "Downlink multi-antenna heterogeneous cellular network with load balancing," *IEEE Trans. Commun.*, vol. 62, no. 11, pp. 4052–4067, Nov. 2014.
20 H. Yang, G. Geraci, and T.Q.S. Quek, "Energy-efficient design of MIMO heterogeneous networks with wireless backhaul," *IEEE Trans. Wireless Commun.*, vol. 15, no. 7, pp. 4914–4927, July 2015.
21 H.H. Yang, J. Lee, and T.Q.S. Quek, "Heterogeneous cellular network with energy harvesting – based D2D communication," *IEEE Trans. Wireless Commun.*, vol. 15, no. 2, pp. 1406–1419, Feb. 2016.

22 H.H. Yang, G. Geraci, T.Q.S. Quek, and J.G. Andrews, "Cell-edge-aware precoding for downlink massive MIMO cellular networks," *IEEE Trans. Signal Process.*, vol. 65, no. 13, pp. 3344–3358, July 2017.
23 G. Zhang, T.Q.S. Quek, M. Kountouris, A. Huang, and H. Shan, "Fundamentals of heterogeneous backhaul design – analysis and optimization," *IEEE Trans. Commun.*, vol. 64, no. 2, pp. 876–889, Feb. 2016.
24 X. Zhang and J.G. Andrews, "Downlink cellular network analysis with multi-slope path loss models," *IEEE Trans. Commun.*, vol. 63, no. 5, pp. 1881–1894, Mar. 2015.
25 M. Ding, P. Wang, D. López-Pérez, G. Mao, and Z. Lin, "Performance impact of LoS and NLoS transmissions in dense cellular networks," *IEEE Trans. Wireless Commun.*, vol. 15, no. 3, pp. 2365–2380, Mar. 2016.
26 H. S. Dhillon, M. Kountouris, and J.G. Andrews, "Downlink MIMO HetNets: modeling, ordering results and performance analysis," *IEEE Trans. Wireless Commun.*, vol. 12, no. 10, pp. 5208–5222, Oct. 2013.
27 T. Bai and R.W. Heath Jr., "Analyzing uplink SIR and rate in massive MIMO systems using stochastic geometry," *IEEE Trans. Commun.*, vol. 64, no. 11, pp. 4592–4606, Nov. 2016.
28 Q. Zhang, H.H. Yang, T.Q.S. Quek, and J. Lee, "Heterogeneous cellular networks with LoS and NLoS transmissions – the role of massive MIMO and small cells," *IEEE Trans. Wireless Commun.*, vol. 16, no. 12, pp. 7996–8010, Dec. 2017.
29 T. Bai and R.W. Heath Jr., "Coverage and rate analysis for millimeter wave cellular networks," *IEEE Trans. Wireless Commun.*, vol. 14, no. 2, pp. 1100–1114, Oct. 2014.
30 M. Haenggi, *Stochastic Geometry for Wireless Networks*. Cambridge University Press, 2012.
31 S. Singh, H.S. Dhillon, and J.G. Andrews, "Offloading in heterogeneous networks: modeling, analysis, and design insights," *IEEE Trans. Wireless Commun.*, vol. 12, no. 5, pp. 2484–2497, May 2013.
32 S. Singh, X. Zhang, and J.G. Andrews, "Joint rate and SINR coverage analysis for decoupled uplink-downlink biased cell associations in HetNets," *IEEE Trans. Wireless Commun.*, vol. 14, no. 10, pp. 5360–5373, Oct. 2015.
33 Spatial Channel Model AHG, "Subsection 3.5.3, Spatial Channel Model Text Description V6.0," Apr. 2003.
34 3GPP, "3GPP TR 36.828 (V11.0.0): Further enhancements to LTE Time Division Duplex (TDD) for Downlink-Uplink (DL-UL) interference management and traffic adaptation," Jan. 2012.
35 H. Yang and T.L. Marzetta, "Performance of conjugate and zero-forcing beamforming in large-scale antenna systems." *IEEE J. Sel. Areas Commun.*, vol. 31, no. 2, pp. 172–179, Feb. 2013.
36 F. Baccelli and B. Błaszczyszyn, *Stochastic Geometry and Wireless Networks. Volume I: Theory*. NOW Publishers, 2009.
37 K.A. Hamdi, "A useful lemma for capacity analysis of fading interference channels," *IEEE Trans. Commun.*, vol. 58, no. 2, pp. 411–416, Feb. 2010.
38 D. Stoyan, W. Kendall, and J. Mecke, *Stochastic Geometry and Its Applications*, 2nd Edition. John Wiley and Sons, 1996.

6

Security for Cell-free Massive MIMO Networks

Tiep M. Hoang[1], *Hien Quoc Ngo*[2], *Trung Q. Duong*[2] *and Hoang D. Tuan*[3]

[1] *Queen's University Belfast, UK*
[2] *School of Electronics, Electrical Engineering and Computer Science, Queen's University Belfast, UK*
[3] *School of Electrical and Data Engineering, University of Technology Sydney, Australia*

Cell-free massive MIMO, where many access points (APs) distributed in a wide area coherently serve many users in the same time/frequency resource, is one of the most promising technologies for the next generation of wireless systems. While many aspects regarding resource allocations, pilot designs, and performance analysis of cell-free massive MIMO have been intensively investigated, physical layer security in cell-free massive MIMO has not been exploited. In this chapter, we first provide some basic concepts of cell-free massive MIMO. Then the security aspect of cell-free massive MIMO under a pilot spoofing attack is investigated. To deal with this attack, we present a simple counterattack scheme, which is based on the power allocation at the APs.

6.1 Introduction

Massive multi-input multi-output (MIMO) is one of the key technologies for next generation wireless networks. Massive MIMO systems can be classified into two types based on the geometric setting of antennas: (i) *collocated* massive MIMO systems and (ii) *distributed* massive MIMO systems. The first type can be described as the use of a large number of antennas in a small area, for example, a system whose transmitter is equipped with a large antenna array, whereas the second type exploits the spatial distribution of antennas over a large area. For the second type, transceivers do not need to be equipped with many antennas to obtain the advanced properties of the massive MIMO technique. Instead, a *virtual* massive MIMO system can be formed once many transceivers, called access points (APs), with a single or several antennas have cooperated with each other over the area of interest. This aspect is the huge advantage of the distributed massive MIMO. In addition, owing to its ability to fully use the space microdiversity and macrodiversity, a distributed massive MIMO system can offer a very high probability of coverage and connectivity. However, a central processing unit (CPU) is required for the cooperation among the APs. Based on these observations, a technique called *cell-free massive MIMO* has been developed recently (see [1]). The terminology *"cell-free"* is used to imply that cellular boundaries may no longer exist, because there is

Ultra-dense Networks for 5G and Beyond: Modelling, Analysis, and Applications,
First Edition. Edited by Trung Q. Duong, Xiaoli Chu and Himal A. Suraweera.

no requirement that a mobile user is only served by a specific AP. In fact, transmitters as well as receivers will not belong to any cell in a cell-free massive MIMO system. Instead, they can be anywhere in the area of interest and each mobile user will be served by all or a group of APs at the same time. As a result, a cell-free massive MIMO system can offer uniformly good quality-of-service to all mobile users. Note that in a cell-free massive MIMO system, simple signal processing such as conjugate beamforming is used. Thus, a huge information exchange between the CPU and the APs is not required. As a result, the deployment of cell-free massive MIMO systems can be scaled up. All of these observations thus support the idea of considering a cell-free massive MIMO system as a replacement for a small-cell system.

Inspired by the above-mentioned advantages, recent studies have started to delve into this new architecture (see [2] to [4]). In order to contribute to the theoretical development, especially in terms of security, this chapter provides another perspective on a cell-free massive MIMO system in the presence of an eavesdropper. Noticeably, the attacks by eavesdroppers can be classified into two types: (i) passively eavesdropping attacks and (ii) actively eavesdropping attacks. When compared with each other, it is shown that active eavesdroppers are more dangerous than passive eavesdroppers (see [5]). That is understandable because of the fact that an active eavesdropper will always seek to overhear as much information as possible, whereas a passive eavesdropper will not do anything other than keeping silent and listening to messages. Focusing on dealing with active eavesdroppers, we will show how they break into and benefit from a cell-free massive MIMO system. In parallel, methods to deal with active attacks will be provided through optimizing power control coefficients at each AP. More specifically, two maximization problems with security constraints will be proposed and resolved. The common goal of these maximization programs is to achieve as much secrecy as possible, while the quality of service (such as the total power constraint and the data rate for each user) still remains guaranteed.

Notation The symbols $[\cdot]^T$, $[\cdot]^*$, and $[\cdot]^\dagger$ denote the transpose operator, conjugate operator, and Hermitian operator, respectively, while $[\cdot]^{-1}$ and $[\cdot]^+$ denote the inverse operator and pseudo-inverse operator, respectively. Vectors and matrices are represented with lowercase boldface and uppercase boldface, respectively. $\mathbf{I}_n$ is the $n \times n$ identity matrix, $\|\cdot\|$ denotes the Euclidean norm, $\mathbb{E}\{\cdot\}$ denotes expectation, and $\mathbf{z} \sim \mathcal{CN}_n(\overline{\mathbf{z}}, \boldsymbol{\Sigma})$ denotes a complex Gaussian vector $\mathbf{z} \in \mathbb{C}^{n\times 1}$ with mean vector $\overline{\mathbf{z}}$ and covariance matrix $\boldsymbol{\Sigma} \in \mathbb{C}^{n\times n}$.

6.2 Cell-free Massive MIMO System Model

A system that consists of M APs and K users is considered. Each AP (as well as each user) is equipped with a single antenna. Let us denote

$$g_{mk} = \sqrt{\beta_{mk}} h_{mk} \sim \mathcal{CN}(0, \beta_{mk})$$

as the channel from the mth AP to the kth user. We assume that the channel is reciprocal, i.e. the channel gain on the uplink equals the channel gain on the downlink. The transmission includes two phases: uplink training for channel estimation and downlink data transmission.

Uplink Training During this phase, all users send their pilot vectors to the APs in order that the APs will be able to estimate the channels. The pilot sent by the kth user is defined as $\mathbf{p}_k \in \mathbb{C}^{T\times 1}$ where T is the length of the training duration. Moreover, all pilots are designed such that $\mathbf{p}_k^\dagger \mathbf{p}_k = \|\mathbf{p}_k\|^2 = 1$, $k \in \{1,\dots,K\}$. At the mth AP, the received pilot vector is given by

$$\mathbf{y}_{p,m}^0 = \sqrt{T\rho_u}\sum_{k=1}^{K} g_{mk}\mathbf{p}_k + \mathbf{w}_m \tag{6.1}$$

where $\rho_u \triangleq P_u/N_0$. Herein, P_u is the average transmit power of each user and N_0 is the average noise power per receive antenna at the mth AP; $\mathbf{w}_m$ is an additive white Gaussian noise (AWGN) vector with $\mathbf{w}_m \sim \mathcal{CN}(\mathbf{0},\mathbf{I})$. Projecting $\mathbf{y}_{p,m}$ along the pilot vector $\mathbf{p}_k^\dagger$, we can obtain the corresponding signal $y_{km}^0 = \mathbf{p}_k^\dagger \mathbf{y}_{p,m}^0$ as

$$y_{km}^0 = \sqrt{T\rho_u}g_{mk} + \sqrt{T\rho_u}\sum_{k'\neq k}^{K} g_{mk}\mathbf{p}_k^\dagger \mathbf{p}_{k'} + \mathbf{p}_k^\dagger \mathbf{w}_m. \tag{6.2}$$

From the above equation, we can use the minimum mean square error (MMSE) method to estimate g_{mk} as follows:[1]

$$\widehat{g_{mk}^0} = c_{mk}y_{km}^0 \tag{6.3}$$

where $c_{mk} = \frac{\sqrt{T\rho_u}\beta_{mk}}{T\rho_u\sum_{k'=1}^{K}\beta_{mk'}|\mathbf{p}_k^\dagger \mathbf{p}_{k'}|^2+1}$.

Downlink Transmisson In this phase, the mth AP uses $\widehat{g_{mk}^0}$ in Eq. (6.3) as the true channel to perform the beamforming technique. First, we denote s_k to be the signal intended for the kth user and P_s the average transmit power for a certain s_k with $k \in \mathcal{K} = \{1, 2, \dots, K\}$. Applying the beamforming technique, the mth AP will transmit the following signal:

$$x_m^0 = \sqrt{P_s}\sum_{k=1}^{K}\sqrt{\eta_{mk}}\widehat{g_{mk}^0}^* s_k \tag{6.4}$$

with s_k being normalized such that $\mathbb{E}\{|s_k|^2\} = 1$. In Eq. (6.4), η_{mk} is the power control coefficient, which corresponds to the downlink channel from the mth AP to the kth user. The received signal at the kth user is then expressed as

$$z_k^0 = \sqrt{\rho_s}\sum_{m=1}^{M} g_{mk}\left(\sum_{k'=1}^{K}\sqrt{\eta_{mk'}}\widehat{g_{mk'}^0}^* s_{k'}\right) + n_k, \tag{6.5}$$

where $\rho_s = P_s/N_0$ and $n_k \sim \mathcal{CN}(0,1)$ is the normalized AWGN at user k. Herein, the average noise power at each user is assumed to be the same as that at each antenna of an AP.

---

1 Let us consider a linear relation $\mathbf{y} = \mathbf{Ph} + \mathbf{n}$ where $\mathbf{y} \in \mathbb{C}^{N_L\times 1}$ is the observed data, $\mathbf{P} \in \mathbb{C}^{N_L\times N_T}$ is a known matrix, $\mathbf{h} \in \mathbb{C}^{N_T\times 1}$ is a random vector obeying $\mathcal{CN}(0,\mathbf{D})$, and $\mathbf{n} \in \mathbb{C}^{N_L\times 1}$ is a noise vector obeying $\mathcal{CN}(0,\boldsymbol{\Omega})$ and being independent of $\mathbf{h}$. The MMSE estimator of $\mathbf{h}$ is $\hat{\mathbf{h}} = (\mathbf{D}^{-1} + \mathbf{P}^\dagger\boldsymbol{\Omega}^{-1}\mathbf{P})^{-1}\mathbf{P}^\dagger\boldsymbol{\Omega}^{-1}\mathbf{y}$. (See 6, Chapters 10 and 11 for more details.) To estimate g_{mk} in Eq. (6.2), we first let $N_L = 1$, $N_T = 1$, $\mathbf{y} = y_{km}^0$, $\mathbf{P} = \sqrt{T\rho_u}$, $\mathbf{h} = g_{mk}$, and $\mathbf{n} = \sqrt{T\rho_u}\sum_{k'\neq k}^{K} g_{mk}\mathbf{p}_k^\dagger\mathbf{p}_{k'} + \mathbf{p}_k^\dagger\mathbf{w}_m$. As a result, we have $\mathbf{D} = \beta_{mk} = \beta_{mk}|\mathbf{p}_k^\dagger\mathbf{p}_k|^2$ and $\boldsymbol{\Omega} = T\rho_u\sum_{k'\neq k}^{K}\beta_{mk}|\mathbf{p}_k^\dagger\mathbf{p}_{k'}|^2 + 1$. Then $\hat{\mathbf{h}} = \widehat{g_{mk}^0}$ can be presented as in Eq. (6.3).

Mutual Information At this stage, we aim to analyze the mutual information $I_k^0(s_k; z_k^0)$ (in nats/s/Hz) between s_k and z_k^0. More specifically, the lower-bound for $I_k^0(s_k; z_k^0)$ will be considered instead of the exact expression. The rationale for the lower-bound analysis is that the lower-bound is at an achievable rate even when the full channel state information (CSI) is unknown to any receiver. In contrast, the exact rate is only obtained in the case where the full CSI is available. The assumption of full CSI, however, is impractical. To find the lower-bound for $I_k^0(s_k; z_k^0)$, we first recall the following lemma.

Lemma 6.1 [7] Let U and V be complex-valued random variables with $U \sim \mathcal{CN}(0, var\{U\})$ and $\mathbb{E}\{|V|^2\} = var\{V\}$. Given that U and V are uncorrelated, the mutual information $I(U; U+V)$ (in nats/s/Hz) between U and $U+V$ is lower-bounded by $\ln(1 + var\{U\}/var\{V\})$. Consequently, the lower-bound SNR can be given by $var\{U\}/var\{V\}$.

Now we rewrite Eq. (6.5) as

$$z_k^0 = \mathrm{DS}_k^0 \times s_k + \underbrace{\mathrm{BU}_k^0 \times s_k + \sum_{k' \neq k}^{K} \mathrm{UI}_{kk'}^0 \times s_{k'} + n_k}_{\text{treated as aggregated noise}}, \tag{6.6}$$

where

$$\mathrm{DS}_k^0 \triangleq \sqrt{\rho_s} \sum_{m=1}^{M} \mathbb{E}\{\sqrt{\eta_{mk}} g_{mk} \hat{g}_{mk}^*\},$$

$$\mathrm{BU}_k^0 \triangleq \sqrt{\rho_s} \sum_{m=1}^{M} (\sqrt{\eta_{mk}} g_{mk} \hat{g}_{mk}^* - \mathbb{E}\{\sqrt{\eta_{mk}} g_{mk} \hat{g}_{mk}^*\}),$$

$$\mathrm{UI}_{kk'}^0 \triangleq \sqrt{\rho_s} \sum_{m=1}^{M} \sqrt{\eta_{mk'}} g_{mk} \hat{g}_{mk'}^*$$

represent the strength of the desired signal s_k, the beamforming gain uncertainty, and the interference caused by the k'th user (with $k' \neq k$), respectively.

More importantly, it is proved that the terms DS_k^0, BU_k^0, $\mathrm{UI}_{kk'}^0$, and n_k in Eq. (6.6) are pair-wisely uncorrelated.[2] Considering the second, third, and fourth terms in Eq. (6.6) as noises, the lower-bound for $I_k^0(s_k; z_k^0)$ can be deduced from Lemma 6.1 as follows:

$$I_k^0(s_k; z_k^0) \geq \ln\left(1 + \frac{|\mathrm{DS}_k^0|^2}{\mathbb{E}\{|\mathrm{BU}_k^0|^2\} + \sum_{k' \neq k}^{K} \mathbb{E}\{|\mathrm{UI}_{kk'}^0|^2\} + 1}\right). \tag{6.7}$$

The explicit expressions of DS_k^0, $\mathbb{E}\{|\mathrm{BU}_k^0|^2\}$ and $\mathbb{E}\{|\mathrm{UI}_{kk'}^0|^2\}$ are presented in Eqs. (6.53), (6.54), and (6.55) in the Appendix. By plugging these results into the RHS of Eq. (6.7), we obtain the explicit expression for the lower-bound.

2 Two random variables X and Y are said to be uncorrelated if their covariance is zero, i.e. $\mathbb{E}\{XY\} - \mathbb{E}\{X\}\mathbb{E}\{Y\} = 0$.

6.3 Cell-free System Model in the presence of an active eavesdropper

We continue to consider the system with M APs and K users. However, we assume that there is an active eavesdropper (Eve) trying to overhear confidential messages. Note that the notations $\mathbf{p}_k$, $\{g_{mk}\}$, $\{\beta_{mk}\}$, $\{\eta_{mk}\}$, T, P_s, P_u, N_0, $\rho_s = P_s/N_0$, $\rho_u = P_u/N_0$, $k \in \{1, \dots, K\}$, and $m \in \{1, \dots, M\}$ still stay the same. Moreover, the assumptions of $\mathbf{w}_m \sim \mathcal{CN}(\mathbf{0}, \mathbf{I})$ and $n_k \sim \mathcal{CN}(0, 1)$ still hold in the rest of this chapter; thus, they will not necessarily be mentioned in the rest of this chapter.

Uplink Training In this phase, Eve sends its pilot sequence $\mathbf{p}_{\text{E}}$ to APs. If Eve aims to overhear the signal s_l intended for user l, she will design $\mathbf{p}_{\text{E}} = \mathbf{p}_l$ to impersonate user l (see [8]). Without the loss of generality, we assume that Eve aims to overhear the signal intended for the first user, i.e. $\mathbf{p}_{\text{E}} = \mathbf{p}_1$. At the mth AP, the received pilot vector is given by

$$\mathbf{y}_{p,m} = \sqrt{T\rho_u} \sum_{k=1}^{K} g_{mk}\mathbf{p}_k + \sqrt{T\rho_{\text{E}}} g_{m\text{E}}\mathbf{p}_1 + \mathbf{w}_m \tag{6.8}$$

where $g_{m\text{E}} \sim \mathcal{CN}(0, \beta_{m\text{E}})$ is the channel between the mth AP and Eve, $\rho_{\text{E}} \triangleq P_{\text{E}}/N_0$ with P_{E} being the average transmit power of Eve. The only difference between $\mathbf{y}_{p,m}$ in Eq. (6.8) and $\mathbf{y}^0_{p,m}$ in Eq. (6.1) lies in the presence of the term $\sqrt{T\rho_{\text{E}}} g_{m\text{E}}\mathbf{p}_1$. The case of $\rho_{\text{E}} = 0$ implies that Eve is a passive eavesdropper and Eq. (6.8) reduces to Eq. (6.1). In contrast, $\rho_{\text{E}} \neq 0$ means that the eavesdropping is active.

To simplify the security problem, we suppose that the length of a pilot vector is larger than or equal to the number of users, i.e. $T \geq K$. This assumption allows us to design K *orthogonal* pilot vectors such that $\mathbf{p}_k^{\dagger}\mathbf{p}_{k'} = 0$ for $k \neq k'$ and $\|\mathbf{p}_k\|^2 = 1$. Projecting $\mathbf{y}_{p,m}$ onto $\mathbf{p}_k^{\dagger}$, we obtain the post-processing signal

$$y_{km} = \mathbf{p}_k^{\dagger}\mathbf{y}_{p,m} = \begin{cases} \sqrt{T\rho_u} g_{mk} + \mathbf{p}_k^{\dagger}\mathbf{w}_m, & k \neq 1, \\ \sqrt{T\rho_u} g_{m1} + \sqrt{T\rho_{\text{E}}} g_{m\text{E}} + \mathbf{p}_1^{\dagger}\mathbf{w}_m, & k = 1. \end{cases} \tag{6.9}$$

This is due to the fact that APs are not aware of an eavesdropping attack until they have found abnormality in the sequence of received signals $\{y_{km}\}$ in Eq. (6.9). Based on such an observation, APs can identify the legitimate user that is being impersonated by Eve. Besides, with the aim of estimating g_{mk} and $g_{m\text{E}}$ from Eq. (6.9), the MMSE method is adopted at the mth AP, i.e. we have

$$\hat{g}_{mk} = \begin{cases} \dfrac{\sqrt{T\rho_u}\beta_{mk}}{T\rho_u\beta_{mk} + 1} y_{km}, & k \neq 1, \\ \dfrac{\sqrt{T\rho_u}\beta_{m1}}{T\rho_u\beta_{m1} + T\rho_{\text{E}}\beta_{m\text{E}} + 1} y_{1m}, & k = 1, \end{cases} \tag{6.10}$$

and

$$\hat{g}_{m\text{E}} = \sqrt{\frac{\rho_{\text{E}}}{\rho_u}\frac{\beta_{m\text{E}}}{\beta_{m1}}}\hat{g}_{m1}. \tag{6.11}$$

Let us denote

$$\gamma_{mk} \triangleq \mathbb{E}\{|\hat{g}_{mk}|^2\} = \begin{cases} \dfrac{T\rho_u\beta_{mk}^2}{T\rho_u\beta_{mk}+1}, & k \neq 1, \\ \dfrac{T\rho_u\beta_{m1}^2}{T\rho_u\beta_{m1}+T\rho_E\beta_{mE}+1}, & k = 1, \end{cases}$$

and $\gamma_{mE} \triangleq \mathbb{E}\{|\hat{g}_{mE}|^2\}$. Using Eq. (6.11), we can also rewrite

$$\gamma_{mE} = \alpha_m\gamma_{m1}$$

where $\alpha_m = (\rho_E\beta_{mE}^2)/(\rho_u\beta_{m1}^2)$. In association with the above, we state the following proposition for later use.

Proposition 6.1 $\hat{g}_{mk}$ and $\hat{g}_{mk'}$ are uncorrelated for $\forall k' \neq k$. At the same time, $\hat{g}_{mE}$ and $\hat{g}_{mk'}$ are uncorrelated for $\forall k' \neq 1$. Furthermore, we have

$$\mathbb{E}\{|\hat{g}_{mk}\hat{g}_{mk'}^*|^2\} = \begin{cases} \gamma_{mk}\gamma_{mk'}, & k' \neq k, \\ 2\gamma_{mk}^2, & k' = k, \end{cases} \tag{6.12}$$

and

$$\mathbb{E}\{|\hat{g}_{mE}\hat{g}_{mk'}^*|^2\} = \begin{cases} \alpha_m\gamma_{m1}\gamma_{mk'}, & k' \neq 1, \\ 2\alpha_m\gamma_{m1}^2, & k' = 1. \end{cases} \tag{6.13}$$

Downlink Transmisson The transmitted signal x_m still remains the same as in Eq. (6.4), i.e. $x_m = \sqrt{P_s}\sum_{k=1}^{K}\sqrt{\eta_{mk}}\hat{g}_{mk}^*s_k$. As such, the received signal at user k and Eve are, respectively, given by

$$z_k = \sqrt{\rho_s}\sum_{m=1}^{M} g_{mk}\left(\sum_{k=1}^{K}\sqrt{\eta_{mk}}\hat{g}_{mk}^*s_k\right) + n_k, \tag{6.14}$$

$$z_E = \sqrt{\rho_s}\sum_{m=1}^{M} g_{mE}\left(\sum_{k=1}^{K}\sqrt{\eta_{mk}}\hat{g}_{mk}^*s_k\right) + n_E, \tag{6.15}$$

where $n_E \sim \mathcal{CN}(0,1)$ is the AWGN at Eve.

Mutual Information: Lower-bound Similar to Subsection 6.2, we also decompose z_k in Eq. (6.14) into four terms, i.e.

$$z_k = \mathrm{DS}_k \times s_k + \underbrace{\mathrm{BU}_k \times s_k + \sum_{k'\neq k}^{K}\mathrm{UI}_{kk'} \times s_{k'} + n_k}_{\text{treated as aggregated noise}}, \tag{6.16}$$

where

$$\mathrm{DS}_k \triangleq \sqrt{\rho_s}\sum_{m=1}^{M}\mathbb{E}\{\sqrt{\eta_{mk}}g_{mk}\hat{g}_{mk}^*\},$$

$$\mathrm{BU}_k \triangleq \sqrt{\rho_s}\sum_{m=1}^{M}(\sqrt{\eta_{mk}}g_{mk}\hat{g}_{mk}^* - \mathbb{E}\{\sqrt{\eta_{mk}}g_{mk}\hat{g}_{mk}^*\}),$$

$$\mathrm{UI}_{kk'} \triangleq \sqrt{\rho_s}\sum_{m=1}^{M}\sqrt{\eta_{mk'}}g_{mk}\hat{g}_{mk'}^*.$$

These terms, DS_k, BU_k, $\mathrm{UI}_{kk'}$ and n_k in Eq. (6.16), can be proved to be pair-wisely uncorrelated. Let $I_k(s_k; z_k)$ (in nats/s/Hz) denote the mutual information between s_k and z_k with $k \in \mathcal{K} = \{1, 2, \ldots, K\}$. Considering the second, third, and fourth terms in Eq. (6.16) as noises, we again apply Lemma 6.1 to obtain the lower-bound for $I_k(s_k; z_k)$, i.e.

$$\begin{aligned} I_k(s_k; z_k) &\geq \ln\left(1 + \frac{|\mathrm{DS}_k|^2}{\mathbb{E}\{|\mathrm{BU}_k|^2\} + \sum_{k' \neq k}^{K}\mathbb{E}\{|\mathrm{UI}_{kk'}|^2\} + 1}\right) \\ &\triangleq \ln(1 + \mathrm{snr}_\mathrm{k}). \end{aligned} \tag{6.17}$$

Although the derivations of terms DS_k, BU_k, $\mathrm{UI}_{kk'}$ are ignored here, they can be totally calculated in the same way as in Subsection 6.2. After some manipulations, we have

$$\mathrm{snr}_\mathrm{k} = \frac{\rho_s\left(\sum_{m=1}^{M}\sqrt{\eta_{mk}}\gamma_{mk}\right)^2}{\rho_s\sum_{k'=1}^{K}\sum_{m=1}^{M}\eta_{mk'}\gamma_{mk'}\beta_{mk} + 1}. \tag{6.18}$$

Mutual Information: Upper-bound We rewrite Eq. (6.15) as

$$z_\mathrm{E} = \mathrm{BU}_{\mathrm{E},1} \times s_1 + \underbrace{\sum_{k' \neq 1}^{K}\mathrm{UI}_{\mathrm{E},k'} \times s_{k'} + n_\mathrm{E}}_{\text{treated as aggregated noise}}. \tag{6.19}$$

where

$$\mathrm{BU}_{\mathrm{E},1} \triangleq \sqrt{\rho_s}\sum_{m=1}^{M}\sqrt{\eta_{m1}}g_{m\mathrm{E}}\hat{g}_{m1}^*,$$

$$\mathrm{UI}_{\mathrm{E},k'} \triangleq \sqrt{\rho_s}\sum_{m=1}^{M}\sqrt{\eta_{mk'}}g_{m\mathrm{E}}\hat{g}_{mk'}^*,$$

respectively, represent the strength of the desired signal s_1 (which Eve aims to overhear) and the interference caused by users $k' \neq k$. We can also prove that the terms $\mathrm{BU}_{\mathrm{E},k}$, $\mathrm{UI}_{\mathrm{E},kk'}$, and n_E in Eq. (6.19) are pair-wisely uncorrelated. We then consider the second and third terms in Eq. (6.19) as noises.

Let $I_{\text{E}}(s_1; z_{\text{E}})$ (in nats/s/Hz) denote the mutual information between s_1 and z_{E}. Then the upper-bound for $I_{\text{E}}(s_k; z_{\text{E}})$ can be formulated as follows:

$$\begin{aligned} I_{\text{E}}(s_1; z_{\text{E}}) &\overset{(a)}{\leq} I_{\text{E}}(s_1; z_{\text{E}} | \{g_{mk}\}_{m,k}, \{\hat{g}_{mk}\}_{m,k}, \{g_{m\text{E}}\}_m) \\ &= \mathbb{E}\left\{ \ln\left(1 + \frac{|\text{BU}_{\text{E},1}|^2}{\sum_{k' \neq 1}^{K} |\text{UI}_{\text{E},k'}|^2 + 1}\right)\right\} \\ &\overset{(b)}{\approx} \ln(1 + \text{snr}_{\text{E}}), \end{aligned} \tag{6.20}$$

where

$$\begin{aligned} \text{snr}_{\text{E}} &= \frac{\mathbb{E}\{|\text{BU}_{\text{E},1}|^2\}}{\sum_{k' \neq 1}^{K} \mathbb{E}\{|\text{UI}_{\text{E},k'}|^2\} + 1} \\ &\overset{(c)}{=} \frac{\rho_s \sum_{m=1}^{M} \eta_{m1}\gamma_{m1} \left(\frac{\rho_{\text{E}}\beta_{m\text{E}}^2}{\rho_u \beta_{m1}^2}\gamma_{m1} + \beta_{m\text{E}}\right)}{\rho_s \sum_{k' \neq 1}^{K} \sum_{m=1}^{M} \eta_{mk'}\gamma_{mk'}\beta_{m\text{E}} + 1}. \end{aligned} \tag{6.21}$$

The RHS of inequality (a) means that Eve has full CSI of every channel. It also implies the worst case in terms of security. Meanwhile, the approximation (b) follows [[9], Lemma 1]. Finally, the derivation of (c) is obtained by using the same methodology as in Section 1.2.

Achievable Secrecy Rate From Eqs. (6.17) and (6.20), we can define the achievable secrecy rate (in nats/s/Hz) and its lower-bound as follows:

$$\Delta \triangleq I_1(s_1; z_1) - I_{\text{E}}(s_1; z_{\text{E}}) \geq \ln((1 + \text{snr}_1)/(1 + \text{snr}_{\text{E}})) \triangleq R_{sec} \tag{6.22}$$

in which the explicit expressions for snr_1 and snr_{E} are presented in Eqs. (6.18) and (6.21), respectively.

In order to facilitate further analysis in the rest of this chapter, we denote $\boldsymbol{\Psi}$ be the matrix in which the (m, k)th entry is $\boldsymbol{\Psi}(m, k) = \sqrt{\eta_{mk}}$. The kth column vector of $\boldsymbol{\Psi}$ is denoted as $\mathbf{u}_k = \boldsymbol{\Psi}(:, k) = [\sqrt{\eta_{1k}}, \sqrt{\eta_{2k}}, \dots, \sqrt{\eta_{Mk}}]^T$. In addition, we also define the following matrices and vectors:

$$\begin{aligned} \mathbf{a}_k &= \sqrt{\rho_s}[\gamma_{1k}, \gamma_{2k}, \dots, \gamma_{Mk}]^T, \\ \mathbf{A}_{kk'} &= \sqrt{\rho_s}\mathsf{diag}(\sqrt{\beta_{1k}\gamma_{1k'}}, \dots, \sqrt{\beta_{Mk}\gamma_{Mk'}}), \\ \mathbf{B}_{\text{E}} &= \sqrt{\rho_s}\mathsf{diag}(\sqrt{\gamma_{11}(\gamma_{1\text{E}} + \beta_{1\text{E}})}, \dots, \sqrt{\gamma_{M1}(\gamma_{M\text{E}} + \beta_{M\text{E}})}), \\ \mathbf{B}_{k'} &= \sqrt{\rho_s}\mathsf{diag}(\sqrt{\beta_{1\text{E}}\gamma_{1k'}}, \dots, \sqrt{\beta_{M\text{E}}\gamma_{Mk'}}) \text{ with } k' \neq 1, \end{aligned}$$

where $\mathsf{diag}(\cdot)$ denotes a diagonal matrix. Finally, Eqs. (6.18) and (6.21) can be rewritten in a more elegant way as follows:

$$\text{snr}_{\text{k}} = \frac{(\mathbf{a}_k^T \mathbf{u}_k)^2}{\varphi_k(\boldsymbol{\Psi})}, \tag{6.23}$$

$$\text{snr}_{\text{E}} = \frac{\|\mathbf{B}_{\text{E}}\mathbf{u}_1\|^2}{\varphi_{\text{E}}(\boldsymbol{\Psi})}, \tag{6.24}$$

where $\varphi_k(\boldsymbol{\Psi}) = \sum_{k'=1}^{K} \|\mathbf{A}_{kk'}\mathbf{u}_{k'}\|^2 + 1$ and $\varphi_{\text{E}}(\boldsymbol{\Psi}) = \sum_{k' \neq 1}^{K} \|\mathbf{B}_{k'}\mathbf{u}_{k'}\|^2 + 1$.

6.4 On Dealing with Eavesdropper

In this section, we aim at designing the power coefficients $\{\eta_{mk}\}$ to maximize the rate of user 1, i.e. $\ln(1+\text{snr}_1)$, subject to security constraints.

Prior to performing this task, the constraint of transmit power at each AP is provided. The power constraint is described as follows:

- Let P_{max} be the maximum transmit power of each AP, i.e. $P_{max} \geq \mathbb{E}\{|x_m|^2\}$. The average transmit power for the mth AP can be given by

$$\mathbb{E}\{|x_m|^2\} = P_s \sum_{k=1}^{K} \eta_{mk}\gamma_{mk}. \tag{6.25}$$

With the power constraint on every AP, we have

$$\sum_{k=1}^{K} \boldsymbol{\Psi}^2(m,k)\gamma_{mk} \leq \frac{\rho_{max}}{\rho_s}, \; m \in \mathcal{M} \tag{6.26}$$

with $\mathcal{M} = \{1,\dots,M\}$. Note that $\rho_{max} = P_{max}/N_0$ is viewed as the maximum possible ratio of the mth AP's average transmit power to the average noise power.

6.4.1 Case 1: Power Coefficients Are Different

When the coefficient $\{\eta_{mk}\}$ varies with m and k, our task is to design the matrix $\boldsymbol{\Psi}$. Hence, we propose the following maximization problem:

$$(\mathbf{P1}) \quad \max_{\boldsymbol{\Psi}} \quad \ln(1+(\mathbf{a}_1^T\mathbf{u}_1)^2/\varphi_1(\boldsymbol{\Psi})) \tag{6.27a}$$

$$\text{s.t.} \quad (6.26), \tag{6.27b}$$

$$\frac{\|\mathbf{B}_{\text{E}}\mathbf{u}_1\|^2}{\varphi_{\text{E}}(\boldsymbol{\Psi})} \leq \theta_{\text{E}}, \tag{6.27c}$$

$$\frac{(\mathbf{a}_k^T\mathbf{u}_k)^2}{\varphi_k(\boldsymbol{\Psi})} \geq \theta_k, \; k \subset \mathcal{K}\backslash\{1\}. \tag{6.27d}$$

We can see that the constraint (6.26) is obviously convex, while (6.27d) is the following second-order cone (SOC) constraint and thus convex:

$$\frac{1}{\sqrt{\theta_k}}\mathbf{a}_k^T\mathbf{u}_k \geq \sqrt{\varphi_k(\boldsymbol{\Psi})}, \; k \in \mathcal{K}\backslash\{1\}. \tag{6.28}$$

Besides, we observe that the objective function of $(\mathbf{P1})$ can be replaced with $(\mathbf{a}_1^T\mathbf{u}_1)^2/\varphi_1(\boldsymbol{\Psi})$. Let $\boldsymbol{\Psi}^{(\vartheta)}$ be a feasible point for $(\mathbf{P1})$. By using the inequality [11]

$$\frac{x^2}{y} \geq 2\frac{\bar{x}}{\bar{y}}x - \frac{\bar{x}^2}{\bar{y}^2}y \quad \forall \quad x>0, y>0, \bar{x}>0, \bar{y}>0, \tag{6.29}$$

we obtain

$$\frac{(\mathbf{a}_1^T\mathbf{u}_1)^2}{\varphi_1(\boldsymbol{\Psi})} \geq f_1^{(\vartheta)}(\boldsymbol{\Psi}) \triangleq a^{(\vartheta)}\mathbf{a}_1^T\mathbf{u}_1 - b^{(\vartheta)}\varphi_1(\boldsymbol{\Psi}) \tag{6.30}$$

with

$$a^{(\vartheta)} = 2\frac{(\mathbf{a}_1^T\mathbf{u}_1^{(\vartheta)})^2}{\varphi_1(\boldsymbol{\Psi}^{(\vartheta)})},\ b^{(\vartheta)} = (a^{(\vartheta)}/2)^2. \tag{6.31}$$

As such, maximizing $(\mathbf{a}_1^T\mathbf{u}_1)^2/\varphi_1(\boldsymbol{\Psi})$ is now equivalent to maximizing $f_1^{(\vartheta)}(\boldsymbol{\Psi})$. Finally, considering the function $\varphi_{\mathrm{E}}(\boldsymbol{\Psi})$ in Eq. (6.27c), we find that it is convex quadratic and thus the non-convex constraint (6.27c) is innerly approximated by the convex quadratic constraint[3]

$$||\mathbf{B}_{\mathrm{E}}\mathbf{u}_1||^2/\theta_{\mathrm{E}} \leq \varphi_{\mathrm{E}}^{(\vartheta)}(\boldsymbol{\Psi}) \tag{6.32}$$

for

$$\varphi_{\mathrm{E}}^{(\vartheta)}(\boldsymbol{\Psi}) \triangleq \sum_{k\neq 1}^{K}\left[\mathbf{u}_k^{(\vartheta)^T}\mathbf{B}_k^2(2\mathbf{u}_k - \mathbf{u}_k^{(\vartheta)})\right] + 1. \tag{6.33}$$

Having the approximations (6.30) and (6.32), at the ϑth iteration we solve the following convex optimization to generate a feasible $\boldsymbol{\Psi}^{(\vartheta+1)}$:

$$\max_{\boldsymbol{\Psi}} f_1^{(\vartheta)}(\boldsymbol{\Psi}) \quad \text{s.t.} \quad (6.26), (6.28), (6.32). \tag{6.34}$$

To find a feasible point for (**P1**) to initialize the above procedure, we address the problem

$$\min_{\boldsymbol{\Psi}} \ ||\mathbf{B}_{\mathrm{E}}\mathbf{u}_1||^2/\theta_{\mathrm{E}} - \varphi_{\mathrm{E}}(\boldsymbol{\Psi}) \quad \text{s.t.} \quad (6.26), (6.28). \tag{6.35}$$

Initialized by any feasible point $\boldsymbol{\Psi}^{(0)}$ for convex constraints (6.26) and (6.28), we iterate the following optimization problem:

$$\min_{\boldsymbol{\Psi}} \ ||\mathbf{B}_{\mathrm{E}}\mathbf{u}_1||^2/\theta_{\mathrm{E}} - \varphi_{\mathrm{E}}^{(\vartheta)}(\boldsymbol{\Psi}) \quad \text{s.t.} \quad (6.26), (6.28), \tag{6.36}$$

until

$$||\mathbf{B}_{\mathrm{E}}\mathbf{u}_1^{(\vartheta)}||^2/\theta_{\mathrm{E}} - \varphi_{\mathrm{E}}(\boldsymbol{\Psi}^{(\vartheta)}) \leq 0, \tag{6.37}$$

so $\boldsymbol{\Psi}^{(\vartheta)}$ is feasible for (**P1**). To sum up, we provide the following algorithm:

Algorithm 6.1 Algorithm for solving (**P1**)

1. **Initialization**: Set $\vartheta = 0$ with a feasible point $\Psi^{(0)}$ for (**P1**).
2. **repeat**
3. Solve (6.34) to obtain the optimal solution $\Psi^{(\vartheta+1)}$.
4. Reset $\vartheta := \vartheta + 1$.
5. **until** Converge.
6. **return** $\Psi^{(\vartheta)}$ as the desired result.

---

3 The right-hand side of Eq. (6.32) is the first-order Taylor approximation of $\varphi_{\mathrm{E}}(\boldsymbol{\Psi})$ near $\boldsymbol{\Psi}^{(\vartheta)}$. With $\varphi_{\mathrm{E}}(\boldsymbol{\Psi})$ being convex, we have $\varphi_{\mathrm{E}}^{(\vartheta)}(\boldsymbol{\Psi}) \leq \varphi_{\mathrm{E}}(\boldsymbol{\Psi})$.

6.4.2 Case 2: Power Coefficients Are the Same

Considering a simple situation with $\eta_{mk} = \eta$ being the same for all m and k, we reconsider the proposed maximization problems for comparison purposes.

Plugging $\eta_{mk} = \eta$ into Eqs. (6.18) to (6.21), we obtain the special expressions for $\mathrm{snr}_{\mathrm{k}}$ and $\mathrm{snr}_{\mathrm{E}}$ as follows:

$$\mathrm{snr}_{\mathrm{k}}|_{\eta_{mk}=\eta} = \frac{\eta\omega_k}{(\eta\breve{\omega}_k + 1)}, \tag{6.38}$$

$$\mathrm{snr}_{\mathrm{E}}|_{\eta_{mk}=\eta} = \frac{\eta\varpi}{\eta\breve{\varpi} + 1}, \tag{6.39}$$

where

$$\omega_k = \rho_s\left(\sum_{m=1}^{M}\gamma_{mk}\right)^2,$$

$$\breve{\omega}_k = \rho_s\sum_{k'=1}^{K}\sum_{m=1}^{M}\gamma_{mk'}\beta_{mk},$$

$$\varpi = \rho_s\sum_{m=1}^{M}\gamma_{m1}\left(\frac{\rho_{\mathrm{E}}\beta_{m\mathrm{E}}^2}{\rho_u\beta_{m1}^2}\gamma_{m1} + \beta_{m\mathrm{E}}\right),$$

$$\breve{\varpi} = \rho_s\sum_{k'\neq 1}^{K}\sum_{m=1}^{M}\gamma_{mk'}\beta_{m\mathrm{E}}.$$

Then the maximization problem (**P1**) reduces to

$$(\underline{\mathrm{P1}})\ \max_{\eta}\ \eta\omega_1/(\eta\breve{\omega}_1 + 1) \tag{6.40a}$$

$$\text{s.t. } \eta \leq \frac{\rho_{max}/\rho_s}{\sum_{k=1}^{K}\gamma_{mk}},\ m \in \mathcal{M}, \tag{6.40b}$$

$$\eta(\varpi - \theta_{\mathrm{E}}\breve{\varpi}) \leq \theta_{\mathrm{E}}, \tag{6.40c}$$

$$\eta(\omega_k - \theta_k\breve{\omega}_k) \geq \theta_k,\ k \in \mathcal{K}\backslash\{1\}. \tag{6.40d}$$

The objective function of $(\underline{\mathrm{P1}})$ increases in η. Hence, maximizing that objective function is equivalent to maximizing η. In other words, we will solve the following problem:

$$(\underline{\mathrm{P1}})\ \max_{\eta}\ \eta \tag{6.41a}$$

$$\text{s.t. (6.40b), (6.40c), (6.40d).} \tag{6.41b}$$

In order for Eq. (6.40d) to be meaningful, we need the condition

$$(\omega_k - \theta_k\breve{\omega}_k) > 0 \Leftrightarrow \theta_k < \omega_k/\breve{\omega}_k \tag{6.42}$$

with $k \in \mathcal{K}\backslash\{1\}$. If θ_k satisfies the above condition, we can infer from both Eqs. (6.40b) and (6.40d) the following:

$$\underbrace{\max_{k\in\mathcal{K}\backslash\{1\}}\left\{\frac{\theta_k}{\omega_k-\theta_k\breve{\omega}_k}\right\}}_{\geq 0}\leq\eta\leq\underbrace{\min_{m\in\mathcal{M}}\left\{\frac{\rho_{max}/\rho_s}{\sum_{k=1}^{K}\gamma_{mk}}\right\}}_{>0}.$$

This also implies another necessary condition as follows:

$$\theta_k<\frac{\omega_k\min_{m\in\mathcal{M}}\left\{\frac{\rho_{max}/\rho_s}{\sum_{k=1}^{K}\gamma_{mk}}\right\}}{1+\breve{\omega}_k\min_{m\in\mathcal{M}}\left\{\frac{\rho_{max}/\rho_s}{\sum_{k=1}^{K}\gamma_{mk}}\right\}} \tag{6.43}$$

for each $k \in \mathcal{K}\backslash\{1\}$. The two conditions (6.42) and (6.43) are now rewritten in the following form:

$$\theta_k<\min\left\{\frac{\omega_k}{\breve{\omega}_k},\frac{\omega_k\min_{m\in\mathcal{M}}\left\{\frac{\rho_{max}/\rho_s}{\sum_{k=1}^{K}\gamma_{mk}}\right\}}{1+\breve{\omega}_k\min_{m\in\mathcal{M}}\left\{\frac{\rho_{max}/\rho_s}{\sum_{k=1}^{K}\gamma_{mk}}\right\}}\right\} \tag{6.44}$$

with $k \in \mathcal{K}\backslash\{1\}$. Once (6.44) has been satisfied, the solution to $(\underline{\text{P1}})$ can be given by

- either

$$\eta^{\star}_{(\underline{\text{P1}})}=\min_{m\in\mathcal{M}}\left\{\frac{\rho_{max}/\rho_s}{\sum_{k=1}^{K}\gamma_{mk}}\right\} \tag{6.45}$$

  for

$$\theta_{\text{E}}\geq\varpi/\breve{\varpi} \tag{6.46}$$

- or

$$\eta^{\star}_{(\underline{\text{P1}})}=\min_{m\in\mathcal{M}}\left\{\frac{\rho_{max}/\rho_s}{\sum_{k=1}^{K}\gamma_{mk}},\frac{\theta_{\text{E}}}{(\varpi-\theta_{\text{E}}\breve{\varpi})}\right\} \tag{6.47}$$

  for

$$\frac{\varpi\max_{k\in\mathcal{K}\backslash\{1\}}\left\{\frac{\theta_k}{\omega_k-\theta_k\breve{\omega}_k}\right\}}{1+\breve{\varpi}\max_{k\in\mathcal{K}\backslash\{1\}}\left\{\frac{\theta_k}{\omega_k-\theta_k\breve{\omega}_k}\right\}}\leq\theta_{\text{E}}<\varpi/\breve{\varpi}. \tag{6.48}$$

6.5 Numerical Results

In this section, the achievable secrecy rate R_{sec} (in nats/s/Hz) will be illustrated for two optimal cases. More particularly, if $\mathbf{\Psi}^{\star}$ is the optimal solution to the problem (**P1**) and $\eta^{\star}$ is the optimal solution to the problem $(\underline{\text{P1}})$, then we will show the secure performance through drawing $R_{sec}(\mathbf{\Psi})$ at $\mathbf{\Psi}=\mathbf{\Psi}^{\star}$ and $R_{sec}(\eta)$ at $\eta=\eta^{\star}$.

Regarding simulation, the Hata-COST231 model is used to imitate the large-scale fading coefficients (see [1] and [10]). Hence, we have

$$\beta_{mk} = 10^{(S+PL(d_{mk}))/10}, \tag{6.49}$$

$$\beta_{m\mathrm{E}} = 10^{(S+PL(d_{m\mathrm{E}}))/10}, \tag{6.50}$$

where $S \sim \mathcal{CN}(0, \sigma_S^2)$ presents the shadowing fading effect with the standard deviation $\sigma_S = 8$ dB and

$$PL(d) = \begin{cases} -139.4 - 35\log_{10}(d) \text{ if } d > 0.05, \\ -119.9 - 20\log_{10}(d) \text{ if } d \in (0.01, 0.05], \\ -79.9 \text{ if } d \leq 0.01 \end{cases} \tag{6.51}$$

represents the path oss in dB with $d \equiv d_{mk}$ (or $d \equiv d_{m\mathrm{E}}$) being the distance in km between the mth AP and user k (or Eve).[4] In addition, the maximum transmit power of each AP is $P_{max} = 1$ W. Meanwhile, the average noise power (in W) is given by

$$N_0 = \text{bandwidth} \times k_B \times T_0 \times \text{noise figure}, \tag{6.52}$$

where $k_B = 1.38 \times 10^{-23}$ (Joule/Kelvin) is the Boltzmann constant and $T_0 = 290$ (Kelvin) is the noise temperature. In all results, the bandwidth is assumed to be 20 MHz, while the noise figure is 9 dB.

In Figure 6.1, we show both $R_{sec}(\boldsymbol{\Psi}^\star)$ and $R_{sec}(\eta^\star)$. For each case, four different subcases of (P_u, P_E) are considered, including $(P_u, P_\mathrm{E}) = (0.3, 0.1)$ W, $(P_u, P_\mathrm{E}) = (0.3, 0.5)$ W,

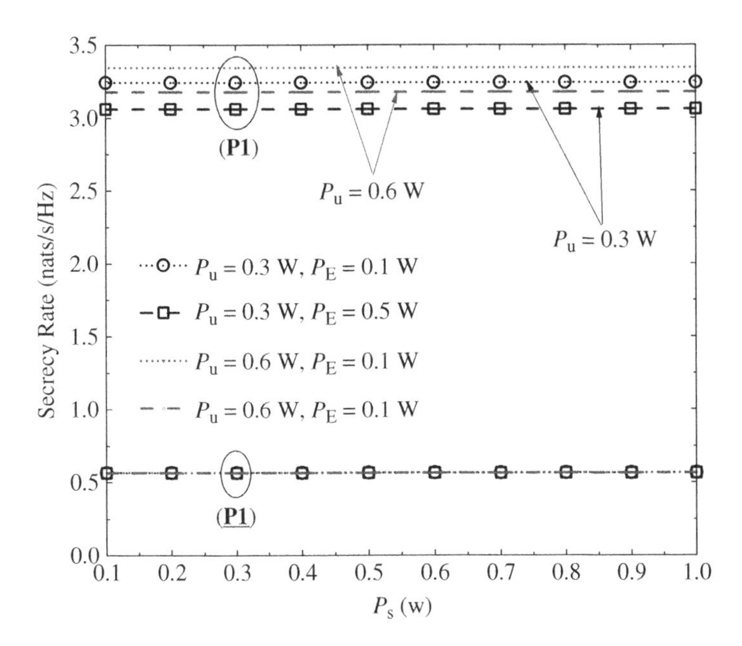

Figure 6.1 Secrecy rate versus P_s with $P_u = \{0.3, 0.6\}$ W, $P_\mathrm{E} = \{0.1, 0.5\}$ W, $M = 50$, $K = 8$, $T = 12$, $\theta_\mathrm{E} = 10^{-4}$, and $\theta_k = 2 \times 10^{-4}$ for $k \in \{2, \ldots, K\}$.

4 Other presentations for $PL(d)$ are also available in the literature. Herein, Eq. (6.51) is suggested for a practical scenario in which the carrier frequency is 1900 MHz, the height of each AP antenna is 20 m, the height of each user antenna (as well as that of Eve antenna) is 1.5 m, and all nodes (APs, users, and Eve) are randomly dispersed over a square of size 1×1 km^2 [[1], Eqs. (6.52) and (6.53)].

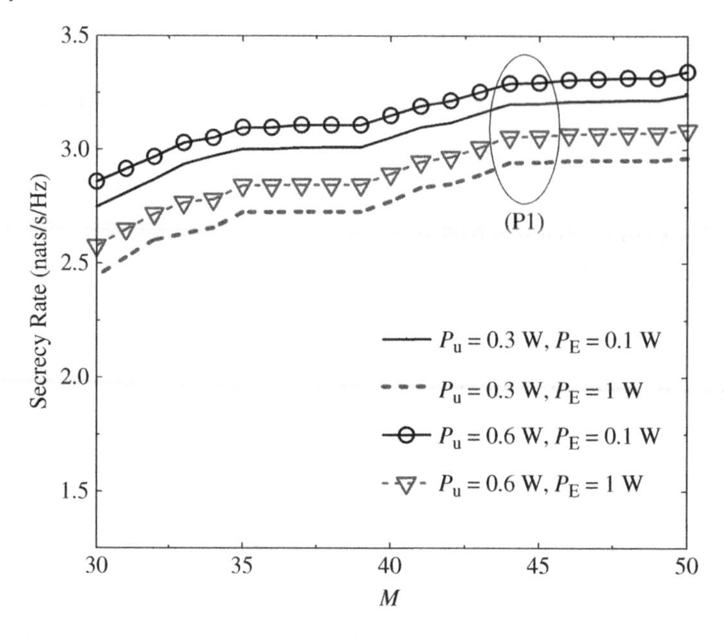

Figure 6.2 Secrecy rate versus M with $P_s = 0.8$ W, $P_u = \{0.3, 0.6\}$ W, $P_E = \{0.1, 1\}$ W, $K = 8$, $T = 12$, $\theta_k = 2 \times 10^{-4}$ for $k \in \{2, \dots, K\}$, and $\theta_E = \theta_k/50$.

$(P_u, P_E) = (0.6, 0.1)$ W, and $(P_u, P_E) = (0.6, 0.5)$ W. It is observed that $R_{sec}(\boldsymbol{\Psi}^\star)$ is significantly higher than $R_{sec}(\eta^\star)$. In fact, the obtained values of $R_{sec}(\eta^\star)$ fall within the interval $(0.553, 0.571)$ nats/s/Hz. As such, the secure performance corresponding to (P1) is much lower than that corresponding to (**P1**). Consequently, it is suggested that (**P1**) should be used to cope with active eavesdropping. Furthermore, the secure performance increases with P_u and reduces with P_E.

In Figure 6.2, we depict only $R_{sec}(\boldsymbol{\Psi}^\star)$ to see its change with M. It is observed that $R_{sec}(\boldsymbol{\Psi}^\star)$ increases with M. This implies that when the number of service APs increases, the cell-free massive MIMO system is improved accordingly. Besides, when considering four subcases $(P_u, P_E) = (0.3, 0.1)$ W, $(P_u, P_E) = (0.3, 1)$ W, $(P_u, P_E) = (0.6, 0.1)$ W, and $(P_u, P_E) = (0.6, 1)$ W, we also reach the same conclusion as above-mentioned, i.e. R_{sec}(**P1**) increases with P_u and decreases with P_E.

6.6 Conclusion

We have described a cell-free massive MIMO network by analyzing two basic processes (including the uplink training and the downlink transmission). Specifically focusing on the security aspect of the cell-free massive MIMO network, we have considered the situation in which Eve impersonates user 1. Then two maximization problems have been proposed and resolved for the purpose of improving the secure performance of the system. Based on numerical results, we have observed that the first case (with a different power allocation to different APs) provides much higher secure performance than the second case (with the same power allocation to all APs). In other words, we have obtained the better performance when the power coefficients are able to vary with m and k.

Appendix

This section presents the derivations of DS_k, $\mathbb{E}\{|\mathrm{BU}_k|^2\}$, and $\mathbb{E}\{|\mathrm{UI}_{kk'}|^2\}$. First, DS_k^0 is simply calculated as

$$\mathrm{DS}_k^0 = \sqrt{\rho_s}\mathbb{E}\left\{\sum_{m=1}^{M}\sqrt{\eta_{mk}}(\widehat{g_{mk}^0} + e_{mk})\widehat{g_{mk}^0}^*\right\} = \sqrt{\rho_s}\sum_{m=1}^{M}\sqrt{\eta_{mk}}\gamma_{mk}, \tag{6.53}$$

where $e_{mk} \triangleq g_{mk} - \widehat{g_{mk}^0}$ is the channel estimation error. Note that e_{mk} is independent of $\widehat{g_{mk}^0}$ and yet we have $e_{mk} \sim \mathcal{CN}(0, \beta_{mk} - \gamma_{mk})$.

Second, $\mathbb{E}\{|\mathrm{BU}_k^0|^2\}$ is calculated as

$$\begin{aligned}
\mathbb{E}\{|\mathrm{BU}_k^0|^2\} &= \sqrt{\rho_s}\sum_{m=1}^{M}\eta_{mk}\mathbb{E}\{|g_{mk}\widehat{g_{mk}^0}^* - \mathbb{E}\{g_{mk}\widehat{g_{mk}^0}^*\}|^2\}\\
&= \sqrt{\rho_s}\sum_{m=1}^{M}\eta_{mk}(\mathbb{E}\{|g_{mk}\widehat{g_{mk}^0}^*|^2\} - |\mathbb{E}\{g_{mk}\widehat{g_{mk}^0}^*\}|^2)\\
&\overset{(a)}{=} \sqrt{\rho_s}\sum_{m=1}^{M}\eta_{mk}(\mathbb{E}\{|e_{mk}\widehat{g_{mk}^0}^*|^2\} + \mathbb{E}\{|\widehat{g_{mk}^0}|^4\} - \gamma_{mk}^2)\\
&= \sqrt{\rho_s}\sum_{m=1}^{M}\eta_{mk}(\gamma_{mk}(\beta_{mk} - \gamma_{mk}) + 2\gamma_{mk}^2 - \gamma_{mk}^2)\\
&= \sqrt{\rho_s}\sum_{m=1}^{M}\eta_{mk}\gamma_{mk}\beta_{mk},
\end{aligned} \tag{6.54}$$

where the equality (a) is obtained by the replacement $g_{mk} = e_{mk} + \widehat{g_{mk}^0}$.

Finally, we derive $\mathbb{E}\{|\mathrm{UI}_{kk'}^0|^2\}$ as follows:

$$\begin{aligned}
\mathbb{E}\{|\mathrm{UI}_{kk'}^0|^2\} &= \rho_s\mathbb{E}\left\{\left|\sum_{m=1}^{M}\sqrt{\eta_{mk'}}c_{mk'}g_{mk}\left(\sqrt{T\rho_u}\sum_{i=1}^{K}g_{mi}\mathbf{p}_{k'}^{\dagger}\mathbf{p}_i + \mathbf{p}_{k'}^{\dagger}\mathbf{w}_m\right)^*\right|^2\right\}\\
&= \rho_s\sum_{m=1}^{M}\eta_{mk'}c_{mk'}^2\beta_{mk} + T\rho_u\rho_s(\mathcal{T}_1 + \mathcal{T}_2),
\end{aligned} \tag{6.55}$$

where

$$c_{mk} = \frac{\sqrt{T\rho_u}\beta_{mk}}{T\rho_u\sum_{k'=1}^{K}\beta_{mk'}|\mathbf{p}_k^{\dagger}\mathbf{p}_{k'}|^2 + 1},$$

$$\mathcal{T}_1 = \mathbb{E}\left\{\left|\sum_{m=1}^{M}\sqrt{\eta_{mk'}}c_{mk'}|g_{mk}|^2\mathbf{p}_{k'}^{\dagger}\mathbf{p}_k\right|^2\right\},$$

$$\mathcal{T}_2 = \mathbb{E}\left\{\left|\sum_{m=1}^{M}\sqrt{\eta_{mk'}}c_{mk'}g_{mk}\left(\sum_{i\neq k}^{K}g_{mi}\mathbf{p}_{k'}^{\dagger}\mathbf{p}_i\right)^*\right|^2\right\}.$$

After some manipulations, $\mathcal{T}_1$ and $\mathcal{T}_2$ can be calculated as

$$\mathcal{T}_1 = 2|\mathbf{p}_{k'}^{\dagger}\mathbf{p}_k|^2 \sum_{m=1}^{M} \eta_{mk'} c_{mk'}^2 \beta_{mk}^2 + |\mathbf{p}_{k'}^{\dagger}\mathbf{p}_k|^2 \sum_{m=1}^{M} \sum_{i\neq m}^{M} \sqrt{\eta_{mk'}\eta_{ik'}} c_{mk'} c_{ik'} \beta_{mk} \beta_{ik},$$

$$\mathcal{T}_2 = \sum_{m=1}^{M} \sum_{i\neq k}^{K} \eta_{mk'} c_{mk'}^2 \beta_{mk} \beta_{mi} |\mathbf{p}_{k'}^{\dagger}\mathbf{p}_i|^2.$$

Plugging $\mathcal{T}_1$ and $\mathcal{T}_2$ into the expression of $\mathbb{E}\{|\mathrm{UI}_{kk'}^0|^2\}$, we have

$$\mathbb{E}\{|\mathrm{UI}_{kk'}^0|^2\} = \rho_s |\mathbf{p}_{k'}^{\dagger}\mathbf{p}_k|^2 \left(\sum_{m=1}^{M} \sqrt{\eta_{mk'}} \gamma_{mk'} \frac{\beta_{mk}}{\beta_{mk'}} \right)^2 + \rho_s \sum_{m=1}^{M} \eta_{mk'} \gamma_{mk'} \beta_{mk}. \tag{6.56}$$

Bibliography

1 H.Q. Ngo, A. Ashikhmin, H. Yang, E.G. Larsson, and T.L. Marzetta, "Cell-free massive MIMO versus small cells," *IEEE Trans. Wirel. Commun.*, vol. 16, pp. 1834–1850, Mar. 2017.

2 S. Buzzi and C.D. Andrea, "Cell-free massive MIMO: user-centric approach," *IEEE Wirel. Commun. Lett.*, vol. 6, pp. 706–709, Dec. 2017.

3 A. Liu and V.K.N. Lau, "Joint BS-user association, power allocation, and user-side interference cancellation in cell-free heterogeneous networks," *IEEE Trans. Sig. Process.*, vol. 65, pp. 335–345, Jan. 2017.

4 L.D. Nguyen, T.Q. Duong, H.Q. Ngo, and K. Tourki, "Energy efficiency in cell-free massive MIMO with zero-forcing precoding design," *IEEE Commun. Lett.*, vol. 21, pp. 1871–1874, Aug. 2017.

5 X. Zhou, B. Maham, and A. Hjorungnes, "Pilot contamination for active eavesdropping," *IEEE Trans. Wirel. Commun.*, vol. 11, pp. 903–907, Mar. 2012.

6 S. Kay, *Fundamentals of Statistical Signal Processing: Estimation Theory*. Prentice-Hall, 1993.

7 T. Yoo and A. Goldsmith, "Capacity and power allocation for fading MIMO channels with channel estimation error," *IEEE Trans. Info. Theo.*, vol. 52, pp. 2203–2214, May 2006.

8 Y. Wu, R. Schober, D.W.K. Ng, C. Xiao, and G. Caire, "Secure massive MIMO transmission in presence of an active eavesdropper," in *Proc. IEEE Int. Conf. Commun. (ICC)*, Jun. 2015.

9 Q. Zhang, S. Jin, K.-K. Wong, H. Zhu, and M. Matthaiou, "Power scaling of uplink massive MIMO systems with arbitrary-rank channel means," *IEEE J. Sel. Top. Sig. Process.*, vol. 8, pp. 966–981, Oct. 2014.

10 T.S. Rappaport, *Wireless Communications: Principles and Practice*, 2nd edn. Prentice Hall PTR, 2002.

11 H. Tuy, *Convex Analysis and Global Optimization*, 2nd edn. Springer, 2016.

7

Massive MIMO for High-performance Ultra-dense Networks in the Unlicensed Spectrum

Adrian Garcia-Rodriguez[1], Giovanni Geraci[2], Lorenzo Galati-Giordano[1] and David López-Pérez[1]

[1] *Nokia Bell Labs, Dublin, Ireland*
[2] *Department of Information and Communication Technologies, Universitat Pompeu Fabra, Spain*

7.1 Introduction

As a result of the ever-increasing wireless data demand, the wireless industry has started to embrace unlicensed spectrum bands to enhance the performance of cellular networks [1–4]. For instance, mobile operators can improve their offered peak data rates via the aggregation of licensed and unlicensed bands, extend the coverage for their users, and/or offload best-effort traffic from the scarce and costly licensed bands below 6 GHz.

Access points (APs) operating in the unlicensed spectrum can complement existing small cell base stations operating in the licensed bands. A variety of technologies have been developed with the purpose of enabling such inter-working in recent years:

1. Long-term evolution (LTE) wireless local area network (WLAN) aggregation (LWA)-like technologies realize licensed unlicensed spectrum aggregation through Wi-Fi APs. Fundamentally, LWA splits protocol data units (PDUs) of a bearer at the packet data convergence protocol (PDCP) layer, transmitting them either through one of the LTE or Wi-Fi radio links by fast switching or through both radio links to enhance capacity. This is achieved by leveraging the dual connectivity framework developed by the Third Generation Partnership Project (3GPP), and specifying a new set of features [5].
2. LTE unlicensed (LTE-U) and licensed assisted access (LAA) [6] rely on a native LTE carrier aggregation approach to operate in the unlicensed spectrum. As a consequence, these approaches always require the presence of a primary carrier component in the licensed band to aggregate unlicensed carrier components at the medium access control (MAC) layer.
3. MulteFire is an industrial standard based on LTE to operate in the unlicensed band [7]. In contrast to LWA, LTE-U, and LAA, MulteFire does not require a licensed carrier anchor, and thus enables stand-alone operation in the unlicensed spectrum.

While the above-mentioned technologies enable operators to access precious frequency resources below 6 GHz, these unlicensed bands are strictly regulated,

Ultra-dense Networks for 5G and Beyond: Modelling, Analysis, and Applications,
First Edition. Edited by Trung Q. Duong, Xiaoli Chu and Himal A. Suraweera.

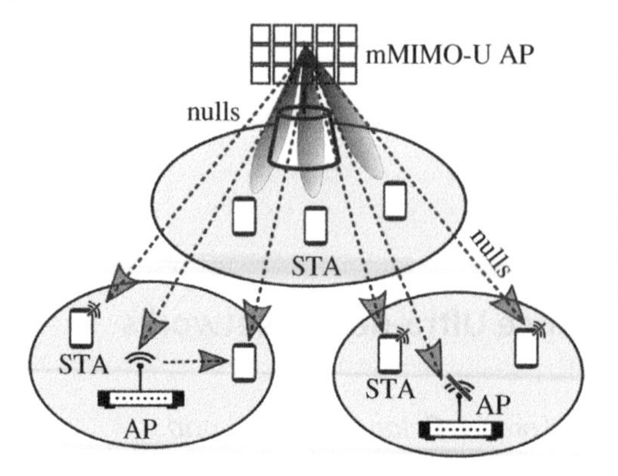

Figure 7.1 Illustration of an mMIMO-U system: an mMIMO-U AP serves a multiplicity of stations (STAs) while spatially mitigating interference towards/from neighboring devices. Colored regions indicate the association area of the APs. Unlicensed devices contend for channel access with all other nodes in the figure; i.e. their coverage area expands the entire topology.

and harmonious coexistence with other incumbent technologies working in the unlicensed spectrum, such as IEEE 802.11x (Wi-Fi), must be guaranteed [8, 9]. Indeed, only solutions implementing listen before talk (LBT) satisfy the strictest regulatory requirements imposed in Europe and Japan [10]. LBT is a channel access procedure based on performing an energy detection phase prior to accessing the unlicensed band. In essence, any device attempting to perform a transmission must evaluate whether the received sum power amounts to less than a regulatory threshold for a given time period [11]. Therefore, LBT only allows the transmission of a single device within a certain coverage area, limiting the potential gains that network densification could bring to communication systems operating in the unlicensed spectrum.

Together with ultra-dense networks, massive MIMO (mMIMO) has also emerged as one of the disruptive technologies for the next-generation wireless communications systems, where APs are envisioned to be equipped with a large number of antennas [12]. This is because mMIMO APs avail of a large number of spatial degrees of freedom (d.o.f.), which allows them to spatially multiplex a large number of users and suppress unintended interference by generating radiation nulls. Indeed, mMIMO could also enable the efficient deployment of ultra-dense networks in the unlicensed spectrum by actively suppressing interference from coexisting neighbors during the mandatory LBT phase [13–16]. This could be achieved by exploiting the spatial awareness offered by large-scale antenna systems, which allows mMIMO APs to design an interference rejection filter that is active during both the LBT and data transmission phases, as illustrated in Figure 7.1. Overall, this enables frequent medium access in ultra-dense networks where a multiplicity of devices contend for channel access. This chapter is dedicated to describe the fundamentals behind mMIMO-unlicensed (mMIMO-U), a technology that, if adopted, could provide significant improvements to existing LAA, MulteFire, and even Wi-Fi systems.

7.2 System Model

This chapter considers a communication system operating in the unlicensed band, where the sets of stations (STAs) and APs are denoted as $\mathcal{S}$ and $\mathcal{A}$, respectively. All

STAs are equipped with a single antenna[1], whereas the ath AP is comprised of $N_a \geq 1$ antennas. STAs associate with the AP that provides the largest average received signal strength (RSS). In this setup, a given AP, $a \in \mathcal{A}$, schedules a subset of its associated STAs $\mathcal{S}_a^{\mathrm{DL}}$ for downlink data transmission at a given time-frequency resource. The maximum number of STAs simultaneously served by the ath AP is defined by a user scheduling algorithm and is given by $M_a \leq N_a$. For ease of formulation, this chapter assumes the same symbol duration independently of the device technology. We also consider that downlink and uplink transmissions might occur simultaneously for devices in different cells, since transmissions share the same frequency band through time division duplexing (TDD) [11]. The subset of APs active at a given time-frequency resource is represented by $\overline{\mathcal{A}}$, whereas $\mathcal{S}_a^{\mathrm{UL}}$ characterizes the subset of STAs transmitting in the ath cell. The signal received by an arbitrary STA s in cell a at a given time-frequency resource can be expressed as

$$y_{as} = \sqrt{P_a}\mathbf{h}_{aas}^{\mathrm{H}}\mathbf{w}_{as}s_{as} + \sqrt{P_a}\sum_{s'\in\mathcal{S}_a^{\mathrm{DL}}\backslash s}\mathbf{h}_{aas}^{\mathrm{H}}\mathbf{w}_{as'}s_{as'} + \sqrt{P_a}\sum_{a'\in\overline{\mathcal{A}}\backslash a}\sum_{s'\in\mathcal{S}_{a'}^{\mathrm{DL}}}\mathbf{h}_{a'as}^{\mathrm{H}}\mathbf{w}_{a's'}s_{a's'} + \sum_{a'\in\mathcal{A}}\sum_{\overline{s}\in\mathcal{S}_{a'}^{\mathrm{UL}}}\sqrt{P_{\overline{s}}}\,g_{a'\overline{s}as}s_{a'\overline{s}}^{\mathrm{UL}} + \epsilon_{as}, \tag{7.1}$$

where:

(i) $P_i \in \mathbb{R}^+$ is the transmission power in watts of the ith AP or STA;
(ii) $\mathbf{h}_{ijk} \in \mathbb{C}^{N_i\times 1}$ is the channel vector between AP i and STA k in cell j;
(iii) $\mathbf{w}_{ik} \in \mathbb{C}^{N_i\times 1}$ is the precoding vector from AP i to STA k in cell i;
(iv) $s_{ik} \in \mathbb{C}$ is the unit-variance data symbol intended for STA k in cell i;
(v) $g_{lijk} \in \mathbb{C}$ is the channel coefficient between STA i in cell l and STA k in cell j;
(vi) $s_{li}^{\mathrm{UL}} \in \mathbb{C}$ is the unit-variance data symbol transmitted by STA i in cell l;
(vii) $\epsilon_{as} \sim \mathcal{CN}(0,\sigma_\epsilon^2)$ is the additive white Gaussian noise (AWGN) component, which follows a complex normal distribution with variance σ_ϵ^2.

The five terms on the right-hand side of Eq. (7.1), represent the useful signal, the intra-cell interference from the serving AP, the inter-cell interference from other APs, the interference from active STAs in other cells, and the thermal noise, respectively.

This chapter adopts a block-fading propagation model, where the communication channels are assumed constant within their respective time/frequency coherence intervals [17]. Moreover, this chapter assumes that the uplink/downlink channel reciprocity holds, as conventionally considered in TDD systems [18]. Moreover, all propagation channels are affected by a slow channel gain (comprising antenna gain, pathloss, and shadowing) and fast fading, i.e.

$$\mathbf{h}_{ijk} = \sqrt{\overline{h}_{ijk}}\widetilde{\mathbf{h}}_{ijk}, \tag{7.2}$$

where $\overline{h}_{ijk} \in \mathbb{R}^+$ represents the slow fading gain, which is assumed constant, and $\widetilde{\mathbf{h}}_{ijk} \in \mathbb{C}^{N_i\times 1}$ contains the fast fading components, which vary at every time/frequency

1 STAs equipped with multiple antennas and implementing spatial multiplexing, beamforming, or active interference suppression techniques could also be considered. This chapter concentrates on single-antenna STAs for ease of exposition, since the fundamentals of the mMIMO-U technology are independent of the specific STA implementation.

coherence interval. The STA-to-STA channels g_{lijk} also follow the above model with their respective slow fading gains and fast fading complex scalar.

The instantaneous signal-to-interference plus noise ratio (SINR) ρ_{as} of STA s located in cell a with perfect channel state information is obtained through an expectation over all transmitted symbols and the noise during a given channel coherence period, and is given by

$$\rho_{as} = \frac{P_a |\mathbf{h}_{aas}^{\mathrm{H}} \mathbf{w}_{as}|^2}{P_a \sum\limits_{s' \in \mathcal{S}_a^{\mathrm{DL}} \setminus s} |\mathbf{h}_{aas}^{\mathrm{H}} \mathbf{w}_{as'}|^2 + P_a \sum\limits_{a' \in \overline{\mathcal{A}} \setminus a} \sum\limits_{s' \in \mathcal{S}_{a'}^{\mathrm{DL}}} |\mathbf{h}_{a'as}^{\mathrm{H}} \mathbf{w}_{a's'}|^2 + \sum\limits_{a' \in \mathcal{A}} \sum\limits_{\overline{s} \in \mathcal{S}_{a'}^{\mathrm{UL}}} P_{\overline{s}} \, |g_{a'\overline{s}as}|^2 + \sigma_\epsilon^2}. \tag{7.3}$$

7.3 Fundamentals of Massive MIMO Unlicensed (mMIMO-U)

This section summarizes the procedures carried out by mMIMO-U systems for downlink data transmissions: (1) channel covariance estimation, (2) enhanced listen before talk, (3) neighboring-node-aware scheduling, (4) acquisition of channel state information, and (5) beamforming with radiation nulls.

7.3.1 Channel Covariance Estimation

In mMIMO-U, the LBT phase is enhanced by placing spatial radiation nulls towards neighboring devices, which include both APs and STAs. To effectively suppress interference from/toward neighboring devices, each AP requires information about their interfering channels. This section describes the blind channel covariance estimation procedure that mMIMO-U APs periodically follow to obtain this information.

The blind channel covariance estimation process occurs when all devices served by a given mMIMO-U AP are silent, e.g. when there are no data packets to transmit or receive. During this period, the mMIMO-U AP processes the received signals from the remaining active APs and STAs located in neighboring cells. Specifically, a given mMIMO-U AP a receives the following signal:

$$\begin{aligned} \mathbf{z}_a[m] = {} & \sqrt{P_a} \sum_{a' \in \overline{\mathcal{A}}[m] \setminus a} \sum_{s' \in \mathcal{S}_{a'}^{\mathrm{DL}}} \mathbf{H}_{a'a} \mathbf{w}_{a's'} s_{a's'}[m] \\ & + \sum_{a' \in \mathcal{A} \setminus a} \sum_{\overline{s} \in \mathcal{S}_{a'}^{\mathrm{UL}}[m]} \sqrt{P_{\overline{s}}} \, \mathbf{h}_{aa'\overline{s}} s_{a'\overline{s}}^{\mathrm{UL}}[m] + \xi_a[m], \end{aligned} \tag{7.4}$$

where m denotes the time-frequency resource index, $\mathbf{H}_{a'a} \in \mathbb{C}^{N_a \times N_{a'}}$ represents the channel between AP a and AP a', and $\xi_a[m] \sim \mathcal{CN}(\mathbf{0}, \sigma_\xi^2 \mathbf{I})$ is the AWGN represented by a circularly symmetric complex normal vector with each entry having a variance σ_ξ^2.

Let $\mathbf{Z}_a \in \mathbb{C}^{N_a \times N_a}$ be the covariance matrix of $\mathbf{z}_a[m]$, i.e.

$$\mathbf{Z}_a = \mathbb{E}[\mathbf{z}_a \mathbf{z}_a^{\mathrm{H}}], \tag{7.5}$$

where the expectation is taken with respect to the active devices ($\overline{\mathcal{A}}[m]$ and $\mathcal{S}_{a'}^{\mathrm{UL}}[m]$), the transmitted symbols ($s_{a's'}[m]$ and $s_{a'\overline{s}}^{\mathrm{UL}}[m]$), and the noise $\xi_a[m]$. The ath AP can thus obtain an estimate $\widehat{\mathbf{Z}}_a$ of $\mathbf{Z}_a$ through averaging over T_{c} symbol intervals as [19]

$$\widehat{\mathbf{Z}}_a = \frac{1}{T_{\mathrm{c}}} \sum_{m=1}^{T_{\mathrm{c}}} \mathbf{z}_a[m]\mathbf{z}_a^{\mathrm{H}}[m], \tag{7.6}$$

where the value of T_{c} must be sufficiently large to ensure that all neighboring devices have been active. The operation in Eq. (7.6) generates a trade-off between improving the quality of the estimate in Eq. (7.6) and reducing the overhead, as further discussed in Section 7.5.1.

Given the channel covariance matrix estimate $\widehat{\mathbf{Z}}_a$, AP a applies a singular value decomposition, obtaining

$$\widehat{\mathbf{Z}}_a = \widehat{\mathbf{U}}_a \widehat{\mathbf{\Lambda}}_a \widehat{\mathbf{U}}_a^{\mathrm{H}}, \tag{7.7}$$

where the columns of $\widehat{\mathbf{U}}_a = [\widehat{\mathbf{u}}_{a1}, \dots, \widehat{\mathbf{u}}_{aN_a}]$ are formed by the eigenvectors of $\widehat{\mathbf{Z}}_a$, which form an orthonormal basis, and

$$\widehat{\mathbf{\Lambda}}_a = \mathrm{diag}(\widehat{\lambda}_{a1}, \dots, \widehat{\lambda}_{aN_a}) \tag{7.8}$$

contains the eigenvalues of $\widehat{\mathbf{Z}}_a$, satisfying $\widehat{\lambda}_{a1} \geq \widehat{\lambda}_{a2} \geq \dots \geq \widehat{\lambda}_{aN_a}$. Strictly speaking, $\widehat{\mathbf{Z}}_a$ is a full-rank matrix, since $\mathbb{E}[\xi_a \xi_a^{\mathrm{H}}] = \mathbf{I}$. However, $\widehat{\mathbf{Z}}_a$ can be generally approximated as a low-rank matrix in many circumstances, e.g. when the number of antennas at the considered mMIMO-U AP exceeds the number of active devices in its neighborhood and/or when the channels toward these neighbor devices are spatially correlated.

From this blind channel covariance estimate, each mMIMO-U AP gains spatial awareness about the wireless transmitters operating in the same coverage area, which may amount to a large number in ultra-dense deployments. This facilitates channel access through the application of interference suppression techniques, as detailed in the following section.

7.3.2 Enhanced Listen Before Talk (eLBT)

To comply with the regulations in the unlicensed band, all APs and STAs must perform an omnidirectional LBT before accessing the channel [10]. Fundamentally, LBT prevents interference in the unlicensed band by allowing only the transmission of a single node within a certain coverage area. With omnidirectional LBT, a transmission opportunity is identified by the ath AP if the sum power received from all active nodes satisfies

$$||\mathbf{z}_a[m]||^2 < \gamma_{\mathrm{LBT}}, \; m = \{1, \dots, M_{\mathrm{LBT}}\}, \tag{7.9}$$

where $\mathbf{z}_a[m]$ is defined in Eq. (7.4), $M_{\mathrm{LBT}} \approx \lceil (\mathrm{DIFS} + n \times \mathrm{slotduration})/T_{\mathrm{OFDM}} \rceil \in \mathbb{Z}^+$ is the LBT duration in number of OFDM symbols for Wi-Fi, and γ_{LBT} is a regulatory threshold. In 802.11ac, $n \in [0, \dots, 1023]$ determines the random number of backoff time slots, the distributed inter-frame space (DIFS) has a duration of $34\mu s$, the slot duration is $9\mu s$, and the OFDM symbol duration with cyclic prefix is $T_{\mathrm{OFDM}} = 4\mu s$ [11][2]. The process in condition (7.9) is conventionally known as energy detection.

2 Although not explicitly characterized here for brevity, we stress that STAs must also follow the LBT procedure when attempting to access the unlicensed spectrum.

In contrast to conventional APs, an mMIMO-U AP a allocates a certain number of spatial d.o.f.s, denoted as D_a, to suppress interference from/toward the dominant directions of the interfering channel subspace. Accordingly, let us now define

$$\hat{\mathbf{\Sigma}}_a \triangleq [\hat{\mathbf{u}}_{a(D_a+1)}, \dots, \hat{\mathbf{u}}_{aN_a}], \tag{7.10}$$

where the columns of $\hat{\mathbf{\Sigma}}_a$ are the eigenvectors corresponding to the $(N_a - D_a)$ smallest eigenvalues of $\hat{\mathbf{Z}}_a$. Similarly, let

$$\hat{\mathbf{\Sigma}}_a^{\perp} \triangleq [\hat{\mathbf{u}}_{a1}, \dots, \hat{\mathbf{u}}_{aD_a}] \tag{7.11}$$

contain the eigenvectors corresponding to the D_a largest eigenvalues of $\hat{\mathbf{Z}}_a$. For a sufficiently large D_a, range$\{\hat{\mathbf{\Sigma}}_a^{\perp}\}$ represents the channel subspace on which AP a receives most of the interference from the neighboring cells. Since channel reciprocity holds in TDD systems, the power transmitted by AP a on range$\{\hat{\mathbf{\Sigma}}_a^{\perp}\}$ also represents the major source of interference towards one or more neighboring devices. These significant interference sources are precisely those that the mMIMO-U AP attempts to suppress. Thus, mMIMO-U enhances the LBT phase by designing a baseband filter that places radiation nulls toward the dominant eigendirections of the interfering channels. When the ath AP listens to the transmissions currently taking place in the unlicensed band, it filters the received signals $\mathbf{z}_a[m]$ with the D_a spatial nulls defined in Eq. (7.10). Unlike Eq. (7.9), an mMIMO-U AP thus obtains a transmission opportunity when

$$\|\hat{\mathbf{\Sigma}}_a^{\mathrm{H}} \mathbf{z}_a[m]\|^2 < \gamma_{\mathrm{LBT}}, \ m = \{1, \dots, M_{\mathrm{LBT}}\} \tag{7.12}$$

holds.

The rationale behind Eq. (7.12) is illustrated in Figure 7.2, and can be described as follows. Since the channel subspace range $\{\hat{\mathbf{\Sigma}}_a^{\perp}\}$ is occupied by neighboring devices, the ath mMIMO-U AP focuses its downlink transmissions on the channel subspace null$\{\hat{\mathbf{\Sigma}}_a^{\perp}\}$ = range$\{\hat{\mathbf{\Sigma}}_a\}$ only. As a direct consequence, the ath mMIMO-U AP must ensure that no concurrent transmissions are detected on range$\{\hat{\mathbf{\Sigma}}_a\}$. This is accomplished by measuring the aggregate power of the received signal filtered through the projection $\hat{\mathbf{\Sigma}}_a$. Provided that a sufficient number of d.o.f.s D_a have been allocated for interference suppression, the condition in (7.9) is met. Therefore, unlike conventional LBT operations, this enhanced LBT (eLBT) phase allows both mMIMO-U APs and their neighboring devices to simultaneously access the unlicensed spectrum.

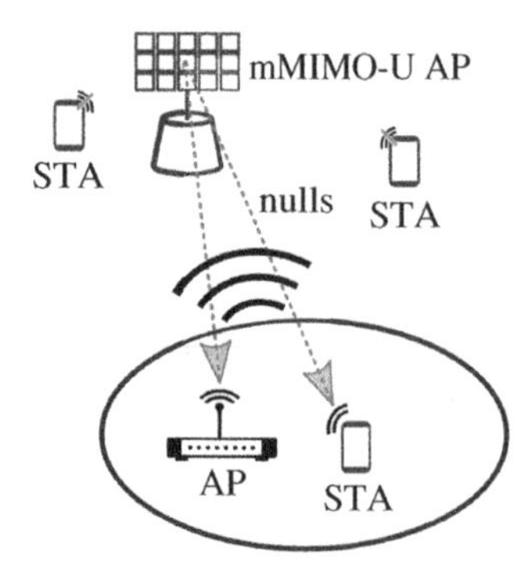

Figure 7.2 An mMIMO-U AP performing eLBT with radiation nulls in place: no concurrent transmissions are detected when a sufficient number of accurate radiation nulls are designed.

Intuitively, since placing radiation nulls is equivalent to dedicating a number of spatial d.o.f. to interference suppression, the remaining $N_a - D_a$ d.o.f. can be dedicated to STA multiplexing. However, this phase must account for the interference concurrently generated by neighboring active devices, as described in the following section.

7.3.3 Neighboring-Node-Aware Scheduling

While the proposed eLBT channel access mechanism creates additional transmission opportunities for mMIMO-U APs, this comes at the expense of increased uncontrolled interference at its served STAs, since the spectrum is reused while other nodes are active. This is especially true for ultra-dense networks, where some STAs are likely to be located near other active nodes, as shown in Figure 7.3. In other words, while the mutual interference between mMIMO-U APs and other coexisting APs with their associated STAs can be suppressed through radiation nulls, the same does not hold for the served STAs. A neighboring-node-aware user selection process is therefore needed, where mMIMO-U APs schedule transmissions toward their STAs depending on their radio proximity to neighboring devices.

In the light of the above, each mMIMO-U AP a ranks its associated STAs according to the metric

$$\mu_{as} = \frac{P_a \bar{h}_{aas}}{P_a \sum_{a' \in \mathcal{A} \setminus a} \bar{h}_{a'as}}, \tag{7.13}$$

which accounts for the average channel gain between the sth associated STA and the serving AP ($\bar{h}_{aas}$), as well as other neighboring APs ($\bar{h}_{a'as}$). These average channel gains vary at a slow rate and they are assumed to be perfectly known by the ath mMIMO-U AP in Eq. (7.13). Intuitively, μ_{as} represents a measure of the average signal-to-interference ratio perceived by the sth STA in cell a during a non-precoded broadcast signal transmission.

In practical implementations, when the mMIMO-U AP adopts a cellular-based technology such as LAA, μ_{as} can be obtained as follows:

- $\bar{h}_{aas}$ can be obtained at the STA through averaging downlink measurements on the cell reference signal (CRS). Specifically, $\bar{h}_{aas}$ is derived from the subtraction in logarithmic scale of the reference signal received power (RSRP) from the reference signal power, which is signaled by the mMIMO-U AP [20, 21].

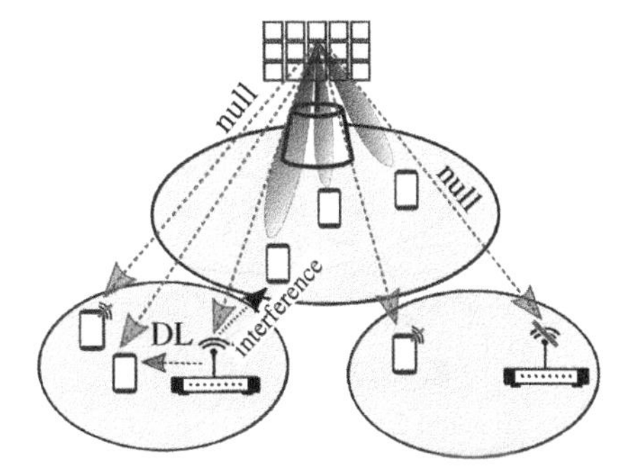

Figure 7.3 Illustration of the uncontrolled interference generated by a neighboring active device.

- For capturing the average channel gains $\overline{h}_{a'as}$ from those neighboring APs that also implement LAA, STAs can follow the same procedure as above but on the CRS transmitted by other APs.
- Instead, if the neighboring APs are Wi-Fi APs, the value of $\overline{h}_{a'as}$ can be obtained through the automatic neighbor relations (ANR) function, where mMIMO-U APs request their STAs to report Wi-Fi measurements that contain the received signal strength indicator (RSSI) from neighboring Wi-Fi APs [22].

Typically, a few dominant terms $\overline{h}_{a'as}$ would be sufficient to provide an accurate estimate of μ_{as}, since a given STA s is unlikely to be near a multitude of APs at the same time due to their limited transmission range.

Regarding the number of scheduled STAs M_a, a variety of criteria can be used to select the pair (M_a, D_a), which must satisfy

$$D_a \leq \min\{N_a - M_a, M_{\text{LBT}}\}. \tag{7.14}$$

The inequality in (7.14) indicates that $D_a + M_a$ should not exceed the spatial d.o.f. N_a available at the AP and that D_a should be upper bounded by the rank M_{LBT} of $\widehat{\mathbf{Z}}_a$ defined in Eq. (7.6). Intuitively, D_a controls the number of d.o.f. used for interference suppression. For instance, D_a could be set as the number of dominant eigenvectors of $\widehat{\mathbf{Z}}_a$, e.g. those containing 99% of the aggregate power in the eigenvalues $\widehat{\mathbf{Z}}_a$.

As a result of the neighbor-aware scheduling procedure, the ath mMIMO-U AP can determine the M_a STAs that are far from neighboring cells, and should be scheduled for transmission using the same spectrum. A number of alternatives can be followed to serve STAs located near other cells:

- They may be served in the same channel when the channel conditions have changed, e.g. because devices have moved or their traffic profile has varied.
- They may be served in the same channel by reverting to conventional LBT, i.e. through discontinuous transmission. Such an approach comes at the expense of a lower spatial reuse, as analyzed in Section 7.4.2.
- They may be served in an alternative unlicensed channel, where they are less affected by transmissions in their vicinity. This approach is depicted in Figure 7.4.
- They may be scheduled for transmission in a licensed band, if this option is available.

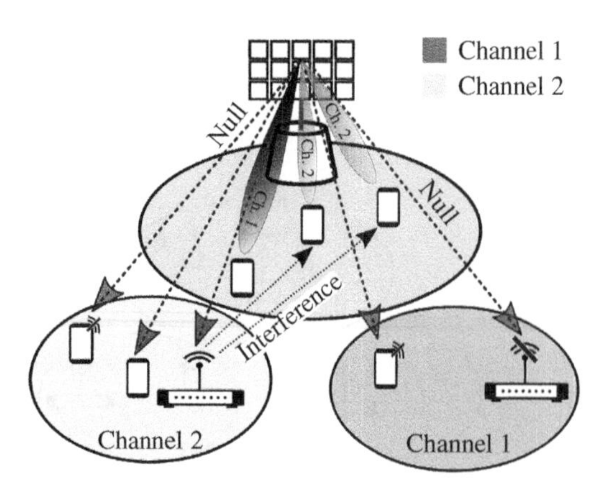

Figure 7.4 Cross-channel user selection for mitigating uncontrolled interference toward STAs served by mMIMO-U APs. Reprinted with permission from [15].

7.3.4 Acquisition of Channel State Information

Once the ath mMIMO-U AP has selected the M_a STAs for data transmission, it should acquire the knowledge of their channels $\mathbf{h}_{aas}$, $s \in \mathcal{S}_a^{\mathrm{DL}}$. Instantaneous channel state information (CSI) from the scheduled STAs can be obtained via pilot signals transmitted during a training phase. During this CSI acquisition period, coexistence between uplink pilots sent by STAs and transmissions from neighboring devices must also be guaranteed, as detailed in the following.

In the channel estimation phase, AP a informs the selected STAs with a request-to-send-pilots (RTSP) message. The RTSP message is transmitted on the subspace range$\{\widehat{\boldsymbol{\Sigma}}_a\}$, such that interference generated at neighboring devices is suppressed. In general, a scheduled STA could receive RTSP messages superimposed with transmissions from active devices from other cells. However, this is unlikely to occur thanks to the STA selection metric in Eq. (7.13), which tends to schedule STAs that are far from other cells. The addressed STAs respond by transmitting back omnidirectional pilot signals [11]. Note that the STA selection metric in Eq. (7.13) ensures that STA pilots do not create significant interference at Wi-Fi devices.[3]

Let the CSI acquisition phase span T_{p} time-frequency resources during each channel coherence interval. The pilot signal transmitted by the sth STA associated to the ath AP is denoted by $\mathbf{v}_{\mathrm{i}_{as}} \in \mathbb{C}^{T_{\mathrm{p}}\times 1}$, where i_{as} represents its chosen codebook entry and all pilots form an orthonormal basis [18, 23]. The pilot signals received at the AP are prone to contamination caused by both concurrent transmissions from neighboring devices and pilot reuse across mMIMO-U cells. The collective received signal during the CSI acquisition phase at the ath AP $\mathbf{Y}_a \in \mathbb{C}^{N_a\times T_{\mathrm{p}}}$ can thus be expressed as

$$\mathbf{Y}_a = \sum_{i\in\mathcal{A}}\sum_{s\in\mathcal{S}_i}\sqrt{P_s}\mathbf{h}_{ais}\mathbf{v}_{\mathrm{i}_{is}}^{\mathrm{T}} + \sum_{a'\in\overline{\mathcal{A}}\backslash a}\sum_{s'\in\mathcal{S}_{a'}^{\mathrm{DL}}}\mathbf{H}_{a'a}\mathbf{x}_{a's'} + \sum_{a'\in\overline{\mathcal{A}}\backslash a}\sum_{\bar{s}\in\mathcal{S}_{a'}^{\mathrm{UL}}}\sqrt{P_{\bar{s}}}\,\mathbf{h}_{aa'\bar{s}}\mathbf{s}_{a'\bar{s}}^{\mathrm{UL}} + \mathbf{N}_a, \tag{7.15}$$

where $\mathbf{x}_{a's'} = [\mathbf{x}_{a's'}[1], \dots, \mathbf{x}_{a's'}[T_{\mathrm{p}}]]$ are the signals transmitted by neighboring active APs, $\mathbf{s}_{a'\bar{s}}^{\mathrm{UL}} = [s_{a'\bar{s}}^{\mathrm{UL}}[1], \dots, s_{a'\bar{s}}^{\mathrm{UL}}[T_{\mathrm{p}}]]$ are the signals conveyed by neighboring active STAs (pilots and/or data), and $\mathbf{N}_a \in \mathbb{C}^{N_a\times T_{\mathrm{p}}}$ is the additive noise received by the ath AP during the pilot transmission stage.

The received signal $\mathbf{Y}_a$ in Eq. (7.15) is first processed at the ath AP by correlating it with the known pilot signal $\mathbf{v}_{\mathrm{i}_{as}}$ from the sth STA, i.e.

$$\widetilde{\mathbf{y}}_a = \mathbf{Y}_a\mathbf{v}_{\mathrm{i}_{as}}^{*}. \tag{7.16}$$

Note that $\widetilde{\mathbf{y}}_a \in \mathbb{C}^{N_a\times 1}$ corresponds to a scaled least-squares estimate of $\mathbf{h}_{aas}$ [18][4]. Subsequently, the mMIMO-U AP projects $\widetilde{\mathbf{y}}_a \in \mathbb{C}^{N_a\times 1}$ on to range$\{\widehat{\boldsymbol{\Sigma}}_a\}$ to reject interference from neighboring transmissions, which results in the channel estimate $\widetilde{\mathbf{h}}_{aas} \in \mathbb{C}^{(N_a-D_a)\times 1}$

3 A scheduled STA might decide to refrain from transmitting its pilot when (a) it deems the channel as busy after LBT or (b) it has been informed of ongoing nearby Wi-Fi transmissions via network allocation vector (NAV) messages [11].

4 While in this chapter we concentrate on the least-squares estimator for ease of exposition, more sophisticated channel estimation methods such as those based on the minimum mean square error criterion could also be applied.

given by

$$\widetilde{\mathbf{h}}_{aas} = \widehat{\boldsymbol{\Sigma}}_a^{\mathrm{H}} \widetilde{\mathbf{y}}_a = \widehat{\boldsymbol{\Sigma}}_a^{\mathrm{H}} \mathbf{Y}_a \mathbf{v}_{\mathrm{i}_{as}}^{*}. \tag{7.17}$$

The above expression illustrates that, from a mathematical perspective, dedicating D_a d.o.f. to radiation nulls is equivalent to using a virtual mMIMO array with $(N_a - D_a)$ antennas in environments with a high number of scatterers.

7.3.5 Beamforming with Radiation Nulls

Thanks to the large number of antennas available, an mMIMO-U AP can spatially multiplex the scheduled S_a STAs in downlink, while simultaneously placing D_a nulls on the channel subspace occupied by the neighboring devices, as depicted in Figure 7.1. The downlink precoder $\widetilde{\mathbf{W}}_a \in \mathbb{C}^{N_a \times M_a}$ at the ath mMIMO-U AP is defined as

$$\widetilde{\mathbf{W}}_a = \widehat{\boldsymbol{\Sigma}}_a \mathbf{W}_a, \tag{7.18}$$

where $\mathbf{W}_a \in \mathbb{C}^{(N_a - D_a) \times M_a}$ can be defined following any standard precoding method such as zero-forcing, i.e.

$$\mathbf{W}_a = \widetilde{\mathbf{H}}_a (\widetilde{\mathbf{H}}_a^{\mathrm{H}} \widetilde{\mathbf{H}}_a)^{-1} \mathbf{D}_a. \tag{7.19}$$

Here, the diagonal matrix $\mathbf{D}_a \in \mathbb{C}^{M_a \times M_a}$ is chosen to satisfy the total power constraint, i.e. such that $\sum_{s \in \mathcal{S}_a^{\mathrm{DL}}} \|\mathbf{w}_{as}\|^2 = P_a$, with $\mathbf{w}_{as}$ denoting the sth column of $\mathbf{W}_a$. In physical terms, the precoder defined in Eq. (7.18) places D_a radiation nulls in range$\{\widehat{\boldsymbol{\Sigma}}_a^{\perp}\}$ by projecting the transmitted signals on to range$\{\widehat{\boldsymbol{\Sigma}}_a^{\perp}\}$.

When operating in the unlicensed spectrum, a tight constraint on the maximum transmit power is imposed [24]. This constraint accounts for the number of spatial d.o.f. used to provide beamforming gain. APs with multiple antennas must thus reduce the radiated power according to the beamforming gain provided to each STA [25], yielding a total transmission power of

$$P_a = P_a^{\max} - 10 \log_{10} \left(\frac{N_a - D_a}{M_a} \right) \text{ dBm}, \tag{7.20}$$

where $P_a^{\max}$ is the maximum transmission power per device specified by regulatory bodies.

7.4 Performance Evaluation

In this section, we present the performance evaluation of mMIMO-U for both outdoor and indoor scenarios.

7.4.1 Outdoor Deployments

We evaluate the performance of outdoor cellular mMIMO-U deployments by considering a wrapped-around hexagonal cellular layout with 19 ten-meter-high sites, three sectors per site, and an inter-site distance of 150 m. This setup is illustrated in Figure 7.5. In this dense network, a cellular operator deploys mMIMO-U APs in the hexagon centers.

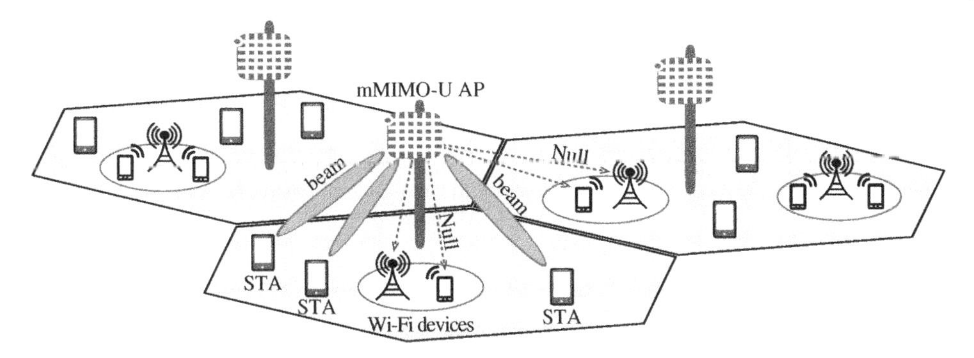

Figure 7.5 The scenario considered for outdoor deployments. Reprinted with permission from [14].

These mMIMO-U APs reuse the unlicensed channel occupied by Wi-Fi hotspots of 10 meter radius and uniformly deployed within each sector. We consider outdoor Wi-Fi hotpots since this case involves no wall penetration losses, making coexistence with APs more challenging. The results are shown for a single 20 MHz channel, and it has been assumed that Wi-Fi devices uniformly distribute their 20 MHz transmissions in the four non-overlapping channels of the U-NII-1 band (5.18 GHz). The detailed simulation parameters can be found in Table 7.1, which are based on [11, 25–29].

7.4.1.1 Cellular/Wi-Fi Coexistence

Figures 7.6 and 7.7 illustrate the coexistence enhancements provided by mMIMO-U APs, where $D_a = 0.75(N_a - M_a)$ d.o.f. are allocated for nulls, with respect to a traditional LBT approach without explicit interference suppression ($D_a = 0$).

In Figure 7.6, the perspective of the Wi-Fi devices is adopted, considering that mMIMO-U APs have gained access to the unlicensed channel. The results are shown for APs with 32, 64, and 128 antenna elements, and implementing both the mMIMO-U APs operations described in this chapter and the conventional omnidirectional LBT. The results shown in Figure 7.6 demonstrate that the aggregate interference at the Wi-Fi devices diminishes when mMIMO-U APs increase their number of antennas, as more d.o.f. are allocated for interference suppression. APs implementing conventional LBT follow a similar trend, which is a direct consequence of the mandatory reduction in the transmission power as per the unlicensed regulations described in Section 7.3.5. Figure 7.6 demonstrates that the aggregate interference falls below γ_{LBT} for mMIMO-U systems with more than 32 antennas. This shows that mMIMO-U enables Wi-Fi devices to access the unlicensed band, while APs are simultaneously transmitting. Instead, Wi-Fi devices are not able to access the channel when APs do not place radiation nulls, since the interference they perceive is above γ_{LBT}.

Figure 7.7 adopts the perspective of APs, assuming that a single Wi-Fi device per hotspot has gained access to the unlicensed channel. Following the specifications of systems such as LAA [6], this figure considers the conservative energy detection threshold value of $\gamma_{\text{LBT}} = -72$ dBm.

The results of Figure 7.7 demonstrate that mMIMO-U APs with a sufficient number of antennas can reuse the same spectrum occupied by Wi-Fi devices. Specifically, the aggregate interference received by an mMIMO-U AP is always below γ_{LBT} for the cases with 64 and 128 antennas. Moreover, it can be observed that the aggregate interference

Table 7.1 Detailed system parameters for the outdoor deployment. Reprinted with permission from [14].

Parameter	Description
Cellular layout	Hexagonal with wrap-around, 19 sites, 3 sectors each, 1 mMIMO-U AP per sector
Inter-site distance	150 m
STAs distribution	Random (P.P.P.), 24 STAs deployed per sector on average
STA association	Based on slow fading gain
STA pilot allocation	Random with reuse 1 ($T_{\mathrm{p}} = 8$)
STA channel estimation	Least-squares estimator
Wi-Fi hotspots	8 single-antenna devices per hotspot: 1 AP and 7 STAs
Carrier frequency	5.18 GHz (U-NII-1)
System bandwidth	20 MHz with 100 resource blocks
LBT regulations	$\gamma_{\mathrm{LBT}} = -62$ dBm
AP precoder	As in Eq. (7.18)
AP antennas	Downtilt:12°, height: 10 m
AP antenna array	Uniform linear, element spacing: $d = 0.5\lambda$
AP antenna pattern	Antennas with 3 dB beamwidth of 65° and 8 dBi max.
mMIMO-U AP max. tx power	$p^{\max} = 30$ dBm
mMIMO-U AP power allocation	Uniform STA power allocation
Wi-Fi AP/STA max. tx power	$p^{\max} = 24/18$ dBm
STA tx pilot power	Fractional uplink power control with $P_0 = -58$ dBm and $\alpha = 0.6$
STA noise figure	9 dB
AP rx sensitivity	−94 dBm
Fast fading	Ricean, distance-dependent K-factor
Shadowing	Lognormal as per 3GPP UMi and 3GPP D2D
Pathloss	3GPP UMi and 3GPP D2D
Thermal noise	−174 dBm/Hz spectral density

is 93% of the time below γ_{LBT} when mMIMO-U APs have 32 antennas, which still entails a high channel access probability for the system. In contrast, Figure 7.7 shows that APs implementing conventional LBT cannot share the same channel used by Wi-Fi devices, since their received interference is consistently higher than γ_{LBT}.

7.4.1.2 Achievable Cellular Data Rates

Figure 7.8 shows the downlink data rates per sector as a function of the number of d.o.f. dedicated to interference suppression at the mMIMO-U APs, which are comprised of 64 antennas. Three scenarios corresponding to 1, 2, and 4 active Wi-Fi hotspots per sector on average are considered. The data rate of the sth STA located in the ath cell,

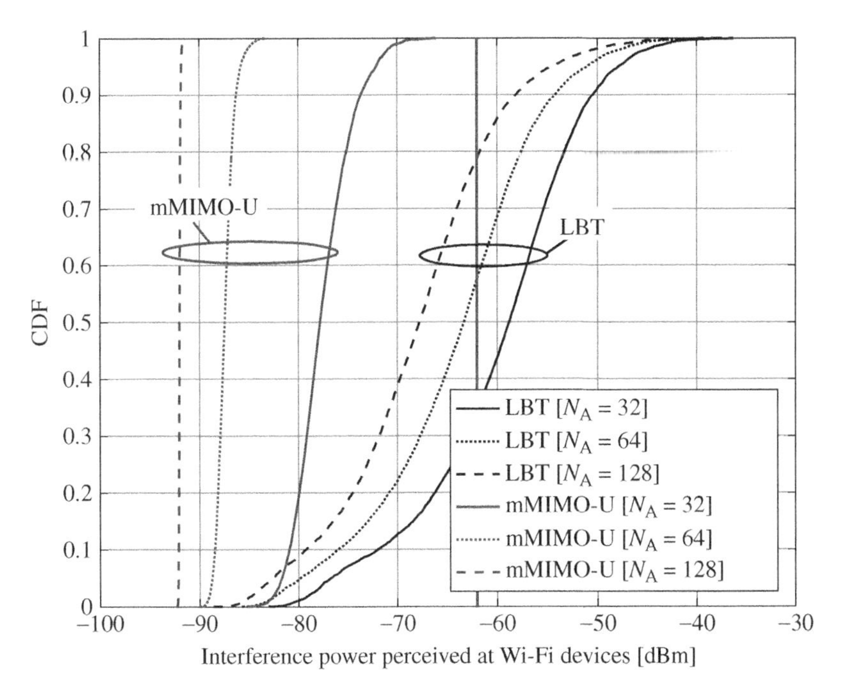

Figure 7.6 Coexistence in the selected unlicensed channel as perceived by Wi-Fi devices in the presence of two active Wi-Fi hotspots per sector.

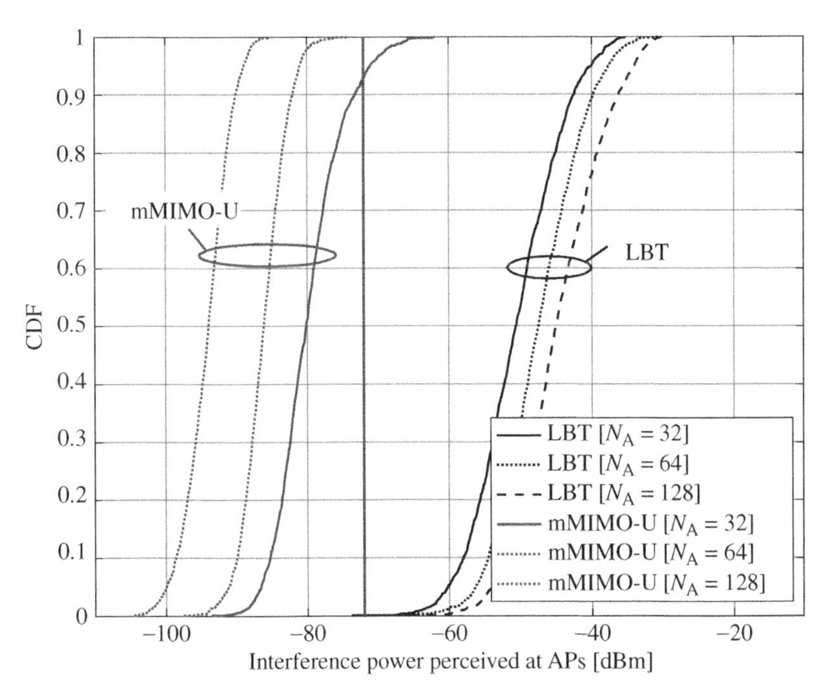

Figure 7.7 Coexistence in the selected unlicensed channel as perceived by APs in the presence of two active Wi-Fi hotspots per sector.

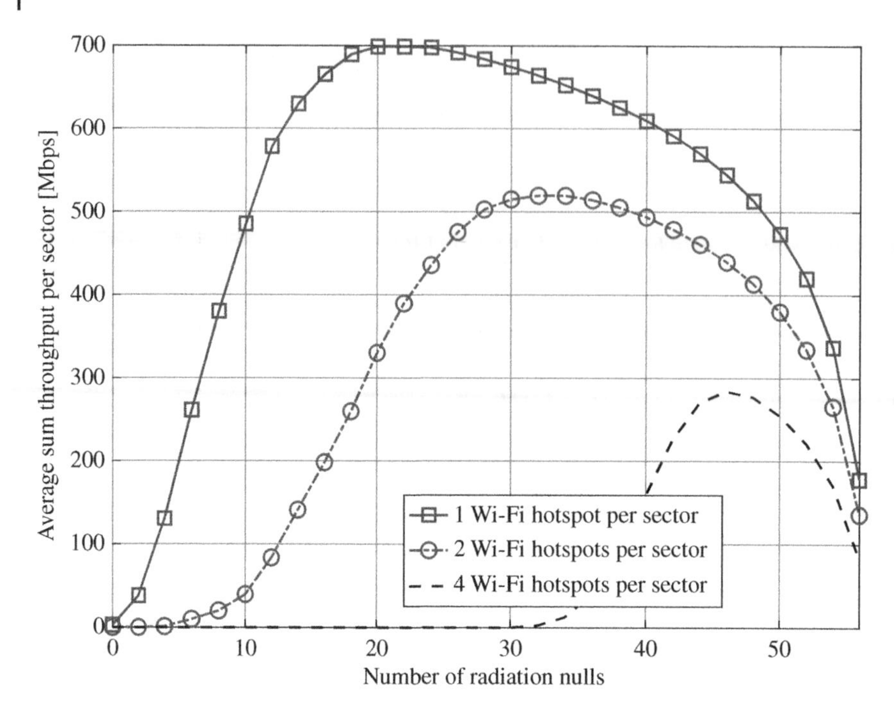

Figure 7.8 Cellular mMIMO-U sum rates per sector versus number of radiation nulls in the presence of 1, 2, and 4 active Wi-Fi hotspots per sector on average. Reprinted with permission from [15].

R_{as}, is calculated by adding up the rates achieved per resource block, which are given by $R_{as} = B_{\mathrm{RB}}\log_2(1 + \rho_{as})$, where ρ_{as} is the SINR per resource block as per Eq. (7.3), and $B_{\mathrm{RB}} = 180$ kHz is the resource block bandwidth [26]. The channel estimation overhead is not considered in these results. In these deployments, the main source of interference toward mMIMO-U STAs are Wi-Fi devices (APs and STAs) located in the same sector.

The results of Figure 7.8 illustrate that a conventional cellular system with omnidirectional LBT ($D = 0$) would not be able to access the channel while Wi-Fi devices are active. Instead, mMIMO-U APs are capable of transmitting information as the number of d.o.f. dedicated to interference suppression increases, since the channel access probability also grows. However, Figure 7.8 also shows that dedicating an excessively large number of d.o.f. for interference suppression is not recommended once mMIMO-U APs are guaranteed channel access, since fewer d.o.f. will be available for spatially separating their associated STAs. Moreover, it can be observed that having to coexist with more Wi-Fi hotspots impacts the attainable mMIMO-U system rates. This is because a larger number of active Wi-Fi devices entail (a) the occupation of a larger number of interfering spatial dimensions with increased power and (b) larger interference from Wi-Fi devices toward mMIMO-U STAs. As a direct consequence, Figure 7.8 illustrates that more radiation nulls must be placed for guaranteeing channel access when the system has to coexist with more Wi-Fi hotspots.

7.4.2 Indoor Deployments

To illustrate the performance of mMIMO-U in ultra-dense indoor deployments, this section considers the single-floor 120 m × 50 m hotspot network depicted in Figure 7.9. This setup is conventionally recommended by the 3GPP for indoor coexistence studies. In this setting, a single operator deploys three Wi-Fi APs on the ceiling of the central corridor to guarantee a full coverage, i.e. a minimum RSS of −82 dBm for all STAs [6]. The three images in Figure 7.9 represent the scenarios considered in this section, namely: (A) single-antenna Wi-Fi, (B) mMIMO Wi-Fi, and (C) mMIMO-U Wi-Fi. Each STA has data packets available with a certain probability P_{tr}. A fraction of such packets are to

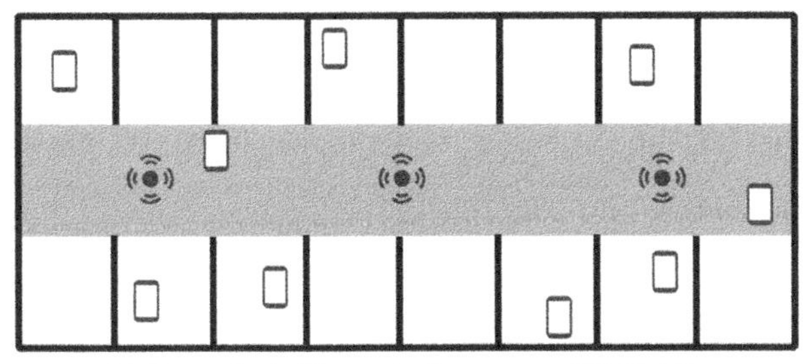

(a) Scenario A: Three single-antenna Wi-Fi APs.

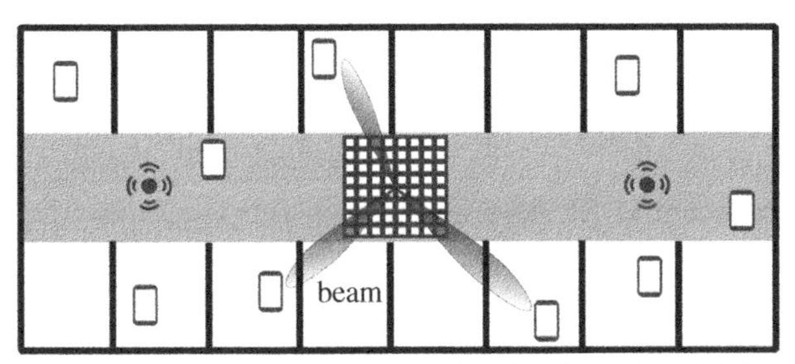

(b) Scenario B: Two single-antenna Wi-Fi APs (left and right) and one mMIMO Wi-Fi AP (center).

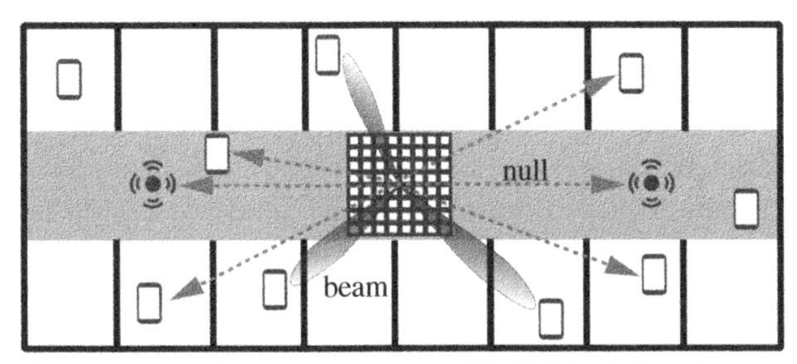

(c) Scenario C: Two single-antenna Wi-Fi APs (left and right) and one mMIMO-U Wi-Fi AP (center).

Figure 7.9 The three scenarios considered for indoor deployments. Reprinted with permission from [16].

be received in downlink from the serving AP when scheduled, whereas the remaining fraction is to be transmitted in uplink toward the serving AP when the channel is available. We consider two different data traffic regimes, i.e. light and heavy traffic, by setting the probability of each user having data packets to transmit/receive to $P_{\text{tr}} = 0.1$ and $P_{\text{tr}} = 1$, respectively. Since all devices follow the Wi-Fi standard, apart from following the mandatory LBT procedure, they also refrain from transmission when a packet preamble from any other node is detected. This preamble identification procedure requires a lower power threshold γ_{preamble} with a minimum SINR of −0.8 dB to deem the channel occupied [11]. Both mMIMO and mMIMO-U APs are each equipped with 36 antennas and schedule a maximum of 4 STAs for multi-user downlink transmission. In the mMIMO-U setup (scenario C), the mMIMO-U AP attempts channel access:

- 40% of the time through eLBT, serving the 40% least interfered STAs with traffic available for downlink transmission as per the criterion defined in Section 7.3.3.
- 60% of the time via conventional LBT operations, serving the remaining 60% most vulnerable STAs.

A detailed list of all system parameters different from those provided in Table 7.1 are provided in Table 7.2, which have been derived from [6, 29, 30].

7.4.2.1 Channel Access Success Rate

Figure 7.10 represents the rate of successful AP channel access under a low traffic load ($P_{\text{tr}} = 0.1$) for the three scenarios considered in Section 7.4.2. It can be observed that the probability of channel access depends on the physical AP location, a direct consequence of the number of interferers that can be perceived. By comparing scenarios A and B, i.e. single-antenna Wi-Fi and mMIMO Wi-Fi, it can be observed that the probability of successful channel access for the central AP does not change. This is because the same LBT procedure is adopted in both scenarios. Figure 7.10 also shows the increased channel access opportunities attained by all APs in scenario C, when the central AP implements mMIMO-U.

Figure 7.11 shows the rate of successful AP channel access in a scenario with a heavy traffic load, i.e. $P_{\text{tr}} = 1$. When compared with Figure 7.10, all success rates are degraded, which is a direct consequence of having to contend with more active nodes for channel access. Interestingly, the outermost APs slightly increase their access rate in scenario B, due to the mandatory power reduction at the central AP as per Eq. (7.20). It is also in this setup where the improvements offered by mMIMO-U are particularly noticeable. Indeed, the central AP increases its channel access success rate by 71% (from 35% to 60%). This is due to the radiation nulls placed during the eLBT phase described in Section 7.3.2. Moreover, it can also be observed that, since the same radiation nulls are used during transmission, mMIMO-U further enhances the channel access success rate of the leftmost and rightmost APs.

7.4.2.2 Downlink User SINR

Figure 7.12 plots the CDF of the SINR per user, as per Eq. (7.3), for the three scenarios considered in Section 7.4.2. The results in Figure 7.12 show that higher SINRs can be achieved in the light traffic scenario due to the smaller number of devices simultaneously transmitting. Scenario A achieves overall larger SINR values than scenarios B and C under heavy traffic load, again because of a reduced number of interfering STAs and

Table 7.2 Detailed system parameters for the indoor deployment. Reprinted with permission from [16].

Parameter	Description
RF	
AP/STA maximum TX power	$P^{\max} = 24/18$ dBm
AP power allocation	Uniform STA power allocation
AP and STA antenna elements	Omnidirectional with 0 dBi gain
Preamble detection	$\gamma_{\text{preamble}} = -82$ dBm with -0.8 dB of minimum SINR
Channel model	
Pathloss and probability of LOS	InH for all links
Shadowing	Lognormal with $\sigma = 3/4$ dB (LOS/NLOS)
Fast fading	Ricean with lognormal K-factor and Rayleigh multi-path
Deployment	
Floor size	120 m $\times$ 50 m
AP positions	Ceiling mounted, equally spaced in central corridor as in Figure 7.9
AP and STA heights	3 and 1.5 meters
STA distribution	30 uniformly deployed STAs
STA association criterion	Strongest received signal
Downlink/uplink traffic fraction	0.8/0.2
Scenario A: single-antenna Wi-Fi	
Number of antennas per AP	1, 1, 1
Maximum number scheduled STAs per AP	1, 1, 1
STA scheduling	Round robin
Scenario B: mMIMO Wi-Fi	
Number of antennas per AP	1, 36 (6 $\times$ 6), 1
Maximum number scheduled STAs per AP	1, 4, 1
Precoder	Zero-forcing
STA scheduling	Round robin
Scenario C: mMIMO-U Wi-Fi	
Number of antennas per AP	1, 36 (6 $\times$ 6), 1
Maximum number scheduled STAs per AP	1, 4, 1
Precoder	Zero-forcing with interference suppression, 24 d.o.f. for nulls
STA scheduling	Round robin with LBT/eLBT user-based selection

APs. This is consistent with Figure 7.11, which shows a lower AP channel access success rate for scenario A. The lowest SINR values in Figure 7.12 correspond to STAs that receive a strong uplink-to-downlink interference from a neighboring active STA. For scenario C, the 5%-worst SINR is bounded thanks to the considered LBT/eLBT joint channel access selection with user scheduling, and amounts to 4 dB. Combined with the

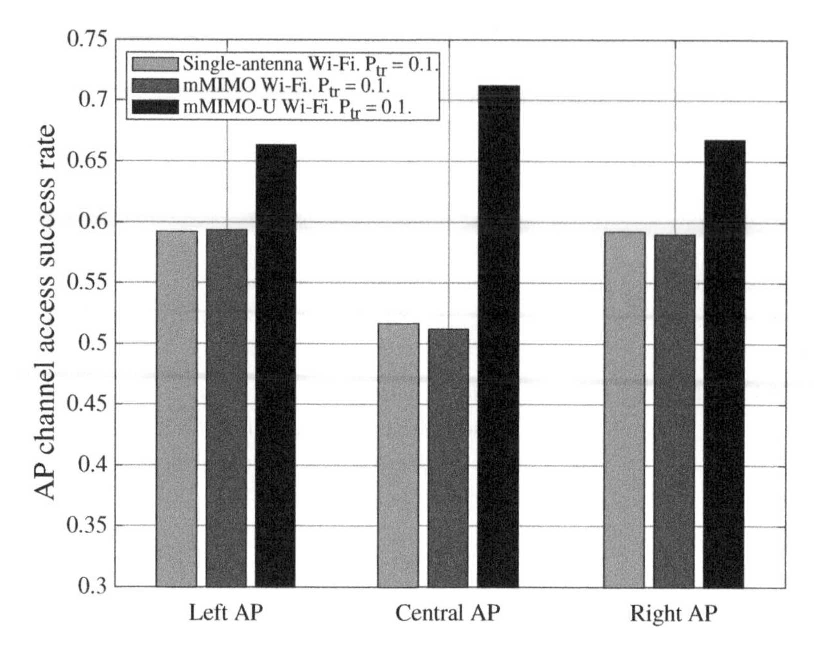

Figure 7.10 Channel access success rate for scenarios A, B, and C with low traffic (P_{tr} = 0.1).

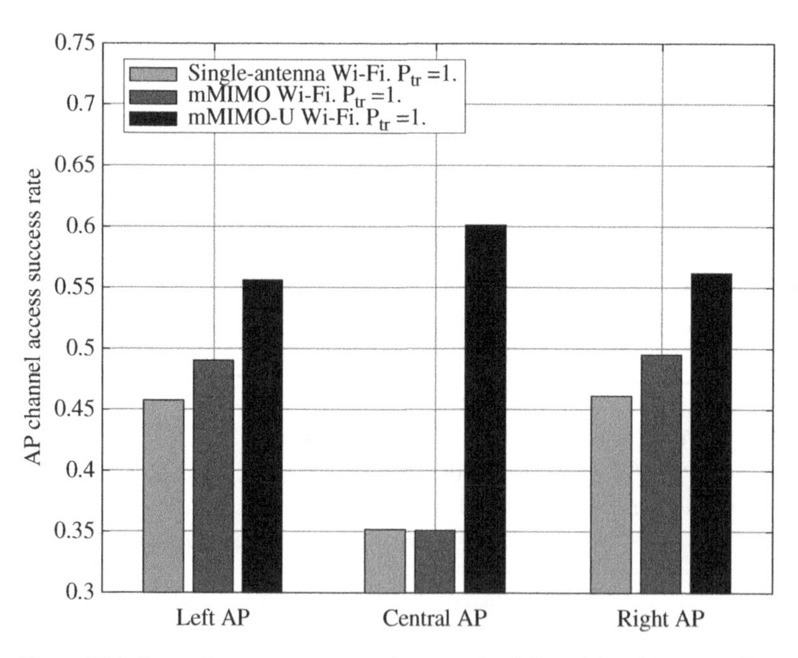

Figure 7.11 Channel access success rate for scenarios A, B, and C with heavy traffic (P_{tr} = 1).

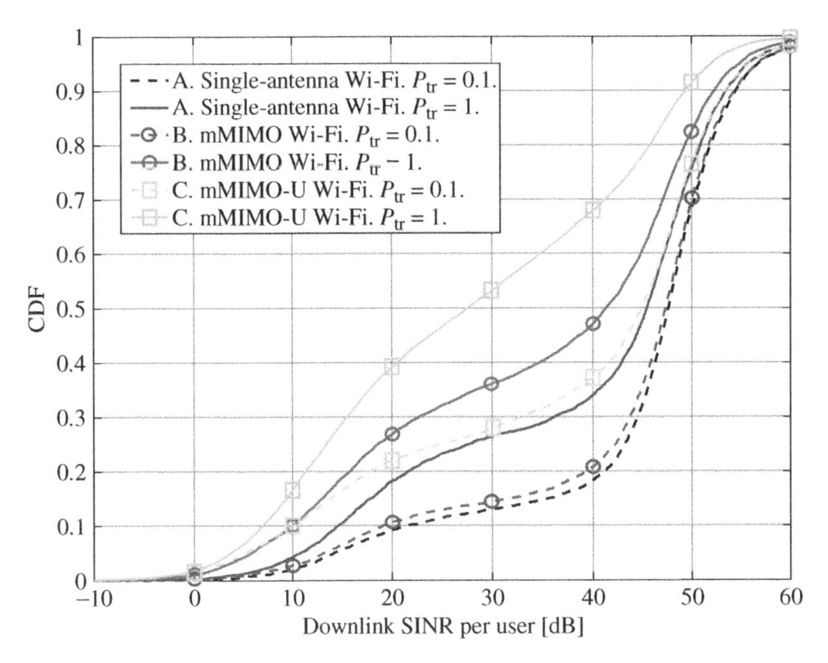

Figure 7.12 Downlink SINR per user for scenarios A, B, and C with $P_{tr} = \{0.1, 1\}$. Reprinted with permission from [16].

increased channel access opportunities provided by mMIMO-U, the SINRs achieved in scenario C yield higher user throughputs, as demonstrated in the following.

7.4.2.3 Downlink Sum Throughput

Figure 7.13 shows the CDF of the downlink sum STA throughput, which is calculated by averaging the SINRs per subcarrier computed as per Eq. (7.3), and mapping the resultant SINRs to a given modulation and coding scheme assuming perfect link adaptation. The Wi-Fi modulation and coding schemes used for communication follow those specified by the 802.11ac standard in [31]. The results shown in this figure account for the service time per STA and for the fraction of time performing uplink/downlink, but do not consider the channel estimation overhead. The former is impacted by the channel access success rate and by the number of STAs sharing the available time resources. As a direct consequence, it can be observed that STAs achieve significantly higher throughputs under light traffic, i.e. for $P_{tr} = 0.1$.

In coherence with the results shown in Figure 7.11, the benefits of mMIMO-U are remarkable under heavy traffic, when a more frequent channel access facilitates scheduling STAs more often. Moreover, the considered LBT/eLBT scheduling mechanism ensures both (i) a guaranteed minimum throughput for STAs located near interfering nodes and (ii) a very large sum throughput for STAs that are more immune to interference. Note that while the former are served after LBT and contribute to the lower part of the CDF, the latter are scheduled after eLBT and are captured in the upper part of the CDF. Overall, mMIMO-U (scenario C) demonstrates 4× gains in the median throughput when compared to the mMIMO Wi-Fi scenario (scenario B), and up to 7× gains with respect to the single-antenna Wi-Fi setup (scenario A).

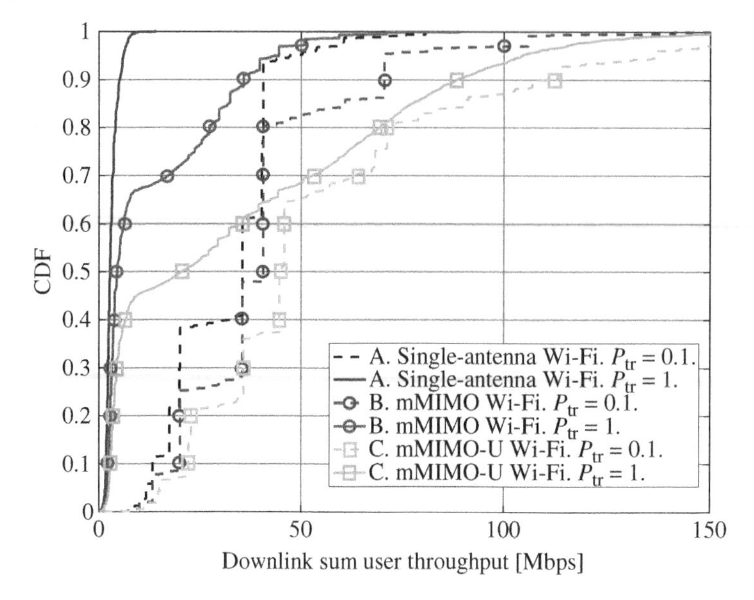

Figure 7.13 Downlink sum STA throughput for scenarios A, B, and C with $P_{tr} = \{0.1, 1\}$. Reprinted with permission from [16].

7.5 Challenges

To realize mMIMO-U and its potential in enhancing spectrum reuse, several challenges must be overcome. This section is devoted to dissecting the main ones.

7.5.1 Wi-Fi Channel Subspace Estimation

The spatial awareness of mMIMO-U relies on performing an accurate estimation of the channel subspace occupied by neighboring Wi-Fi devices, in order to correctly place radiation nulls and achieve additional spectrum reuse in ultra-dense unlicensed networks [13, 14]. In practice, a channel covariance estimate can be obtained during a silent phase, as described in Section 7.3.1. It is clear that silent phases incur an overhead, and an inherent tradeoff exists between improving the quality of the covariance estimate and limiting such effective loss of data transmission time. To reduce the overhead, signal samples acquired during the mandatory eLBT phase can be stored and reused within a validity time. An AP can then undergo additional silent phases only when the number of available samples is deemed insufficient, as illustrated in Figure 7.14. Repeated failures of the eLBT phase may indicate that (a) the required number of samples has been underestimated or (b) that the network density is too high and the mMIMO-U AP does not have enough spatial d.o.f. to suppress interference towards all the neighboring Wi-Fi devices.

7.5.2 Uplink Transmission

While eLBT and radiation nulls can be used to fully exploit the potential of mMIMO-U in the downlink, appropriate procedures should be defined for the uplink, where

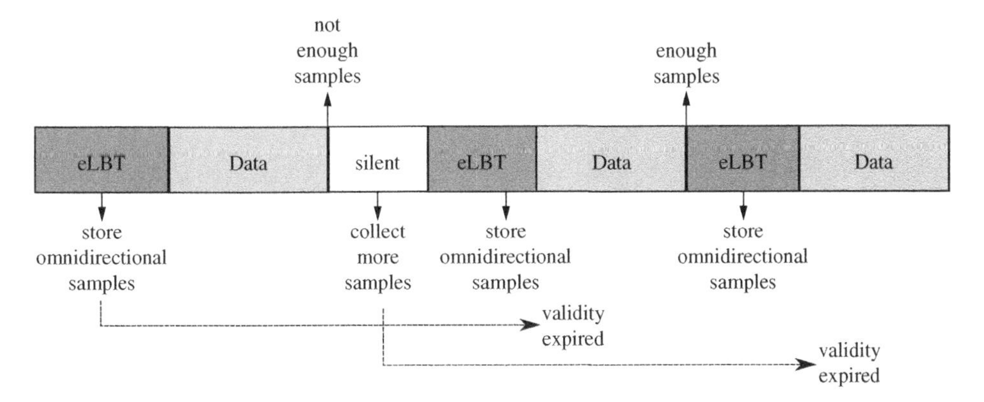

Figure 7.14 Adaptive allocation of silent phases for optimizing the channel covariance matrix acquisition.

collisions between STA-generated pilot or data signals and concurrent Wi-Fi transmissions must be prevented.

As in conventional mMIMO operations, STA uplink pilots are required at every AP-STA channel coherence interval in order to perform spatial multiplexing. Similar to the CSI acquisition procedure described in Section 7.3.4, uplink data must be transmitted in a synchronized fashion for the AP to perform spatial de-multiplexing. This might require APs to gain access to the medium via an LBT contention phase with no nulls in place and to reserve it through omnidirectional transmissions. The channel reservation guarantees that the scheduled STAs can transmit their uplink pilots and data in a conventional multi-user MIMO fashion.

7.5.3 Hidden Terminals

Consider an AP-to-STA downlink-only transmission, as shown in Figure 7.15. The lack of traffic from the STA's side might impede the mMIMO-U AP to estimate its channel covariance matrix, causing a radiation null to be placed only toward/from the AP. As a result, the mMIMO-U AP might access the channel during the Wi-Fi downlink transmission, thus disrupting it. This circumstance is more likely to occur in ultra-dense networks due to the large number of neighboring Wi-Fi devices.

The above hidden terminal problem highlights the need to perform channel covariance estimation sufficiently often, capturing the MAC ACKs or even TCP ACKs, if available, sent by the Wi-Fi STA potentially affected. A complementary approach to alleviate this issue consists in periodically reverting to conventional LBT operations with discontinuous transmission. A more effective solution may be attained by implementing a network listening mode capability at the mMIMO-U. The network listening mode allows decoding of Wi-Fi packet headers and a per-device channel covariance estimation to be performed, maintaining a list of tracked devices. Detecting a downlink-only Wi-Fi transmission whose recipient is not on the list informs the mMIMO-U AP that one or more Wi-Fi devices are hidden and that the spectrum should only be accessed through conventional LBT.

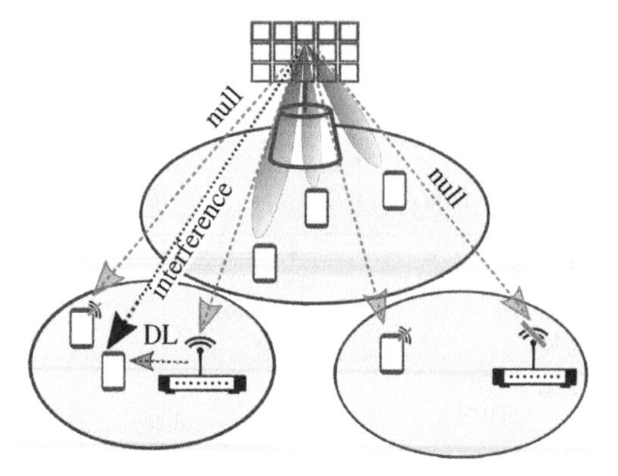

Figure 7.15 Illustration of the hidden terminal problem in mMIMO-U.

7.6 Conclusion

This chapter has described the fundamentals behind the mMIMO-U technology, which targets solving the spectrum crunch of ultra-dense networks operating in the unlicensed spectrum. This objective is achieved by leveraging the spatial interference suppression capabilities of multi-antenna systems. The results included in this chapter demonstrate that mMIMO-U is capable of significantly boosting the performance of unlicensed systems deployed both outdoors and indoors. Moreover, this chapter has identified the main use cases and challenges that this novel technology faces. Overall, the substantial enhancements attainable over existing alternatives make mMIMO-U a key solution for future unlicensed ultra-dense network deployments.

Bibliography

1 3GPP RP-131723, "Discussion paper on unlicensed spectrum integration to IMT systems." Dec. 2013.

2 Qualcomm News, "Introducing MulteFire: LTE-like performance with Wi-Fi-like simplicity," June 2015.

3 Nokia Executive Summary, "The MulteFire opportunity for unlicensed spectrum," 2015.

4 R. Zhang, M. Wang, L.X. Cai, Z. Zheng, X. Shen, and L.L. Xie, "LTE-unlicensed: the future of spectrum aggregation for cellular networks," *IEEE Wireless Communications*, vol. 22, no. 3, pp. 150–159, June 2015.

5 3GPP TS 36.360, "Evolved universal terrestrial radio access (E-UTRA); LTE-WLAN aggregation adaptation protocol (LWAAP) specification," v.13.0.0. Mar. 2016.

6 3GPP TR 36.889, "Feasibility study on licensed-assisted access to unlicensed spectrum," v.13.0.0, Jan. 2015.

7 MulteFire Alliance, MulteFire Release 1.0 technical paper: "A new way to wireless," Jan. 2017 [Online]. Available at: https://www.multefire.org/specification/release-1-0-technical-paper-download/. Accessed on: 11 Oct. 2017.

8 H. Zhang, X. Chu, W. Guo, and S. Wang, "Coexistence of Wi-Fi and heterogeneous small cell networks sharing unlicensed spectrum," *IEEE Comms. Mag.*, vol. 53, no. 3, pp. 158–164, Mar. 2015.

9 A. Mukherjee, J.F. Cheng, S. Falahati, L. Falconetti, A. Furuskr, B. Godana, D.H. Kang, H. Koorapaty, D. Larsson, and Y. Yang, "System architecture and coexistence evaluation of licensed-assisted access LTE with IEEE 802.11," in *Proc. IEEE Int. Conf. on Comm. Workshop (ICCW)*, pp. 2350–2355, June 2015.
10 3GPP RP 140808, "Review of regulatory requirements for unlicensed spectrum," June 2014.
11 E. Perahia and R. Stacey, *Next Generation Wireless LANs: 802.11n and 802.11ac.* Cambridge University Press, June 2013.
12 F. Rusek, D. Persson, B.K. Lau, E.G. Larsson, T.L. Marzetta, O. Edfors, and F. Tufvesson, "Scaling up MIMO: opportunities and challenges with very large arrays," *IEEE Signal Process. Mag.*, vol. 30, no. 2, pp. 40–60, Oct. 2013.
13 G. Geraci, A. Garcia Rodriguez, D. López-Pérez, A. Bonfante, L. Galati Giordano, and H. Claussen, "Enhancing coexistence in the unlicensed band with massive MIMO," in *Proc. IEEE Int. Conf. on Comm. (ICC)*, May 2017.
14 G. Geraci, A. Garcia-Rodriguez, D. López-Pérez, A. Bonfante, L. Galati Giordano, and H. Claussen, "Operating massive MIMO in unlicensed bands for enhanced coexistence and spatial reuse," *IEEE J. Sel. Areas Commun.*, vol. 35, no. 6, pp. 1282–1293, June 2017.
15 A. Garcia-Rodriguez, G. Geraci, L. Galati Giordano, A. Bonfante, M. Ding, and D. López-Pérez, "Massive MIMO unlicensed: A new approach to dynamic spectrum access," in *IEEE Commun. Mag.*, vol. 56, no. 6, pp. 186–192, Jun. 2018
16 A. Garcia-Rodriguez, G. Geraci, D. Lopez-Perez, L.G. Giordano, M. Ding, and H. Claussen, "Massive MIMO unlicensed for high-performance indoor networks," in *IEEE Globecom Workshops (GC Wkshps)*, pp. 1–6, Dec. 2017.
17 M. Medard and D.N.C. Tse, "Spreading in block-fading channels," in *Proc. Asilomar Conf. on Signals, Systems, and Computers*, vol. 2, pp. 1598–1602, Oct. 2000.
18 J. Jose, A. Ashikhmin, T.L. Marzetta, and S. Vishwanath, "Pilot contamination and precoding in multi-cell TDD systems," *IEEE Trans. Wireless Commun.*, vol. 10, no. 8, pp. 2640–2651, Aug. 2011.
19 J. Hoydis, K. Hosseini, S. Ten Brink, and M. Debbah, "Making smart use of excess antennas: massive MIMO, small cells, and TDD," *Bell Labs Tech. J.*, vol. 18, no. 2, pp. 5–21, Sept. 2013.
20 3GPP TS 36.201, "LTE; evolved universal terrestrial radio access (E-UTRA); LTE physical layer (Release 10)," June 2011.
21 3GPP TS 36.213, "LTE; evolved universal terrestrial radio access (E-UTRA); physical layer procedures (Release 10)," June 2011.
22 3GPP TR 36.300, "Evolved universal terrestrial radio access (E-UTRA) and evolved universal terrestrial radio access network (E-UTRAN); overall description," v.14.0.0, Sept. 2016.
23 H.Q. Ngo, E.G. Larsson, and T.L. Marzetta, "The multicell multiuser MIMO uplink with very large antenna arrays and a finite-dimensional channel," *IEEE Trans. Commun.*, vol. 61, no. 6, pp. 2350–2361, June 2013.
24 FCC 662911, "Emissions testing of transmitters with multiple outputs in the same band," Oct. 2013.
25 FCC 14-30, "Revision of part 15 of the commission's rules to permit unlicensed national information infrastructure (U-NII) devices in the 5 GHz band," Apr. 2014.

26 3GPP TR 36.814, "Further advancements for E-UTRA physical layer aspects," v.9.0.0, Mar. 2013.

27 3GPP TR 36.873, "Study for 3D channel model for LTE (Release 12)," June 2015.

28 C. Ubeda Castellanos, D.L. Villa, C. Rosa, K.I. Pedersen, F.D. Calabrese, P.H. Michaelsen, and J. Michel, "Performance of uplink fractional power control in UTRAN LTE," in *Proc. IEEE Veh. Tech. Conference (VTC)*, pp. 2517–2521, May 2008.

29 3GPP TR 36.843, "Study on LTE device to device proximity services; radio aspects," v.12.0.0, Mar. 2014.

30 Matthew Gast, "*802.11 Wireless Networks: The Definitive Guide*," O'Reilly Media, Inc., 2005.

31 IEEE. "Standard for Information Technology — Telecommunications and information exchange between systems local and metropolitan area networks. Part 11: Wireless LAN medium access control (MAC) and physical layer (PHY) specifications," 2016.

8

Energy Efficiency Optimization for Dense Networks

Quang-Doanh Vu[1], *Markku Juntti*[1], *Een-Kee Hong*[2] *and Le-Nam Tran*[3]

[1] *Centre for Wireless Communications, University of Oulu, Finland*
[2] *College of Electronics and Information, Kyung Hee University, South Korea*
[3] *School of Electrical and Electronic Engineering, University College Dublin, Ireland*

8.1 Introduction

The two major challenges for future wireless networks are the ever-growing data traffic and the increasing energy consumption [1]. Both are the natural result of the popularity of wireless communications. In particular, energy consumption in wireless networks needs to be satisfactorily dealt with for sustainable economic growth. A dense network paradigm is introduced in order to solve the problem of exponential growth of data demand in the communications industry [2]. Moreover, network densification is also expected to be the key architecture for future wireless networks where wireless connectivity is pervasive. By bringing base stations near to the end users, dense small cell deployment has the potential of using energy in an efficient manner, since the required transmit power for combating the path loss is reduced. However, dense small cell deployment has its own cost. Using more base stations (BSs) means that the large number of hardware elements may lead to a sharp increase in the circuit power consumption. Moreover, with full frequency reuse and the small cells being close together, the interference becomes complicated [3]. If not properly managed, this will become a performance bottleneck in dense networks.

In general, efficiency is defined as the magnitude to which the least amount of resources is used to achieve a certain target. Thus, efficiency is often expressed as the ratio of the obtained output to the resource consumption. For data communications, the outcome is usually the amount of received (or transmitted) information and the resource is the required power. Consequently, the definition of energy efficiency (EE) in data communications is given by [4]

$$\mathcal{E}_{\text{eff}} = \frac{\int i(t)dt}{\int p(t)dt} \tag{8.1}$$

where $i(t)$ and $p(t)$ are the reliably decoded information and consumed power at a given time t, respectively. Thus, energy efficiency is measured in *bits (or nats) per Joule.*

Ultra-dense Networks for 5G and Beyond: Modelling, Analysis, and Applications,
First Edition. Edited by Trung Q. Duong, Xiaoli Chu and Himal A. Suraweera.

Commonly, the channels are supposed to be flat and quasi-static; then (8.1) simplifies to [5]

$$\mathcal{E}_{\text{eff}}^{\text{static}} = \frac{R}{P} \tag{8.2}$$

where R is the total data rate and P is the total power consumed in the corresponding signaling interval. The EE definition in (8.2) is mainly used in this chapter.

By the definition, the EE optimization involves fractional programs. In some special cases, e.g. quasi-convex problems, a fractional program can be solved efficiently using, for example, Dinkelbach's algorithms or Charnes Cooper's transformation [6–8]. For others, the fractional program of interest is *truly non-convex* in the sense that there is no equivalent convex reformulation and globally optimal solutions cannot be found in polynomial time. In such situations, to find a locally optimal solution with low complexity is more practically useful.

In dense networks, there are possibly hundreds of nodes in a small geographic area. Thus, it is practically impossible and inefficient to coordinate the operation of all nodes with some kind of centralized mechanisms since the networks need to be reorganized frequently. The signaling for gathering relevant information may be overwhelming. Consequently, it is certain that self-organization including self-optimization is necessary [3]. Thus, it is important to design algorithms that can be implemented in a decentralized manner.

This chapter is organized as follows. In the first part, we discuss the fractional programs and the methods that can be used to solve these, including Dinkelbach's methods, parameter-free equivalent transformation, and successive convex approximation. The alternating direction method of multipliers (ADMM) is also introduced as a powerful approach for parallel optimization algorithms. In the second part, we illustrate the applications of these tools in two specific scenarios of dense networks.

8.2 Energy Efficiency Optimization Tools

8.2.1 Fractional Programming

An energy efficiency maximization problem can be generally formulated as a fractional program

$$\underset{\mathbf{x}\in\mathcal{K}}{\text{maximize}}\, f(\mathbf{x}) \tag{8.3}$$

where $\mathcal{K} \in \mathbb{R}^n$ represents the design constraints and $f : \mathbb{R}^n \to \mathbb{R}$ can take one of the following forms:

$$f(\mathbf{x}) \triangleq \frac{\sum_{i=1}^{p} g_i(\mathbf{x})}{\sum_{i=1}^{q} h_i(\mathbf{x})}, \tag{8.4}$$

$$f(\mathbf{x}) \triangleq \sum_{i=1}^{q} \frac{g_i(\mathbf{x})}{h_i(\mathbf{x})}, \tag{8.5}$$

or

$$f(\mathbf{x}) \triangleq \min_{1\le i\le q} \left\{ \frac{g_i(\mathbf{x})}{h_i(\mathbf{x})} \right\} \tag{8.6}$$

with $g_i : \mathbb{R}^n \to \mathbb{R}_+$, $h_i : \mathbb{R}^n \to \mathbb{R}_{++}$ for all $i, p \ge 1$, and $q \ge 1$. Specifically, $g_i(\cdot)$ and $h_i(\cdot)$ stand for the expression of the data rate and power consumption. As we shall see, the assumptions $g_i(\cdot) \ge 0$ and $h_i(\cdot) > 0$ for all i are automatically satisfied in the EE problems to be considered in this chapter since the consumed power is always positive and the achievable rate is always non-negative. The cost function in (8.4) involves a single ratio, which arises from the problem of maximizing EE of the overall network [9]. The cost function in (8.5) is the sum of multiple ratios, which appears in the problem of weighted sum energy efficiency [10]. The objective in (8.6) results in a max–min fractional program, which occurs when the EE fairness among the nodes is concerned [11]. Generally, the EE maximization problems are not convex even if $\mathcal{K}$ is convex. The difficulty in dealing with different classes of fractional programs is discussed below.

- For the objective in (8.4) and $\mathcal{K}$ is convex, g_i is concave, and $h_i(\mathbf{x})$ is convex on $\mathcal{K}$, the resulting problem is called the concave fractional program (CFP), which is a class of quasi-concave programs. Dinkelbach's method or Charnes Cooper's transformation can be used to solve this type of problem efficiently.
- For the objective in (8.6) and $\mathcal{K}$ is convex, g_i is concave, and $h_i(\mathbf{x})$ is convex on $\mathcal{K}$, the obtained program is called the max–min fractional program (MMFP). This problem is considered as a *generalized* convex program for which an efficient solution is also possible [6].
- For the remaining cases, efficient optimal solutions for (8.3) remain open. Alternatively, one is more interested in finding the local solution for these programs.

8.2.2 Concave Fractional Programs

We provide two well-known approaches to solving a CFP. Let $g(\mathbf{x}) = \sum_{i=1}^{p} g_i(\mathbf{x})$ and $h(\mathbf{x}) = \sum_{i=1}^{q} h_i(\mathbf{x})$. Then the CFP is rewritten as

$$\max_{\mathbf{x}\in\mathcal{K}} \quad f(\mathbf{x}) = \frac{g(\mathbf{x})}{h(\mathbf{x})}. \tag{8.7}$$

Note that the concavity of $g(\mathbf{x})$ and convexity of $h(\mathbf{x})$ are easily justified from that of the individual functions.

8.2.2.1 Parameterized Approach

The parameterized approach for solving problem (8.7) is to solve a series of concave programs. In particular, let us consider the following parametric problem:

$$\mathcal{P}(\alpha) = \max_{\mathbf{x}\in\mathcal{K}} \{g(\mathbf{x}) - \alpha h(\mathbf{x})\} \tag{8.8}$$

where $\alpha > 0$. The relationship between CFP (8.7) and (8.8) is as follows [7].

Lemma 8.1 A point $\mathbf{x}^*$ is an optimal solution of (8.7) if and only if $\mathcal{P}(g(\mathbf{x}^*)/h(\mathbf{x}^*)) = 0$. In addition, there is unique α^* such that $\mathcal{P}(\alpha^*) = 0$.

Thus one can easily obtain $\mathbf{x}^*$ when α^* is given. Based on the relationship and the fact that $\mathcal{P}(\alpha)$ is strictly monotonic decreasing, Dinkelbach developed an iterative procedure whose main steps are outlined in Algorithm 8.1 for solving the CFP [7]. In each iteration, a concave problem is solved and the value of α is updated. It is worth mentioning that the properties of sequence $\{\alpha^{(l)}\}_l$ obtained by Dinkelback's algorithm are:

- $\{\alpha^{(l)}\}_l$ is an increasing sequence.
- $\lim_{l\to\infty}\alpha^{(l)} = \alpha^*$.

Algorithm 8.1 Dinkelbach's procedure for solving CFP

1. **Initialization:** $\mathbf{x}^{(0)} \in \mathcal{K}$ (or set $\alpha^{(0)} = 0$); $l = 0$; error tolerant $\epsilon > 0$
2. **repeat**
3. $\quad \alpha^{(l)} := \dfrac{g(\mathbf{x}^{(l)})}{h(\mathbf{x}^{(l)})}$
4. $\quad \mathbf{x}^{(l+1)} := \arg\max_{\mathbf{x}\in\mathcal{K}} g(\mathbf{x}) - \alpha^{(l)}h(\mathbf{x})$
5. $\quad l := l + 1$
6. **until** $\left(g(\mathbf{x}^{(l)}) - \alpha^{(l-1)}h(\mathbf{x}^{(l)})\right) < \epsilon$
7. **Output:** $\mathbf{x}^{(l)}$

8.2.2.2 Parameter-free Approach

Another well-known approach solving the CFP is to transform it into an equivalent convex problem. Here the equivalence means that the optimal solution of the former can be obtained from that of the latter. Such a transformation was first introduced in [12] for linear CFP. The technique was then extended to nonlinear CFP in [8]. Particularly, CFP (8.7) is equivalent to the following problem:

$$\underset{(\mathbf{y},\mu)\in\tilde{\mathcal{K}}}{\text{maximize}}\ \mu g\left(\frac{\mathbf{y}}{\mu}\right) \tag{8.9}$$

where

$$\tilde{\mathcal{K}} = \left\{\mathbf{y} \in \mathbb{R}^n, \mu | \mu > 0, \frac{\mathbf{y}}{\mu} \in \mathcal{K}, \mu h\left(\frac{\mathbf{y}}{\mu}\right) \leq 1\right\}. \tag{8.10}$$

Clearly, (8.9) is a convex program (concave maximization to be precise) since $\mu g(\mathbf{y}/\mu)$ and $\mu h(\mathbf{y}/\mu)$ are concave and convex, respectively. Note that the perspective functions preserve the convexity of the original ones. The relationship between CFP (8.7) and (8.9) is stated in the following lemma [8].

Lemma 8.2 Suppose $(\mathbf{y}^*, \mu^*)$ to be an optimal solution of (8.9); then $(\mathbf{y}^*/\mu^*)$ is the optimal of CFP (8.7). Conversely, suppose $\mathbf{x}^*$ to be an optimal solution of CFP (8.7); then $(\mathbf{x}^*/h(\mathbf{x}^*), 1/h(\mathbf{x}^*))$ is the optimal of (8.9).

The above lemma means that once an optimal of one of the two problems is known, we can easily obtain that of the other. Notice that when h is linear, we can write $\mu h(\mathbf{y}/\mu) = 1$.

There exist other approaches for dealing with the CFP. The interested reader is referred to [6] for further details.

8.2.3 Max–Min Fractional Programs

We now discuss details on the MMFP. In fact, this class of problem can also be solved via a parameterized approach, where the iterative procedure is an extension of Algorithm 8.1 [13]. The main steps are outlined in Algorithm 8.2. Similar to Algorithm 8.1, the sequence $\{\alpha^{(l)}\}_l$ returned by Algorithm 8.2 is also increasing and converges to an optimal solution of the MMFP. However, the convergence rate of Algorithm 8.2 (normally a linear rate) is slower than that of Algorithm 8.1 (superlinear convergence). Some methods for improving the convergence rate for the MMFP can be found in [14] and the references therein.

The subproblem at Step 4 in Algorithm 8.2 is a convex program since the pointwise minimum of concave functions is concave. Currently, some solvers and parsers do not accept the pointwise operations. This practical issue can be simply overcome via the epigraph of the problem, i.e. by introducing a new variable β we can rewirte the problem as $\underset{(\mathbf{x},\beta)\in\hat{\mathcal{K}}^{(l)}}{\text{maximize}}\ \beta$ where $\hat{\mathcal{K}}^{(l)} = \{\mathbf{x} \in \mathcal{K}, \beta | g_i(\mathbf{x}) - \alpha^{(l)} h_i(\mathbf{x}) \geq \beta, i = 1, ..., q\}$.

Algorithm 8.2 Generalized Dinkelbach's procedure for solving MMFP

1. **Initialization:** $\mathbf{x}^{(0)} \in \mathcal{K},\ l = 0$, error tolerant $\epsilon > 0$
2. **repeat**
3. $\quad \alpha^{(l)} := \min_{1\leq i\leq q} \left\{ \dfrac{g_i(\mathbf{x}^{(l)})}{h_i(\mathbf{x}^{(l)})} \right\}$
4. $\quad \mathbf{x}^{(l+1)} := \arg\max_{\mathbf{x}\in\mathcal{K}} \min_{1\leq i\leq q} \{g_i(\mathbf{x}) - \alpha^{(l)} h_i(\mathbf{x})\}$
5. $\quad l := l + 1$
6. **until** $\min_{1\leq i\leq q} \left\{g_i(\mathbf{x}^{(l)}) - \alpha^{(l-1)} h_i(\mathbf{x}^{(l)})\right\} < \epsilon$
7. **Output:** $\mathbf{x}^{(l)}$

8.2.4 Generalized Non-convex Fractional Programs

For the remaining cases, (8.3) has no hidden convexity so it is generally difficult to find an optimal solution. Here we present an efficient local optimization framework called successive convex approximation (SCA), which is widely applied to solve non-convex problem in wireless communications.

By a slight abuse of notation, let us consider the general optimization problem

$$\begin{aligned} &\underset{\mathbf{x}\in\mathbb{R}^n}{\text{maximize}}\ f(\mathbf{x}) \\ &\text{subject to}\ g_i(\mathbf{x}) \leq 0,\ i = 1, \dots, p, \\ &\qquad\qquad h_k(\mathbf{x}) \leq 0, k = 1, \dots, q, \end{aligned} \tag{8.11}$$

where $f(\mathbf{x}) : \mathbb{R}^n \to \mathbb{R}$ and $g_i(\mathbf{x}) : \mathbb{R}^n \to \mathbb{R}$, $i = 1, \dots, p$, are differentiable convex functions, $h_k(\mathbf{x}) : \mathbb{R}^n \to \mathbb{R}$, $k = 1, \dots, q$ are differentiable non-convex functions, and $\mathcal{K} \triangleq \{\mathbf{x} \in \mathbb{R}^n | g_i(\mathbf{x}) \leq 0,\ i = 1, \dots, p,\ h_k(\mathbf{x}) \leq 0,\ k = 1, \dots, q\}$ is a compact set. Clearly, the *non-convex part* of the problem is due to the last q constraints. To locally solve problem (8.11), an iterative procedure outlined in Algorithm 8.3 was first introduced in [15], and thoroughly studied later in [16]. The idea is to successively approximate the non-convex set by its inner one via replacing $\{h_k(\mathbf{x})\}_k$ by their convex upper bounds. Particularly, let us denote by $\overline{\mathbf{x}}$ a feasible point of (8.11). Let $\overline{\mathbf{y}}_k \triangleq \overline{h}_k(\overline{\mathbf{x}}) : \mathbb{R}^n \to \mathbb{R}^{m_k}$, $m_k \geq 1$ and $\{\tilde{h}_k(\mathbf{x};\overline{\mathbf{y}}_k) : \mathbb{R}^n \to \mathbb{R}\}$ be the convex upper bounds functions having the following properties:

$$\begin{aligned} h_k(\mathbf{x}) &\leq \tilde{h}_k(\mathbf{x};\overline{\mathbf{y}}), \ \forall \mathbf{x} \in \mathcal{K}, \\ h_k(\overline{\mathbf{x}}) &= \tilde{h}_k(\overline{\mathbf{x}};\overline{\mathbf{y}}), \\ \nabla_{\mathbf{x}} h_k(\mathbf{x}) &= \nabla_{\mathbf{x}} \tilde{h}_k(\mathbf{x};\overline{\mathbf{y}}). \end{aligned} \tag{8.12}$$

The first and the second properties guarantee that the sequence of objective values obtained by the algorithm is non-decreasing while the second and the third properties ensure that the convergent points satisfy the necessary optimality (Karush-Kuhn-Tucker) conditions of (8.11).

Algorithm 8.3 General procedure of successive convex approximation method

1. **Initialization**: generate random point $\mathbf{x}^{(0)} \in \mathcal{K}$. Set $l = 0$
2. **repeat**
3. Determine $\overline{\mathbf{y}}_k^{(l)} \triangleq \overline{h}_k(\mathbf{x}^{(l)})$ for all k
4. Solve
$$\begin{aligned} \underset{\mathbf{x}\in\mathbb{R}^n}{\text{maximize}} \quad & f(\mathbf{x}) \\ \text{subject to} \quad & g_i(\mathbf{x}) \leq 0, \ i = 1, \dots, p, \\ & \tilde{h}_k(\mathbf{x};\overline{\mathbf{y}}_k^{(l)}) \leq 0, \ k = 1, \dots, q \end{aligned}$$
and denote the obtained solution by $\mathbf{x}^{(l+1)}$
5. Update $l := l + 1$
6. **until** Convergence
7. **Output**: $\mathbf{x}^{(l)}$

There are several convergence results of the SCA method [15–17]. Herein we briefly summarize these results for the sake of completeness. First, if $f(\mathbf{x})$ is coercive on the feasible set or the feasible set is bounded, then $\{f(\mathbf{x}^{(l)})\}_l$ converges to a finite value since the sequence $\{f(\mathbf{x}^{(l)})\}_{l=1}^{\infty}$ is non-increasing. Moreover, if $f(\mathbf{x})$ is strongly convex, the convergence of the iterates $\{\mathbf{x}^{(l)}\}_{l=0}^{\infty}$ is guaranteed [16, Theorem 1]. This means that, when $f(\mathbf{x})$ is not strongly convex, Algorithm 8.3 is not guaranteed to output a Karush-Kuhn-Tucker solution. To overcome this issue, we can add a *proximal term* [17], i.e. the objective function at each iteration is modified into $f(\mathbf{x}) + \lambda||\mathbf{x} - \mathbf{x}^{(l)}||_2^2$, where λ is some positive scalar.

8.2.5 Alternating Direction Method of Multipliers for Distributed Implementation

One of the popular optimization techniques for developing decentralized algorithms in wireless communications design is the dual decomposition method. The ADMM is a combination of the dual descent method and the multiplier method [18], being amenable to distributed implementations especially for large scale systems. A good introduction of ADMM can be found in [18]. To be self-contained we provide a brief review of ADMM that is necessary for the next section. Let us consider the convex problem

$$\begin{aligned} &\underset{\mathbf{x}\in\mathcal{K}_1,\mathbf{y}\in\mathcal{K}_2}{\text{minimize}}\ f(\mathbf{x})+h(\mathbf{y}) \\ &\text{subject to}\ \mathbf{A}\mathbf{x}+\mathbf{B}\mathbf{y}=\mathbf{c}, \end{aligned} \tag{8.13}$$

where $\mathcal{K}_1 \in \mathbb{R}^{n_1}$ and $\mathcal{K}_2 \in \mathbb{R}^{n_2}$ are convex sets, $\mathbf{A} \in \mathbb{R}^{m\times n_1}$, $\mathbf{B} \in \mathbb{R}^{m\times n_2}$, and $\mathbf{c} \in \mathbb{R}^{m}$; $f : \mathbb{R}^{n_1} \rightarrow \mathbb{R}$ and $h : \mathbb{R}^{n_2} \rightarrow \mathbb{R}$ are convex. Different from the dual decomposition method, the ADMM works on the augmented Lagrangian function given by

$$L_{\mathrm{A}}(\mathbf{x},\mathbf{y};\boldsymbol{\mu}) = f(\mathbf{x}) + h(\mathbf{y}) + \boldsymbol{\mu}^{\mathrm{T}}(\mathbf{A}\mathbf{x}+\mathbf{B}\mathbf{y}-\mathbf{c}) + \frac{d}{2}\|\mathbf{A}\mathbf{x}+\mathbf{B}\mathbf{y}-\mathbf{c}\|_2^2 \tag{8.14}$$

where $\boldsymbol{\mu}$ is the Lagrange multiplier (or dual variable) vector and $d > 0$ is the penalty parameter.

The ADMM is an iterative procedure where the variables are successively updated. In particular, the following steps take place at the lth iteration:

$$\begin{aligned} \mathbf{x}^{(l)} &:= \underset{\mathbf{x}\in\mathcal{K}_1}{\text{minimize}}\ L_{\mathrm{A}}(\mathbf{x},\mathbf{y}^{(l-1)},\boldsymbol{\mu}^{(l-1)}), \\ \mathbf{y}^{(l)} &:= \underset{\mathbf{y}\in\mathcal{K}_2}{\text{minimize}}\ L_{\mathrm{A}}(\mathbf{x}^{(l)},\mathbf{y},\boldsymbol{\mu}^{(l-1)}), \\ \boldsymbol{\mu}^{(l)} &:= \boldsymbol{\mu}^{(l-1)} + d(\mathbf{A}\mathbf{x}^{(l)}+\mathbf{B}\mathbf{y}^{(l)}-\mathbf{c}). \end{aligned} \tag{8.15}$$

Although the update in (8.15) is quite similar to that in the dual decomposition method, the ADMM converges with quite mild conditions. In particular, the ADMM does not require $f(\mathbf{x})$ and $h(\mathbf{y})$ to be strictly convex as the strong convexity of the objective is automatically achieved via the penalty term $(d/2)\|\mathbf{A}\mathbf{x}+\mathbf{B}\mathbf{y}-\mathbf{c}\|_2^2$ included in the augmented Lagrangian. We will see that this property is important for developing distribution solutions to EE problems. We refer the interested reader to [18] for technical details about the convergence of the ADMM.

It is worth discussing the complexity of the provided optimization tools. The tools for both tractable and intractable fractional problems only require solving convex programs. The fact is that there exist many efficient *dimension-free* algorithms, e.g. the first-order methods [19]. In addition, the ADMM separates a large problem into smaller subproblems. In some cases, solutions of the subproblems can be expressed by explicit analytical formulas. Therefore, the provided tools are quite efficient for large-scale problems, and thus are well suited to ultra dense networks.

8.3 Energy Efficiency Optimization for Dense Networks: Case Studies[1]

8.3.1 Multiple Radio Access Technologies

In the first case study, we apply the provided optimization tools to optimize the energy efficiency of a dense network where multiple radio access technologies (RATs) coexist. These include wireless local area network (WLAN), wireless metropolitan area network

1 The system models and simulation setups presented here are merely for illustrating the use of the provided optimization tools. In order to evaluate the performances of ultra-dense networks, more sophisticated system models and simulation setups should be considered.

(WMAN), and the emerging long-term evolution (LTE) as examples. Due to the evolution of the smart-phone, there are more and more portable devices equipped with multiple wireless interfaces. In an ultra-dense network, it is reasonable to assume that there are many geographic areas covered by multiple radio access networks (RANs). When subscribers enter an overlapped coverage area, their devices can be configured to receive data from multiple RANs [6, 21].

We investigate resource allocation in a multi-RAT scenario where users can simultaneously receive data from multiple RANs using multiple air interfaces. The target is optimally assigning the bandwidth and power to each user-RAN connection so as to maximize EE of the entire network subject to user-specific quality of service (QoS) requirements as well as the available resource budgets. Our main focus includes: first, the EE maximization (EEmax) problem is formulated, and then an iterative algorithm for solving the problem based on the Dinkelbach method is presented. To facilitate a distributed implementation, we transform the EEmax problem into an equivalent convex program by applying the Charnes-Cooper transformation. Finally, we resort to the ADMM to propose a decentralized algorithm. The contents here are built on [22], where *the uplink* transmission is considered.

8.3.1.1 System Model and Energy Efficiency Maximization Problem

We consider a region covered by a set of B wireless access network base stations (BSs) denoted by $\mathcal{B} = \{1, 2, ..., B\}$. In this region, there exist a set of U multi-homing users, denoted by $\mathcal{U} = \{1, 2, ..., U\}$, which are able to receive data from multiple network BSs simultaneously. An example of the system model is sketched in Figure 8.1. We assume that all BSs and users are equipped with a single antenna. The objective is to allocate the BSs' power and the available bandwidth for each user-BS link to maximize the EE

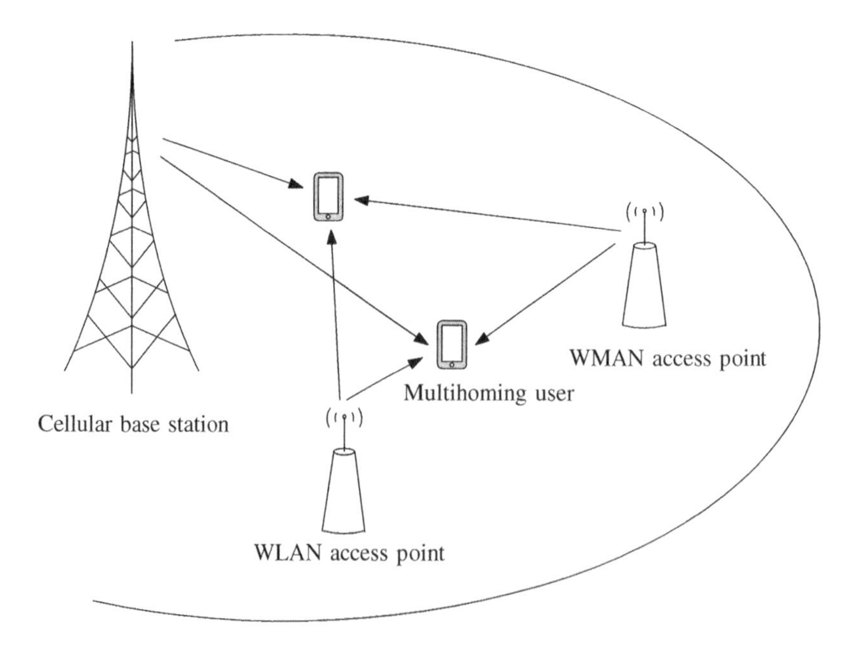

Figure 8.1 An example of multi-RAT systems.

of the entire network. For the connection between user u and BS b, let us denote by β_{ub} and ρ_{ub} the allocated bandwidth and power, respectively. We assume that the channels are flat and the channel coefficients remain constant during a transmission time interval [23, 24].Then the data rate transmitted over the connection is

$$r_{ub} = a_b \beta_{ub} \log\left(1 + \frac{g_{ub}\rho_{ub}}{\beta_{ub}}\right) \tag{8.16}$$

where $a_b \in (0, 1)$ is the network efficiency depending on the network [23], $g_{ub} \triangleq |h_{ub}|^2/N_0$, N_0 is the power spectral density of the Gaussian background noise, and h_{ub} is the channel coefficient. Each user may be assigned by the corresponding network provider a certain degree of priority. Let us denote by m_{ub} the priority of user u in the network of BS b. Then the overall weighted sum rate transmitted of the entire network is

$$R = \sum_{b\in\mathcal{B}} \sum_{u\in\mathcal{U}} m_{ub} r_{ub}. \tag{8.17}$$

Let p_b^{ct} and p_u^{cr} be the circuit power for activating BS b and user u, respectively. In addition, the data transmission between user u and BS b requires an amount of circuit power ρ_{ub}^{ca} for activating the processing blocks (e.g. mixers, filters). Then the total power consumed in the network is

$$P = p^{\mathrm{c}} + \sum_{b\in\mathcal{B}} \frac{1}{\lambda_b}\left(\sum_{u\in\mathcal{U}} \rho_{ub}\right) \tag{8.18}$$

where $p^{\mathrm{c}} = \sum_{b\in\mathcal{B}}\left(p_b^{\mathrm{ct}} + \sum_{u\in\mathcal{U}} p_{ub}^{\mathrm{ca}}\right) + \sum_{u\in\mathcal{U}} p_u^{\mathrm{cr}}$ is constant and $\lambda_b \in (0, 1)$ is the power amplifier's efficiency at BS b.

The design problem is to find optimal values for $\boldsymbol{\rho} \triangleq \{\rho_{ub}\}_{u,b}$ and $\boldsymbol{\beta} \triangleq \{\beta_{ub}\}_{u,b}$ to maximize the EE of the whole network, which is stated as

$$\underset{\boldsymbol{\rho}\geq 0, \boldsymbol{\beta}\geq 0}{\text{maximize}} \ \mathcal{E}(\boldsymbol{\rho}, \boldsymbol{\beta}) \triangleq \frac{R}{P} \tag{8.19a}$$

$$\text{subject to} \ \sum_{u\in\mathcal{U}} \rho_{ub} \leq \overline{P}_b, \forall b \in \mathcal{B}, \tag{8.19b}$$

$$\sum_{u\in\mathcal{U}} \beta_{ub} \leq \overline{W}_b, \forall b \in \mathcal{B}, \tag{8.19c}$$

$$\sum_{b\in\mathcal{B}} r_{ub} \geq \overline{R}_u, \forall u \in \mathcal{U}, \tag{8.19d}$$

where $\overline{P}_b$ and $\overline{W}_b$ are the maximum power budget and bandwidth of BS b, respectively; $\overline{R}_u$ is the predefined data rate threshold for user u.

We assume that there exists a proper scheduler guaranteeing problem (8.19) that is feasible [25–28]. We further assume that the system has enough resources such that the data rate of all users can be larger than thresholds; i.e. there exist $\boldsymbol{\rho}$ and $\boldsymbol{\beta}$ such that the constraints hold with strict inequalities.

The rate function (8.16) is the perspective function of the concave function $a_b \log_2(1 + g_{ub}\rho_{ub})$, and thus it is jointly concave with respect to β_{ub} and ρ_{ub}. In addition, the feasible set of (8.19) is convex. Hence problem (8.19) is a CFP that can be optimally solved by using paramterized or parameterized-free approaches.

8.3.1.2 Solution via Parameterized Approach

We first solve (8.19) using the procedure in Algorithm 8.1. Let us define the function $f_\alpha(\boldsymbol{\rho}, \boldsymbol{\beta}) \triangleq R - \alpha P$ for a fixed $\alpha > 0$ and arrive at the parameterized problem

$$\underset{\boldsymbol{\rho}\geq 0, \boldsymbol{\beta}\geq 0}{\text{maximize}} \quad f_\alpha(\boldsymbol{\rho}, \boldsymbol{\beta}) \tag{8.20a}$$

$$\text{subject to} \quad \sum_{u\in\mathcal{U}} \rho_{ub} \leq \overline{P}_b, \forall b \in \mathcal{B}, \tag{8.20b}$$

$$\sum_{u\in\mathcal{U}} \beta_{ub} \leq \overline{W}_b, \forall b \in \mathcal{B}, \tag{8.20c}$$

$$\sum_{b\in\mathcal{B}} r_{ub} \geq \overline{R}_u, \forall u \in \mathcal{U}. \tag{8.20d}$$

In each iteration, problem (8.20) is solved and parameter α is updated as $\alpha := \mathcal{E}(\boldsymbol{\rho}, \boldsymbol{\beta})$.

There are some interesting insights into the energy-efficient resource allocation inferred from (8.20). Recall that (8.19) is assumed to be strictly feasible; thus strong duality holds for problem (8.19) by Slater's condition and the duality gap is zero for (8.20) [29]. By maneuvering the Lagrangian function of (8.20), we can express the optimal transmit power and allocated bandwidth for a given α as

$$\rho^*_{ub} = \beta^*_{ub}\left[\frac{a_b(m_{ub} + \mu^*_u)}{(\nu^*_b + \alpha)} - \frac{1}{g_{ub}}\right]^+ \tag{8.21}$$

where $[x]^+ \triangleq \max\{0, x\}$, $\{\nu^*_b\}_{b\in\mathcal{B}} \geq 0$, and $\{\mu^*_u\}_{u\in\mathcal{U}} \geq 0$ are the optimal Lagrangian multipliers corresponding to (8.20b) and (8.20d), respectively. We can observe from (8.21) that user u receives data from BS b if the channel gain g_{ub} is good enough. Moreover, power ρ^*_{ub} increases with a_b and m_{ub}, which intuitively means that more power should be allocated to a user having higher network efficiency and/or priority.

8.3.1.3 Solution via Parameter-free Approach

We can use the Charnes-Cooper transformation to translate (8.19) into the equivalent convex problem given as

$$\underset{\tilde{\boldsymbol{\rho}}\geq 0,\tilde{\boldsymbol{\beta}}\geq 0,\mu>0}{\text{maximize}} \quad \tilde{\mathcal{E}}(\tilde{\boldsymbol{\rho}}, \tilde{\boldsymbol{\beta}}) \triangleq \sum_{b\in\mathcal{B}}\sum_{u\in\mathcal{U}} m_{ub} a_b \tilde{\beta}_{ub} \log\left(1 + \frac{g_{ub}\tilde{\rho}_{ub}}{\tilde{\beta}_{ub}}\right) \tag{8.22a}$$

$$\text{subject to} \quad \mu p^{\mathrm{c}} + \sum_{b\in\mathcal{B}} \frac{1}{\lambda_b} \sum_{u\in\mathcal{U}} \tilde{\rho}_{ub} = 1, \tag{8.22b}$$

$$\sum_{u\in\mathcal{U}} \tilde{\rho}_{ub} \leq \mu \overline{P}_b, \forall b \in \mathcal{B}, \tag{8.22c}$$

$$\sum_{u\in\mathcal{U}} \tilde{\beta}_{ub} \leq \mu \overline{W}_b, \forall b \in \mathcal{B}, \tag{8.22d}$$

$$\sum_{b\in\mathcal{B}} a_b \tilde{\beta}_{ub} \log\left(1 + \frac{g_{ub}\tilde{\rho}_{ub}}{\tilde{\beta}_{ub}}\right) \geq \mu \overline{R}_u, \forall u \in \mathcal{U}. \tag{8.22e}$$

After solving (8.22) and obtaining the optimal solution $(\tilde{\rho}^*, \tilde{\beta}^*, \mu^*)$, we follow the result in Lemma 8.2 to determine the optimal of (8.19) as

$$\rho^* = \frac{\tilde{\rho}^*}{\mu^*}, \beta^* = \frac{\tilde{\beta}^*}{\mu^*}. \tag{8.23}$$

8.3.1.4 Distributed Implementation

The parameterized approach is not suitable for decentralized implementation since it requires updating parameter α, which is only done when the global data rate and global consumed power are collected. Thus we focus on the convex equivalent program (8.22). The current form of (8.22) is not amenable to applying the ADMM. Hence, we first translate it into a proper formulation. Let $\tilde{\rho}_b \triangleq \{\tilde{\rho}_{ub}\}_{u\in\mathcal{U}}$ and $\tilde{\beta}_b \triangleq \{\tilde{\beta}_{ub}\}_{u\in\mathcal{U}}$. For the connection between user u and BS b, $\forall u, b$, we introduce two local variables t_{ub} and t'_{ub} and a global variable ν_{ub}. We also introduce local variables $\tilde{\mu}_b$ for all $b \in \mathcal{B}$ and $\tilde{\mu}'_u$ for all $u \in \mathcal{U}$. With these new variables, we equivalently rewrite (8.22) as

$$\underset{\substack{\tilde{\rho},\tilde{\beta},\mu>0 \\ \mathbf{t},\mathbf{t}',\mathbf{v}\geq 0,\tilde{\mu},\tilde{\mu}'}}{\text{minimize}} - \sum_{b\in\mathcal{B}} \mathbf{m}_b^{\mathrm{T}} \mathbf{t}_b \tag{8.24a}$$

$$\text{subject to } \mu p^{\mathrm{c}} + \sum_{b\in\mathcal{B}} \frac{1}{\lambda_b} \mathbf{1}^{\mathrm{T}} \tilde{\rho}_b = 1, \tag{8.24b}$$

$$\mathbf{1}^{\mathrm{T}} \tilde{\rho}_b \leq \tilde{\mu}_b \overline{P}_b, \forall b \in \mathcal{B}, \tag{8.24c}$$

$$\mathbf{1}^{\mathrm{T}} \tilde{\beta}_b \leq \tilde{\mu}_b \overline{W}_b, \forall b \in \mathcal{B}, \tag{8.24d}$$

$$R_{ub}(\tilde{\rho}_b, \tilde{\beta}_b) \geq t_{ub}, \forall b \in \mathcal{B}, u \in \mathcal{U}, \tag{8.24e}$$

$$\mathbf{1}^{\mathrm{T}} \mathbf{t}'_u \geq \tilde{\mu}'_u \overline{R}_u, \forall u \in \mathcal{U}, \tag{8.24f}$$

$$t_{ub} = \nu_{ub}, \forall b \in \mathcal{B}, u \in \mathcal{U}, \tag{8.24g}$$

$$t'_{ub} = \nu_{ub}, \forall b \in \mathcal{B}, u \in \mathcal{U}, \tag{8.24h}$$

$$\tilde{\mu}_b = \mu, \forall b \in \mathcal{B}, \tag{8.24i}$$

$$\tilde{\mu}'_u = \mu, \forall u \in \mathcal{U}, \tag{8.24j}$$

where $\mathbf{m}_b \triangleq \{m_{ub}\}_{u\in\mathcal{U}}$, $\mathbf{t}_b \triangleq \{t_{ub}\}_{u\in\mathcal{U}}$, $\mathbf{t}'_u \triangleq \{t'_{ub}\}_{b\in\mathcal{B}}$, and $R_{ub}(\tilde{\rho}_b, \tilde{\beta}_b) \triangleq a_b \tilde{\beta}_{ub} \log(1 + g_{ub}\tilde{\rho}_{ub}/\tilde{\beta}_{ub})$. The purpose of introducing $\tilde{\boldsymbol{\mu}} \triangleq \{\tilde{\mu}_b\}$ is to make the constraints in (8.24c) and (8.24d) handled *locally* at the BSs. Similarly, introducing $\tilde{\boldsymbol{\mu}}' \triangleq \{\tilde{\mu}'_u\}$ is to make the constraints in (8.24f) handled *locally* at the users. The global variables $\{\nu_{ub}\}$ are introduced to keep the corresponding local versions of data rate at the BSs and the users to be equal.

For the ease of presentation, let $\boldsymbol{\theta}_b \triangleq [\tilde{\mu}_b, \mathbf{t}_b^{\mathrm{T}}]^{\mathrm{T}}$, $\boldsymbol{\theta}'_u \triangleq [\tilde{\mu}'_u, \mathbf{t}_u'^{\mathrm{T}}]^{\mathrm{T}}$, $\mathbf{v}_b \triangleq \{v_{ub}\}_{u\in\mathcal{U}}$, $\mathbf{v}'_u \triangleq \{v_{ub}\}_{b\in\mathcal{B}}$, $\mathbf{z}_b \triangleq [\mu, \mathbf{v}_b^{\mathrm{T}}]^{\mathrm{T}}$, and $\mathbf{z}'_u \triangleq [\mu, \mathbf{v}_u'^{\mathrm{T}}]^{\mathrm{T}}$. We note that $\{\mathbf{z}_b\}$ and $\{\mathbf{z}'_u\}$ are derived from *the same set* of variables $\{v_{ub}\}$ and μ. For each BS b we define a local feasible set $\mathcal{K}_b$ as

$$\mathcal{K}_b = \{(\tilde{\mu}_b, \tilde{\rho}_b, \tilde{\boldsymbol{\beta}}_b, \mathbf{t}_b) | \mathbf{1}^{\mathrm{T}}\tilde{\boldsymbol{\beta}}_b \le \tilde{\mu}_b \overline{W}_b, \mathbf{1}^{\mathrm{T}}\tilde{\rho}_b \le \tilde{\mu}_b \overline{P}_b, R_{ub}(\tilde{\rho}_b, \tilde{\boldsymbol{\beta}}_b) \ge t_{ub}, \forall u \in \mathcal{U}\}. \tag{8.25}$$

Similarly, for each user u we define a local feasible set $\mathcal{K}'_u$ as

$$\mathcal{K}'_u = \{(\tilde{\mu'}_u, \mathbf{t}'_u) | \mathbf{1}^{\mathrm{T}}\mathbf{t}'_u \ge \tilde{\mu}'_u \overline{R}_u, \}. \tag{8.26}$$

Then we can rewrite (8.24) in a more compact form as follows:

$$\underset{\substack{\{\tilde{\rho}_b\},\{\tilde{\boldsymbol{\beta}}_b\},\mu>0\\ \{\mathbf{t}_b\},\{\mathbf{t}'_u\},\mathbf{v}\ge 0,\tilde{\mu}_b,\tilde{\mu}'_u,\mathbf{x}}}{\text{minimize}} \quad -\sum_{b\in\mathcal{B}} \mathbf{m}_b^{\mathrm{T}}\mathbf{t}_b \tag{8.27a}$$

$$\text{subject to} \quad \mathbf{g}^{\mathrm{T}}\boldsymbol{\theta}_b - \frac{1}{B} = x_b, \forall b \in \mathcal{B}, \tag{8.27b}$$

$$\sum_{b\in\mathcal{B}} x_b = 0, \tag{8.27c}$$

$$\boldsymbol{\theta}_b = \mathbf{z}_b, \forall b \in \mathcal{B}, \tag{8.27d}$$

$$\boldsymbol{\theta}'_u = \mathbf{z}'_u, \forall u \in \mathcal{U}, \tag{8.27e}$$

$$(\tilde{\mu}_b, \tilde{\rho}_b, \tilde{\boldsymbol{\beta}}_b, \mathbf{t}_b) \in \mathcal{K}_b, \forall b \in \mathcal{B}, \tag{8.27f}$$

$$(\tilde{\mu'}_u, \mathbf{t}'_u) \in \mathcal{K}'_u, \forall u \in \mathcal{U}, \tag{8.27g}$$

where $\mathbf{g} \triangleq [p^{\mathrm{c}}/B, 1/\lambda_b \mathbf{1}^{\mathrm{T}}, \mathbf{0}^{\mathrm{T}}]^{\mathrm{T}}$. Here (8.24b) is equivalently rewritten into the two constraints (8.27b) and (8.27c) due to the equality in (8.24i).

We now form the partial augmented Lagrangian function of (8.27) given by

$$\begin{aligned} L_{\mathrm{A}}(\boldsymbol{\theta}, \boldsymbol{\theta}', \mathbf{z}, \mathbf{x}; \boldsymbol{\tau}, \boldsymbol{\zeta}, \boldsymbol{\zeta}', \xi) = &-\sum_{b\in\mathcal{B}} \mathbf{m}_b^{\mathrm{T}}\mathbf{t}_b + \sum_{b\in\mathcal{B}} \tau_b \left(\tau_b \left(\mathbf{g}^{\mathrm{T}}\boldsymbol{\theta}_b - x_b - \frac{1}{B}\right)\right. \\ &+ \frac{d}{2}\left(\mathbf{g}^{\mathrm{T}}\boldsymbol{\theta}_b - x_b - \frac{1}{B}\right)^2 + \boldsymbol{\zeta}_b^{\mathrm{T}}(\boldsymbol{\theta}_b - \mathbf{z}_b) \\ &\left. + \frac{d}{2}(\boldsymbol{\theta}_b - \mathbf{z}_b) + \frac{d}{2}\|\boldsymbol{\theta}_b - \mathbf{z}_b\|^2\right) \\ &+ \sum_{u\in\mathcal{U}} \left(\boldsymbol{\zeta}_u'^{\mathrm{T}}(\boldsymbol{\theta}'_u - \mathbf{z}'_u) + \frac{d}{2}\|\boldsymbol{\theta}'_u - \mathbf{z}'_u\|^2\right) \\ &+ \xi \sum_{b\in\mathcal{B}} x_b + \frac{d}{2}\left(\sum_{b\in\mathcal{B}} x_b\right)^2 \end{aligned} \tag{8.28}$$

where $d > 0$ is the penalty parameter and ξ, $\boldsymbol{\tau} \triangleq \{\tau_b\}_{b\in\mathcal{B}}$, $\boldsymbol{\zeta} \triangleq \{\boldsymbol{\zeta}_b\}_{b\in\mathcal{B}}$, $\boldsymbol{\zeta}' \triangleq \{\boldsymbol{\zeta}'_u\}_{u\in\mathcal{U}}$ are the Lagrangian multipliers.

In what follows, we detail the variable update at iteration $(l+1)$ based on the ADMM to solve (8.27). First $\{\boldsymbol{\theta}_b\}_{b\in\mathcal{B}}$ and $\{\boldsymbol{\theta}'_u\}_{u\in\mathcal{U}}$ are updated as

$$(\boldsymbol{\theta}^{(l+1)}, \boldsymbol{\theta}'^{(l+1)}) = \arg\min_{\boldsymbol{\theta},\boldsymbol{\theta}'} L_{\mathrm{A}}(\boldsymbol{\theta}, \boldsymbol{\theta}', \mathbf{z}^{(l)}, \mathbf{x}^{(l)}; \boldsymbol{\tau}^{(l)}, \boldsymbol{\zeta}^{(l)}, \boldsymbol{\zeta}'^{(l)}, \xi^{(l)}). \tag{8.29}$$

The update in Eq. (8.29) can be carried out independently at each BS and user. Particularly, BS b solves the following convex subproblem:

$$\begin{aligned}\underset{(\tilde{\mu}_b,\tilde{\rho}_b,\tilde{\beta}_b,\mathbf{t}_b)\in\mathcal{K}_b}{\text{minimize}} \; & -\mathbf{m}_b^{\mathrm{T}}\mathbf{t}_b + \tau_b^{(l)}\left(\mathbf{g}^{\mathrm{T}}\boldsymbol{\theta}_b - x_b^{(l)} - \frac{1}{B}\right) \\ & + \frac{d}{2}\left(\mathbf{g}^{\mathrm{T}}\boldsymbol{\theta}_b - x_b^{(l)} - \frac{1}{B}\right)^2 + \boldsymbol{\zeta}_b^{(l)\mathrm{T}}(\boldsymbol{\theta}_b - \mathbf{z}_b^{(l)}) + \frac{d}{2}\|\boldsymbol{\theta}_b - \mathbf{z}_b^{(l)}\|^2\end{aligned} \tag{8.30}$$

and user u solves the following quadratic program:

$$\underset{(\tilde{\mu}'_u,\mathbf{t}'_u)\in\mathcal{K}'_u}{\text{minimize}} \; \boldsymbol{\zeta}_u'^{(l)\mathrm{T}}(\boldsymbol{\theta}'_u - \mathbf{z}'_u) + \frac{d}{2}\|\boldsymbol{\theta}'_u - \mathbf{z}'_u\|^2. \tag{8.31}$$

Next, the global variables $\mathbf{x}$ and $\mathbf{z}$ are updated as

$$\begin{aligned}\underset{\mathbf{x}}{\text{minimize}} \; & \xi^{(l)}\sum_{b\in\mathcal{B}} x_b + \frac{d}{2}\left(\sum_{b\in\mathcal{B}} x_b\right)^2 \\ & + \sum_{b\in\mathcal{B}}\left(\tau_b^{(l)}\left(\mathbf{g}^{\mathrm{T}}\boldsymbol{\theta}_b^{(l+1)} - x_b - \frac{1}{B}\right) + \frac{d}{2}\left(\mathbf{g}^{\mathrm{T}}\boldsymbol{\theta}_b^{(l+1)} - x_b - \frac{1}{B}\right)^2\right)\end{aligned} \tag{8.32}$$

and

$$\begin{aligned}\underset{\mathbf{z}}{\text{minimize}} \; & \sum_{b\in\mathcal{B}}\left(\boldsymbol{\zeta}_b^{(l)\mathrm{T}}(\boldsymbol{\theta}_b^{(l+1)} - \mathbf{z}_b) + \frac{d}{2}\|\boldsymbol{\theta}_b^{(l+1)} - \mathbf{z}_b\|_2^2\right) \\ & + \sum_{u\in\mathcal{U}}\left(\boldsymbol{\zeta}_u'^{(l)\mathrm{T}}(\boldsymbol{\theta}_u'^{(l+1)} - \mathbf{z}'_u) + \frac{d}{2}\|\boldsymbol{\theta}_u'^{(l+1)} - \mathbf{z}'_u\|_2^2\right).\end{aligned} \tag{8.33}$$

Problem (8.32) admits a closed-form solution given as

$$x_b^{(l+1)} := X_b^{(l)} - \frac{B\overline{X}^{(l)}}{B+1} \tag{8.34}$$

where

$$X_b^{(l)} = \mathbf{g}^{\mathrm{T}}\boldsymbol{\theta}_b^{(l+1)} - \frac{1}{B} - \frac{\xi^{(l)} - \tau_b^{(l)}}{d} \quad \text{and} \quad \overline{X}^{(l)} = \frac{\sum_{b\in\mathcal{B}} X_b^{(l)}}{B}.$$

Thus, for updating $x_b^{(l+1)}$, we can perform an average consensus algorithm among the BSs to compute $\overline{X}^{(l)}$ [30]. We turn to updating $\mathbf{z}$ in (8.33). For decomposing (8.33), let us define $\zeta_{b,\mu} \triangleq [1\ \mathbf{0}]\boldsymbol{\zeta}_b$, $\zeta'_{u,\mu} \triangleq [1\ \mathbf{0}]\boldsymbol{\zeta}'_u$, $\boldsymbol{\zeta}_{b,\mathbf{t}} \triangleq [\mathbf{0}\ \mathbf{I}_{|\zeta_b-1|}]\boldsymbol{\zeta}_b$. Then (8.33) is rewritten as

$$\begin{aligned}\underset{(\mu,\mathbf{v}_b)}{\text{minimize}} \; & \sum_{b\in\mathcal{B}} \zeta_{b,\mu}^{(l)}(\mu_b^{(l+1)} - \mu) + \frac{d}{2}\|\mu_b^{(l+1)} - \mu\|_2^2 + \boldsymbol{\zeta}_{b,\mathbf{t}}^{(l)\,\mathrm{T}}(\mathbf{t}_b^{(l+1)} - \mathbf{v}_b) \\ & + \frac{d}{2}\|\mathbf{t}_b^{(l+1)} - \mathbf{v}_b\|_2^2 + (\tilde{\boldsymbol{\zeta}}'_b)^{(l)^{\mathrm{T}}}((\tilde{\mathbf{t}}'_b)^{(l+1)} - \mathbf{v}_b) + \frac{d}{2}\|(\tilde{\mathbf{t}}'_b)^{(l+1)} - \mathbf{v}_b\|_2^2 \\ & \sum_{u\in\mathcal{U}} \zeta'^{(l)}_{u,\mu}(\mu_u'^{(l+1)} - \mu) + \frac{d}{2}\|\mu_u'^{(l+1)} - \mu\|_2^2\end{aligned} \tag{8.35}$$

Algorithm 8.4 Decentralized algorithm solving (8.22)

1. **Initialization:** Set $\epsilon^{ADMM} > 0$, $l = 0$ and choose initial values for $\mathbf{x}^{(0)}, \mathbf{z}^{(0)}, \mathbf{z}'^{(0)}, \xi^{(0)}, \boldsymbol{\tau}^{(0)}, \boldsymbol{\zeta}^{(0)}, \boldsymbol{\zeta}'^{(0)}$
2. **repeat**
3. **for** $b \in \mathcal{B}$ **and** $u \in \mathcal{U}$ **do**
4. BS b updates $\boldsymbol{\theta}_b$ using (8.30); user u updates $\boldsymbol{\theta}'_u$ using (8.31)
5. BS b calculates X_b, exchanges this value with other BSs to compute $\overline{X}^{(l)}$, and then updates x_b as in (8.34)
6. BS b receives $\boldsymbol{\theta}'_{u'}$, $\boldsymbol{\zeta}'_{u'}$ from user u' for all $u' \in \mathcal{U}$, and forms $(\tilde{\mathbf{t}}'_b)^{(l+1)}$,$(\tilde{\boldsymbol{\zeta}}'_b)^{(l)}$ and $\phi'_{u'}$. Then BS b exchanges ϕ_b, ϕ'_u with other BSs
7. BS b updates $\mathbf{v}_b$ using (8.36), and μ using (8.37)
8. User u receives $v_{ub'}$ and μ from BS b' all $b' \in \mathcal{B}$ to form $\mathbf{z}'_u$
9. BS b updates ξ, τ_b, $\boldsymbol{\zeta}_b$ while user u updates $\boldsymbol{\zeta}'_u$.
10. **end for**
11. $l := l + 1$
12. **until** $\|\mathbf{t} - \mathbf{v}\|_2 \leq \epsilon^{ADMM}$

where $\tilde{\mathbf{t}}'_b \triangleq \{t'_{ub}\}_{u \in \mathcal{B}}$ and $\tilde{\boldsymbol{\zeta}}'_b$ are the corresponding elements in $\{\boldsymbol{\zeta}'_u\}_u$. As the objective in (8.35) is a quadratic function, then the update is simple as

$$\mathbf{v}_b^{(l+1)} := 0.5\left(\mathbf{t}_b^{(l+1)} + (\tilde{\mathbf{t}}'_b)^{(l+1)} + \frac{\boldsymbol{\zeta}_{b,\mathbf{t}}^{(l)} + (\tilde{\boldsymbol{\zeta}}'_b)^{(l)}}{d}\right), \tag{8.36}$$

$$\mu^{(l+1)} := \frac{\sum_{b \in \mathcal{B}} \phi_b^{(l)} + \sum_{u \in \mathcal{U}} \phi_u'^{(l)}}{d(B + U)} \tag{8.37}$$

where $\phi_b^{(l)} \triangleq (\zeta_{b,\mu}^{(l)} + c\mu_b^{(l+1)})$ and $\phi_u'^{(l)} \triangleq (\zeta_{u,\mu}'^{(l)} + d\mu_u'^{(l+1)})$. Updating $\mathbf{z}$ can be done independently at each BS as follows. BS b receives information from its connected users and forms $(\tilde{\mathbf{t}}'_b)^{(l+1)}$and $(\tilde{\boldsymbol{\zeta}}'_b)^{(l)}$ to update $\mathbf{v}_b^{(l+1)}$. The update of μ can be implemented by means of consensus among all BSs and users by exchanging ϕ_b, ϕ'_u. After a consensus is reached, $\mu^{(l+1)}$ is recovered by dividing the consensus value by d. After the $\mathbf{z}$-update is finished at all BSs, the users gather the required information to form $\mathbf{z}'^{(l+1)}$.

The last step of the ADMM is to update the Lagrangian multipliers as

$$\xi^{(l+1)} = \xi^{(l)} + \frac{dB\overline{X}^{(l)}}{B+1}, \tag{8.38}$$

$$\tau_b^{(l+1)} = \tau_b^{(l)} + d\left(\mathbf{g}^{\mathrm{T}}\boldsymbol{\theta}_b^{(l+1)} - x_b^{(l+1)} - \frac{1}{B}\right), \tag{8.39}$$

$$\boldsymbol{\zeta}_b^{(l+1)} = \boldsymbol{\zeta}_b^{(l)} + d(\boldsymbol{\theta}_b^{(l+1)} - \mathbf{z}_b^{(l+1)}), \tag{8.40}$$

$$\boldsymbol{\zeta}_u'^{(l+1)} = \boldsymbol{\zeta}_u'^{(l)} + d(\boldsymbol{\theta}_u'^{(l+1)} - \mathbf{z}_u'^{(l+1)}). \tag{8.41}$$

Since $\overline{X}^{(l)}$ is available at all BSs, ξ can be updated without requiring extra gathered information. Updating τ_b and $\boldsymbol{\zeta}_b$ is carried out at BS b, while user u updates $\boldsymbol{\zeta}'_u$. In summary, the decentralized procedure is outlined in Algorithm 8.4.

8.3.1.5 Numerical Examples

We consider a simple simulation model as follows. In the coverage area with radius 300 m centered at origin, there are four BSs ($B = 4$) placed at (200 m, 0), (0, 200 m), (−200 m, 0) and (0, −200 m). There are $U = 7$ users whose positions are randomly generated inside the area. We use the modified Okumura-Hata urban model for pathloss, which is given as [23]

$$PL(\text{dB}) = \begin{cases} 122 + 38\log(x) & \text{if } x \geq 0.05 \text{ km}, \\ 122 + 38\log(0.05) & \text{otherwise}, \end{cases} \tag{8.42}$$

where x is the distance in kilometers. The standard deviation of lognormal distribution shadowing is 8 dB. The noise power density is $N_0 = -174$ dBm/Hz. The maximum bandwidth at the BSs are $\overline{W}_1 = 2.4$ MHz, $\overline{W}_2 = 1.2$ MHz, $\overline{W}_3 = 1.8$ MHz, and $\overline{W}_4 = 2.5$ MHz. The maximum transmission power at the BSs are $\overline{P}_1 = 35$ dBm, $\overline{P}_2 = 32$ dBm, $\overline{P}_3 = 26$ dBm, and $\overline{P}_4 = 37$ dBm. The power amplifier's efficiency at the BSs are $\lambda_1 = 0.95$, $\lambda_2 = 0.91$, $\lambda_3 = 0.88$, and $\lambda_4 = 0.9$. Without loss of generality, we set $a_b = 1, \forall b \in \mathcal{B}$, and $m_{ub} = 1, \forall u \in \mathcal{U}, b \in \mathcal{B}$. We set $p_u^{\text{cr}} = p^{\text{cr}}$, $p_b^{\text{ct}} = p^{\text{ct}}$, and $p_{ub}^{\text{ca}} = p^{\text{ca}}$, $\forall u \in \mathcal{U}, b \in \mathcal{B}$. The QoSs for the users are $\overline{\mathbf{R}} = [2, 1.4, 1.1, 2.2, 3.1, 2.5, 3.7]$ Mbits/s, where $\overline{\mathbf{R}} \triangleq \{\overline{R}_u\}$.

Figure 8.2 plots the convergence behavior of the Dinkelbach algorithm over two random channel realizations. We can observe that, with superlinear convergence, the algorithm outputs the optimal solutions after a few iterations.

Figure 8.3 shows the convergence performance of Algorithm 8.4 over a random channel realization with two different values of penalty parameter d. Specifically, Figure 8.3a plots the objective function and Figure 8.3b plots the term $\|\mathbf{t} - \mathbf{t}'\|_2$, which represents the difference between the local versions of data rate at the users and the BSs. The expected observation is that the iterative procedure converges to the centralized solution with all cases of d. We can also see that the convergence rate depends on d.

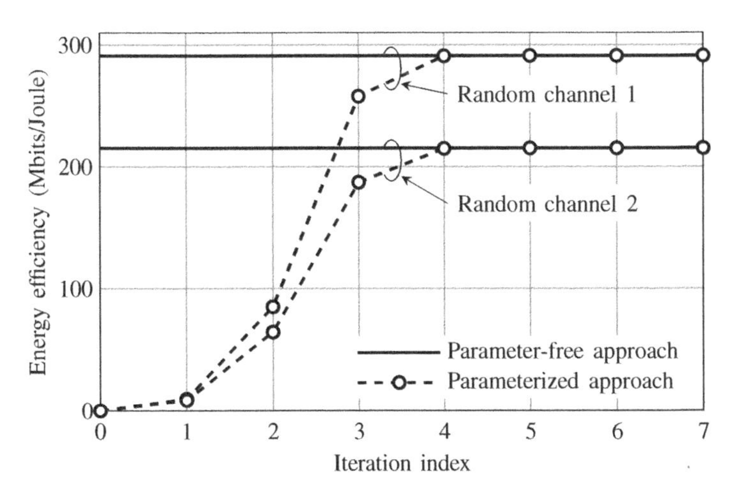

Figure 8.2 Convergence behavior of Dinkelbach's procedure for two random channel realizations. The circuit power parameters are taken as $p^{\text{ct}} = p^{\text{cr}} = 10$ dBm and $p^{\text{ca}} = 5$ dBm.

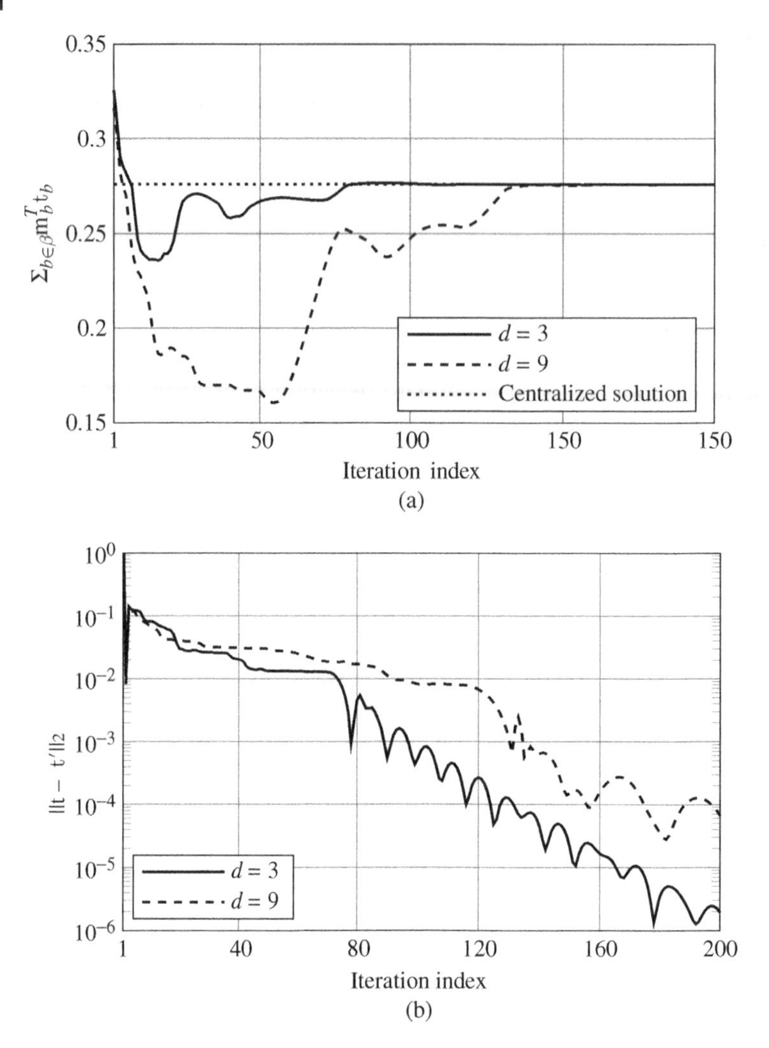

Figure 8.3 Convergence behavior of the ADMM procedure over a random channel. The circuit power parameters are taken as $p^{ct} = p^{cr} = 10$ dBm and $p^{ca} = 5$ dBm. (a) Convergence on objective value. (b) The gap between local versions of data rate at the BSs and the users.

We study the average EE performance in Figure 8.4. For comparison purposes, we also consider the schemes sum rate maximization (SRmax) and power minimization (Powermin), which are mathematically stated as

$$\text{SRmax} \triangleq \max \left\{ \sum_{b\in\mathcal{B}} \sum_{u\in\mathcal{U}} m_{ub} r_{ub} | (8.19\text{b}) \text{ and } (8.19\text{c}) \right\} \tag{8.43}$$

and

$$\text{Powermin} \triangleq \min \left\{ \sum_{b\in\mathcal{B}} \frac{\sum_{u\in\mathcal{U}} \rho_{ub}}{\lambda_b} | (8.19\text{b}) \text{ and } (8.19\text{c}) \right\}. \tag{8.44}$$

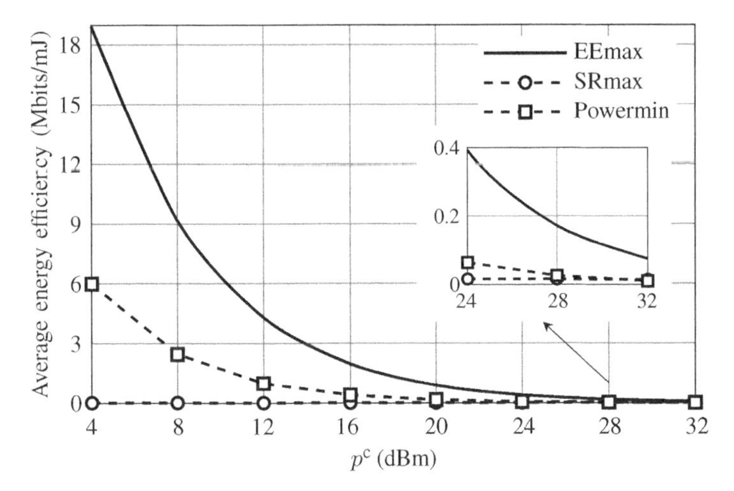

Figure 8.4 Average energy efficiency performance of EEmax, SRmax, and Powermin with different total circuit powers p^c.

As expected, EEmax always outperforms the others in terms of EE. The other observation is that the EE performances of all schemes reduce when p^c increases. For SRmax and Powermin, p^c has no impact on the optimal solution of (8.43) and (8.44); thus when p^c increases, the total consumption powers increase while the sum rates do not change.

8.3.2 Dense Small Cell Networks

In the second application, we use the provided optimization tools in designing EE beamforming vectors for a dense network scenario where macrocell and small cell BSs cooperate in transmitting data to users. The group of BSs using the same bandwidth to simultaneously serve multiple users, i.e. interference channels, are considered. Here we are interested in non-conherent joint transmission in which the information for a specific user is encoded independently at the BSs [31]. This transmit technique requires less strict synchronization compared to the joint transmission coordinated multi-point (or coherent transmission). Thus it is easier to implement in practice.

The EEmax problem is cast as a generalized non-convex fractional program and therefore we use SCA to locally solve the problem. Specifically, the main works are as follows. We first transform the problem into a formulation amenable to the SCA. We then present how to arrive at the convex approximation subproblem solved at each iteration of the SCA procedure. Finally, we implement the solution in a decentralized manner with the ADMM.

8.3.2.1 System Model

We consider a region covered by a macrocell BS and a set of B small cell BSs denoted by $\mathcal{B} = \{1, 2, ..., B\}$, as shown in Figure 8.5. For notational convenience, we refer to the macro BS as BS 0 and let $\overline{\mathcal{B}} = \mathcal{B} \cup \{0\}$. Let K_b denote the number of antennas equipped at BS b. In the region, there is a set of U single-antenna users, denoted by $\mathcal{U} = \{1, 2, ..., U\}$, simultaneously served by the BSs under the same frequency band. We assume that linear precoding is used at all BSs and the users receive information from all BSs under

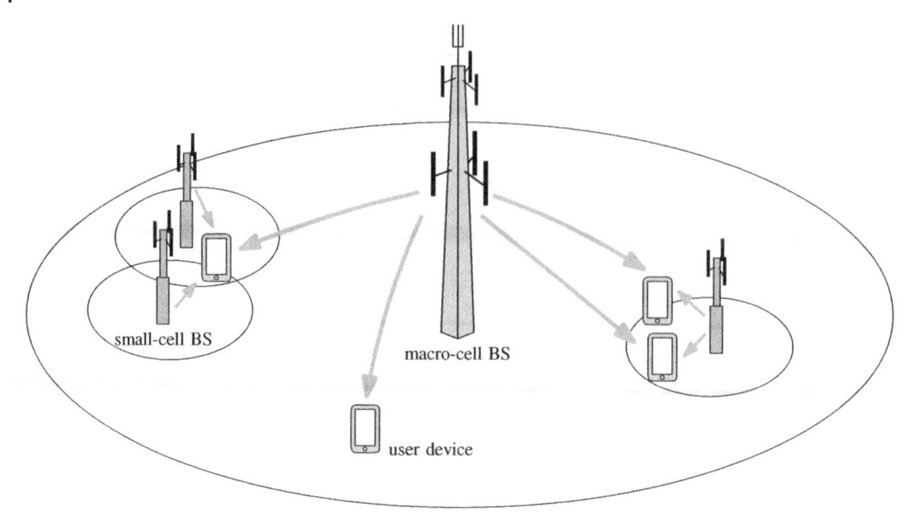

Figure 8.5 An example of dense small cell systems.

non-conherent transmission [31]. Particularly, let $\mathbf{h}_{ub} \in \mathbb{C}^{1\times K_b}$(row vector) denote the channels between user u and BS b and let x_{ub} and $\mathbf{w}_{ub} \in \mathbb{C}^{K_b\times 1}$ be the normalized symbol and the beamforming vector at BS b for user u, respectively. Suppose the channels to be flat. The received signal at user u is

$$r_u = \mathbf{h}_{u0}\mathbf{w}_{u0}x_{u0} + \underbrace{\sum_{b\in\mathcal{B}} \mathbf{h}_{ub}\mathbf{w}_{ub}x_{ub}}_{\text{from small cell BSs}} + \underbrace{\sum_{b\in\overline{\mathcal{B}}}\sum_{m\in\mathcal{U}\setminus u} \mathbf{h}_{ub}\mathbf{w}_{mb}x_{mb}}_{\text{interference}} + n_u \tag{8.45}$$

where $n_u \backsim \mathcal{N}(0,\sigma_u^2)$ is the additive white Gaussian noise at user u. We further assume that users treat the signal of other users as noise. The effective signal-to-interference-plus-noise ratio (SINR) at user u is

$$\gamma_u = \frac{|\mathbf{h}_{u0}\mathbf{w}_{u0}|^2 + \sum_{b\in\mathcal{B}} |\mathbf{h}_{ub}\mathbf{w}_{ub}|^2}{\sum_{b\in\overline{\mathcal{B}}}\sum_{m\in\mathcal{U}\setminus u} |\mathbf{h}_{ub}\mathbf{w}_{mb}|^2 + \sigma_u^2}. \tag{8.46}$$

We note that Eq. (8.46) is due to the non-cohenrent transmission, which does not require phase synchronization between the BSs. For notational convenience, let $\mathbf{w}_u \triangleq [\mathbf{w}_{u0}^{\mathrm{T}}, \mathbf{w}_{u1}^{\mathrm{T}}, ..., \mathbf{w}_{uB}^{\mathrm{T}}]^{\mathrm{T}} \in \mathbb{C}^{\sum_{b\in\overline{\mathcal{B}}}K_b\times 1}$ and $\overline{\mathbf{H}}_u \triangleq \mathrm{blkdiag}\{\mathbf{H}_{u0}, \mathbf{H}_{u1}, ..., \mathbf{H}_{uB}\}$, where $\mathbf{H}_{ub} \triangleq \mathbf{h}_{ub}^{\mathrm{H}}\mathbf{h}_{ub}$. Then the SINR can be rewritten as

$$\gamma_u = \frac{\mathbf{w}_u^{\mathrm{H}}\overline{\mathbf{H}}_u\mathbf{w}_u}{\sum_{m\in\mathcal{U}\setminus u} \mathbf{w}_m^{\mathrm{H}}\overline{\mathbf{H}}_u\mathbf{w}_m + \sigma_u^2}. \tag{8.47}$$

The power consumption model similar to that in Section 8.3.1 is considered. Specifically, let us denote by p_b^{cirTx} and p_u^{cirRx} the circuit power of the idle mode of BS b and user u, respectively. Also, let p_{ub}^{cirCo} denote the extra circuit consumed power when BS b

transmits data to user u. Then the total circuit power is

$$P^{\text{cir}} = \sum_{b\in\overline{\mathcal{B}}} P_b^{\text{cirTx}} + \sum_{u\in\mathcal{U}} P_u^{\text{cirRx}} + \sum_{b\in\overline{\mathcal{B}}} \sum_{u\in\mathcal{U}} P_{ub}^{\text{cirCo}}. \tag{8.48}$$

The total consumption power is

$$P^{\text{Total}} = \sum_{b\in\overline{\mathcal{B}}} \frac{1}{\lambda_b} \sum_{u\in\mathcal{U}} \mathbf{w}_u^{\text{H}} \mathbf{E}_b \mathbf{w}_u + P^{\text{cir}} \tag{8.49}$$

where $\lambda_b \in (0,1)$ is the amplifier's efficiency at BS b and $\mathbf{E}_b = \text{blkdiag}\{\mathbf{0}_{\sum_{b'=0}^{b-1} K_{b'}}, \mathbf{I}_{K_b}, \mathbf{0}_{\sum_{b'=b+1}^{B} K_{b'}}\}$.

The aim is to design beamforming vectors $\{\mathbf{w}_u\}_u$ so that the overall EE is maximized under the constraints of the transmit power budget at the BSs and QoSs for the users. Mathematically, the interest problem reads

$$\underset{\{\mathbf{w}_u\}}{\text{maximize}} \ \frac{\sum_{u\in\mathcal{U}} a_u \log(1+\gamma_u)}{P^{\text{Total}}} \tag{8.50a}$$

$$\text{subject to} \ \log(1+\gamma_u) \ge \overline{R}_u, \forall u \in \mathcal{U}, \tag{8.50b}$$

$$\sum_{u\in\mathcal{U}} \mathbf{w}_u^{\text{H}} \mathbf{E}_b \mathbf{w}_u \le P_b, \forall b \in \mathcal{B}, \tag{8.50c}$$

where coefficient a_u represents the priority of user u, $\overline{R}_u$ is the predefined data rate threshold representing the QoS for user u, and P_b is the maximum total transmit power at BS b. Problem (8.50) is a generalized non-convex fractional program due to the SINR terms $\{\gamma_k\}$.

8.3.2.2 Centralized Solution via Successive Convex Approximation

In what follows, we present how to locally solve (8.50) using the SCA framework. First, we transform the problem into an equivalent formulation as follows:

$$\underset{\{\mathbf{w}_u\},\{t_u\},\{\mu_u\},\{\pi_b\}}{\text{maximize}} \ \frac{\sum_{u\in\mathcal{U}} a_u t_u^2}{\sum_{b\in\overline{\mathcal{B}}} \frac{\pi_b}{\lambda_b} + P^{\text{cir}}} \tag{8.51a}$$

$$\text{subject to} \ \frac{\mathbf{w}_u^{\text{H}} \overline{\mathbf{H}}_u \mathbf{w}_u}{\sum_{m\in\mathcal{U}\setminus u} \mathbf{w}_m^{\text{H}} \overline{\mathbf{H}}_u \mathbf{w}_m + \sigma_u^2} \ge \mu_u, \forall u \in \mathcal{U}, \tag{8.51b}$$

$$\log(1+\mu_u) \ge t_u^2, \forall u \in \mathcal{U}, \tag{8.51c}$$

$$t_u \ge \sqrt{\overline{R}_u}, \forall u \in \mathcal{U}, \tag{8.51d}$$

$$\sum_{u\in\mathcal{U}} \mathbf{w}_u^{\text{H}} \mathbf{E}_b \mathbf{w}_u \le \pi_b, \forall b \in \overline{\mathcal{B}}, \tag{8.51e}$$

$$\pi_b \le P_b, \forall b \in \overline{\mathcal{B}}, \tag{8.51f}$$

where $\{t_u\}$, $\{\mu_u\}$, and $\{\pi_b\}$ are newly introduced slack variables. Problem (8.51) is equivalent to (8.50) in the sense of the optimal set; it is easy to justify that the constraints in (8.51b), (8.51c), and (8.51e) are active at the optimal points. We can see that the non-convex parts lie in (8.51a) and (8.51b).

The objective in problem (8.51) is a quadratic-over-linear function that is convex. A common approach arriving at a concave approximate of the function is to use the first-order Taylor series. Particularly, let $(\{\mathbf{w}_u^{(l)}\}, \{t_u^{(l)}\}, \{\mu_u^{(l)}\}, \{\pi_b^{(l)}\})$ be a feasible point of (8.51). A concave approximate of the objective function is

$$\sum_{u\in\mathcal{U}} a_u \left(\frac{2t_u^{(l)}}{\left(\sum_{b\in\overline{\mathcal{B}}}\pi_b^{(l)}/\lambda_b + P^{\text{cir}}\right)} t_u - \frac{(t_u^{(l)})^2}{\left(\sum_{b\in\overline{\mathcal{B}}}\pi_b^{(l)}/\lambda_b + P^{\text{cir}}\right)^2} \left(\sum_{b\in\overline{\mathcal{B}}} \frac{\pi_b}{\lambda_b} \right) \right). \tag{8.52}$$

We now turn to the non-convex constraints in (8.51b), which can be rewritten for each u as

$$\frac{\mathbf{w}_u^{\text{H}}\overline{\mathbf{H}}_u\mathbf{w}_u}{\mu_u} \geq \sum_{m\in\mathcal{U}\backslash u} \mathbf{w}_m^{\text{H}}\overline{\mathbf{H}}_u\mathbf{w}_m + \sigma_u^2. \tag{8.53}$$

Again, the left side of the inequality is a quadratic-over-linear function. Then an approximate of (8.51b) is

$$\frac{2\text{Re}\{(\mathbf{w}_u^{(l)})^{\text{H}}\overline{\mathbf{H}}_u\mathbf{w}_u\}}{\mu_u^{(l)}} - \frac{(\mathbf{w}_u^{(l)})^{\text{H}}\overline{\mathbf{H}}_u\mathbf{w}_u^{(l)}}{(\mu_u^{(l)})^2}\mu_u \geq \sum_{m\in\mathcal{U}\backslash u} \mathbf{w}_m^{\text{H}}\overline{\mathbf{H}}_u\mathbf{w}_m + \sigma_u^2, \tag{8.54}$$

where $\text{Re}\{\cdot\}$ denotes the real part. It is easy to justify that (8.52) and (8.54) satisfy the three conditions posted in (8.12). Finally, the convex approximation subproblem solved at the $(l+1)$th iteration of the SCA procedure is

$$\underset{\{\mathbf{w}_u\},\{t_u\},\{\mu_u\},\{\pi_b\}}{\text{maximize}} \quad \sum_{u\in\mathcal{U}} A_u^{(l)} t_u - \sum_{b\in\overline{\mathcal{B}}} B_b^{(l)} \pi_b \tag{8.55a}$$

$$\text{subject to} \quad \frac{2\text{Re}\{(\mathbf{w}_u^{(l)})^{\text{H}}\overline{\mathbf{H}}_u\mathbf{w}_u\}}{\mu_u^{(l)}} - \frac{(\mathbf{w}_u^{(l)})^{\text{H}}\overline{\mathbf{H}}_u\mathbf{w}_u^{(l)}}{(\mu_u^{(l)})^2}\mu_u \geq \sum_{m\in\mathcal{U}\backslash u} \mathbf{w}_m^{\text{H}}\overline{\mathbf{H}}_u\mathbf{w}_m + \sigma_u^2, \forall u \in \mathcal{U}, \tag{8.55b}$$

$$\log(1+\mu_u) \geq t_u^2, \forall u \in \mathcal{U}, \tag{8.55c}$$

$$t_u \geq \sqrt{\overline{R}_u}, \forall u \in \mathcal{U}, \tag{8.55d}$$

$$\sum_{u\in\mathcal{U}} \mathbf{w}_u^{\text{H}}\mathbf{E}_b\mathbf{w}_u \leq \pi_b, \forall b \in \overline{\mathcal{B}}, \tag{8.55e}$$

$$\pi_b \leq P_b, \forall b \in \overline{\mathcal{B}}, \tag{8.55e}$$

where

$$A_u^{(l)} = \frac{2a_u t_u^{(l)}}{\left(\sum_{b\in\overline{\mathcal{B}}}\frac{\pi_b^{(l)}}{\lambda_b} + P^{\text{cir}}\right)} \quad \text{and} \quad B_b^{(l)} = \left(\frac{\sum_{u\in\mathcal{U}} a_u (t_u^{(l)})^2}{\lambda_b\left(\sum_{b\in\overline{\mathcal{B}}}\frac{\pi_b^{(l)}}{\lambda_b} + P^{\text{cir}}\right)^2} \right).$$

A feasible point is required to start the SCA algorithm. For (8.51), this requirement might be a challenge due to the QoS constraints. Here, we show a simple heuristic method to overcome this practical issue. By introducing non-negative slack variables $\{\varphi_u\}$, we arrive at a regularization formulation of (8.51) given as

$$\underset{\{\mathbf{w}_u\},\{t_u\},\{\mu_u\},\{\pi_b\},\{\varphi_u\}}{\text{maximize}} \quad \frac{\sum_{u\in\mathcal{U}} a_u t_u^2}{\sum_{b\in\overline{\mathcal{B}}} \frac{\pi_b}{\lambda_b} + P^{\text{cir}}} - \beta \sum_{u\in\mathcal{U}} \varphi_u \tag{8.56a}$$

$$\text{subject to } \; t_u + \varphi_u \geq \sqrt{\bar{R}_u}, \forall u \in \mathcal{U}, \tag{8.56b}$$

$$\text{(8.51b), (8.51c), (8.51e), (8.51f)} \tag{8.56c}$$

where parameter $\beta > 0$. A feasible point of (8.51) is easy to find as follows. We first randomly generate beamforming vectors $\{\mathbf{w}_u^{(0)}\}_u$ so that (8.50c) is satisfied. Then, $\mu_u^{(0)}$, $t_u^{(0)}$, and $\pi_b^{(0)}$ are determined according to (8.51b), (8.51c), and (8.51e), respectively. Constraint (8.56b) is automatically satisfied with sufficient large $\varphi_u^{(0)}$. By using the SCA to solve (8.51), it is expected that $\varphi_u \to 0$ for all u due to the term $\beta \sum_{u\in\mathcal{U}} \varphi_u$ at the objective. Once $\varphi_u = 0$ for all u, we yield a feasible point of (8.51) which could be used to initialize the SCA algorithm solving (8.51).

8.3.2.3 Distributed Implementation

We now implement the SCA solution in a decentralized manner. The core idea is to use ADMM to solve the convex approximation subproblem (8.55). For the purpose, we first equivalently rewrite (8.55) as

$$\underset{\substack{\{\mathbf{w}_{ub}\},\{t_u\},\{\mu_u\},\\ \{\pi_b\},\{y_{ub}\},\{\tilde{y}_{ub}\},\\ \{\tilde{\mu}_{ub}\},\{s_{ub}\},\{z_u\}}}{\text{maximize}} \quad \sum_{u\in\mathcal{U}} A_u^{(l)} t_u - \sum_{b\in\overline{\mathcal{B}}} B_b^{(l)} \pi_b \tag{8.57a}$$

$$\text{subject to } \; f_{ub}^{(l)}(\{\mathbf{w}_{ub}\}_u, \tilde{\mu}_{ub}; \mathbf{w}_{ub}^{(l)}, \mu_u^{(l)}) \geq \tilde{y}_{ub}, \forall u \in \mathcal{U}, b \in \overline{\mathcal{B}}, \tag{8.57b}$$

$$\sum_{b\in\overline{\mathcal{B}}} y_{ub} - \sigma_u^2 \geq 0, \forall u \in \mathcal{U}, \tag{8.57c}$$

$$\log(1+\mu_u) \geq t_u^2, \; \forall u \in \mathcal{U}, \tag{8.57d}$$

$$t_u \geq \sqrt{\bar{R}_u}, \forall u \in \mathcal{U}, \tag{8.57e}$$

$$\sum_{u\in\mathcal{U}} \mathbf{w}_{ub}^{\text{H}} \mathbf{w}_{ub} \leq \pi_b, \forall b \in \overline{\mathcal{B}}, \tag{8.57f}$$

$$\pi_b \leq P_b, \; \forall b \in \overline{\mathcal{B}}, \tag{8.57g}$$

$$\tilde{y}_{ub} = s_{ub}, \forall u \in \mathcal{U}, b \in \overline{\mathcal{B}}, \tag{8.57h}$$

$$y_{ub} = s_{ub}, \forall u \in \mathcal{U}, b \in \overline{\mathcal{B}}, \tag{8.57i}$$

$$\tilde{\mu}_{ub} = z_u, \forall u \in \mathcal{U}, b \in \overline{\mathcal{B}}, \tag{8.57j}$$

$$\mu_u = z_u, \forall u \in \mathcal{U}, \tag{8.57k}$$

where $\{y_{ub}\}$, $\{\tilde{y}_{ub}\}$, $\{\tilde{\mu}_{ub}\}$, $\{s_{ub}\}$, and $\{z_u\}$ are newly introduced variables:

$$f_{ub}^{(l)}(\{\mathbf{w}_{ub}\}_u, \tilde{\mu}_{ub}; \mathbf{w}_{ub}^{(l)}, \mu_u^{(l)}) \triangleq \frac{2\mathrm{Re}\{(\mathbf{w}_{ub}^{(l)})^{\mathrm{H}}\mathbf{H}_{ub}\mathbf{w}_{ub}\}}{\mu_u^{(l)}} - \frac{(\mathbf{w}_{ub}^{(l)})^{\mathrm{H}}\mathbf{H}_{ub}\mathbf{w}_{ub}^{(l)}}{(\mu_u^{(l)})^2}\tilde{\mu}_{ub} - \sum_{m\in\mathcal{U}\backslash u} \mathbf{w}_{mb}^{\mathrm{H}}\mathbf{H}_{ub}\mathbf{w}_{mb}.$$

The introduction of $\{y_{ub}\}$, $\{\tilde{y}_{ub}\}$, and $\{\tilde{\mu}_{ub}\}$ is to decompose (8.55b) into (8.57b) and (8.57c), which will be handled locally at BS b and user u, respectively. The introduction of global variables $\{s_{ub}\}$ and $\{z_u\}$ is to guarantee that the corresponding local versions of the variables are equal to the others via the constraints in (8.57h) to (8.57k). Particularly, problem (8.57) can be handled distributively as follows.

For each user $u \in \mathcal{U}$, let us denote by $\mathbf{v}_u \triangleq \{t_u, \mu_u, \{y_{ub}\}_b\}$ the local variables and define the local feasible set as

$$\mathcal{K}_u = \{\mathbf{v}_u | \sum_{b\in\overline{\mathcal{B}}} y_{ub} - \sigma_u^2 \geq 0, \log(1+\mu_u) \geq t_u^2, t_u \geq \sqrt{\bar{R}_u}\}. \tag{8.58}$$

Similarly, for each BS $b \in \overline{\mathcal{B}}$, let $\tilde{\mathbf{v}}_b \triangleq (\{\mathbf{w}_{ub}\}_u, \{\tilde{\mu}_{ub}\}_u, \{\tilde{y}_{ub}\}_u, \pi_b)$ be the local variables, and the local feasible set is defined as

$$\tilde{\mathcal{K}}_b = \{\tilde{\mathbf{v}}_b | f_{ub}^{(l)}(\{\mathbf{w}_{ub}\}_u, \tilde{\mu}_{ub}; \mathbf{w}_{ub}^{(l)}, \mu_u^{(l)}) \geq \tilde{y}_{ub} \forall u \in \mathcal{U}, \sum_{u\in\mathcal{U}} \mathbf{w}_{ub}^{\mathrm{H}}\mathbf{w}_{ub} \leq \pi_b \leq P_b\}. \tag{8.59}$$

With these definitions, the problem (8.57) is rewritten as

$$\underset{\{\tilde{\mathbf{v}}_b\},\{\mathbf{v}_u\},\{s_{ub}\},\{z_u\}}{\text{minimize}} \quad -\sum_{u\in\mathcal{U}} A_u^{(l)} t_u + \sum_{b\in\overline{\mathcal{B}}} B_b^{(l)} \pi_b \tag{8.60a}$$

$$\text{subject to} \quad \tilde{\mathbf{v}}_b \in \tilde{\mathcal{K}}_b, \forall b \in \overline{\mathcal{B}}, \tag{8.60b}$$

$$\mathbf{v}_u \in \mathcal{K}_u, \forall u \in \mathcal{U}, \tag{8.60c}$$

$$\tilde{\boldsymbol{\theta}}_b = \tilde{\boldsymbol{\phi}}_b, \forall b \in \overline{\mathcal{B}}, \tag{8.60d}$$

$$\boldsymbol{\theta}_u = \boldsymbol{\phi}_u, \forall u \in \mathcal{U}, \tag{8.60e}$$

where $\tilde{\boldsymbol{\theta}}_b \triangleq \{\{\tilde{y}_{ub}\}_u, \{\tilde{\mu}_{ub}\}_u\}$, $\boldsymbol{\theta}_u \triangleq \{\{y_{ub}\}_b, \mu_u\}$ and $\tilde{\boldsymbol{\phi}}_b$ and $\boldsymbol{\phi}_u$ are the rearranged vectors from the same set of variables $(\{s_{ub}\}, \{z_u\})$. The augmented Lagrangian function of (8.60) is

$$
\begin{aligned}
&L_{\mathrm{A}}(\{\tilde{\mathbf{v}}_b\},\{\mathbf{v}_u\},\{s_{ub}\},\{z_u\};\{\boldsymbol{\xi}_u\},\{\boldsymbol{\rho}_b\})\\
&\quad=\sum_{u\in\mathcal{U}}\left(\xi_u^{\mathrm{T}}(\theta_u-\phi_u)+\frac{d}{2}||\theta_u-\phi_u||_2^2-A_u^{(l)}t_u\right)\\
&\quad+\sum_{b\in\overline{\mathcal{B}}}\left(B_b^{(l)}x_b+\rho_b^{\mathrm{T}}(\tilde{\theta}_b-\tilde{\phi}_b)+\frac{d}{2}||\tilde{\theta}_b-\tilde{\phi}_b||_2^2\right)
\end{aligned}\tag{8.61}
$$

where $\{\xi_u\}$ and $\{\rho_b\}$ are the Lagrangian multipliers and d is the penalty parameter.

In what follows, we present the variable updates at iteration $(k+1)$ of the ADMM procedure. The global variables $\{s_{ub}\}$ and $\{z_u\}$ are updated via solving the following problem extracted from (8.61)

$$
\begin{aligned}
\underset{\{s_{ub}\},\{z_u\}}{\text{minimize}}\ &\sum_{b\in\overline{\mathcal{B}}}\sum_{u\in\mathcal{U}}\left([\rho_b^{(k)}]_{s_{ub}}(\tilde{y}_{ub}^{(k)}-s_{ub})+\frac{d}{2}(\tilde{y}_{ub}^{(k)}-s_{ub})^2\right.\\
&\left.+[\xi_u^{(k)}]_{s_{ub}}(y_{ub}^{(k)}-s_{ub})+\frac{d}{2}(y_{ub}^{(k)}-s_{ub})^2\right)\\
&+\sum_{u\in\mathcal{U}}\left(\sum_{b\in\overline{\mathcal{B}}}\left([\rho_b^{(k)}]_{z_u}(\tilde{\mu}_{ub}^{(k)}-z_u)+\frac{d}{2}(\tilde{\mu}_{ub}^{(k)}-z_u)^2\right)\right.\\
&\left.+[\xi_u^{(k)}]_{z_u}(\mu_u^{(k)}-z_u)+\frac{d}{2}(\mu_u^{(k)}-z_u)^2\right),
\end{aligned}\tag{8.62}
$$

where $[\rho_b^{(k)}]_{s_{ub}}$ is the element in $\rho_b^{(k)}$ corresponding to constraint $\tilde{y}_{ub}=s_{ub}$; a similar definition is applied to $[\rho_b^{(k)}]_{z_u}$, $[\xi_u^{(k)}]_{s_{ub}}$, and $[\xi_u^{(k)}]_{z_u}$. Problem (8.62) has the closed-form solution as follows:

$$
s_{ub}^{(k+1)}=\frac{([\rho_b^{(k)}]_{s_{ub}}+d\tilde{y}_{ub}^{(k)})+([\xi_u^{(k)}]_{s_{ub}}+dy_{ub}^{(k)})}{2d},\tag{8.63}
$$

$$
z_u^{(k+1)}=\frac{([\xi_u^{(k)}]_{z_u}+d\mu_u^{(k)})+\sum_{b\in\overline{\mathcal{B}}}([\rho_b^{(k)}]_{z_u}+d\tilde{\mu}_{ub}^{(k)})}{(B+2)d}.\tag{8.64}
$$

Updating $s_{ub}^{(k+1)}$ can be done at BS b or user u; if BS b updates $s_{ub}^{(k+1)}$, it requires the term $[\xi_u^{(k)}]_{s_{ub}}+dy_{ub}^{(k)}$ from user u. Variable $z_u^{(k+1)}$ can be updated at user u; for this, user u requires the term $[\rho_b^{(k)}]_{z_u}+d\tilde{\mu}_{ub}^{(k)}$ from BS b for all $b\in\overline{\mathcal{B}}$.

BS b updates its local variables $\tilde{\mathbf{v}}_b$ by solving the following QCQP problem:

$$
\underset{\tilde{\mathbf{v}}_b\in\tilde{\mathcal{K}}_b}{\text{minimize}}\ B_b^{(l)}x_b+(\rho_b^{(k)})^{\mathrm{T}}(\tilde{\theta}_b-\tilde{\phi}_b^{(k+1)})+\frac{d}{2}||\tilde{\theta}_b-\tilde{\phi}_b^{(k+1)}||_2^2.\tag{8.65}
$$

Similarly, user u updates local variable $\mathbf{v}_u$ by solving the convex problem:

$$
\underset{\mathbf{v}_u\in\mathcal{K}_u}{\text{minimize}}\ (\xi_u^{(k)})^{\mathrm{T}}(\theta_u-\phi_u^{(k+1)})+\frac{d}{2}||\theta_u-\phi_u^{(k+1)}||_2^2-A_u^{(l)}t_u.\tag{8.66}
$$

Thus, for the local variable update, user u receives $s_{ub}^{(k+1)}$ from BS b, for all b, to form $\phi_u^{(k+1)}$, and BS b receives $z_u^{(k+1)}$ from user u, for all u, to form $\tilde{\phi}_b^{(k+1)}$.

Algorithm 8.5 Decentralized procedure solving (8.50)

1. **Initialization**: Set $\epsilon^{\mathrm{ADMM}} > 0, \epsilon^{\mathrm{SCA}} > 0, l := 0$ and $k := 0$, then choose initial values for $(\{\mathbf{w}_u^{(0)}\}, \{t_u^{(0)}\}, \{\mu_u^{(0)}\}, \{\pi_b^{(0)}\})$and $(\tilde{\mathbf{v}}_b^{(0)}, \mathbf{v}_u^{(0)}; \boldsymbol{\rho}_b^{(0)}, \boldsymbol{\xi}_u^{(0)})$
2. **repeat** {Outer loop (SCA procedure)}
3. BS b receives $t_u^{(l)}$ from user $u \in \mathcal{U}$ to determine $B_b^{(l)}$; user u receives $\pi_b^{(l)}$ from BS $b \in \overline{\mathcal{B}}$ to determine $A_u^{(l)}$
4. **repeat** {Inner loop (ADMM procedure)}
5. **for** $b \in \overline{\mathcal{B}}$ **and** $u \in \mathcal{U}$ **do**
6. BS b updates $s_{ub}^{(k+1)}$ using (8.63); user u updates $z_u^{(k+1)}$ using (8.64)
7. BS b updates $\tilde{\mathbf{v}}_b^{(k+1)}$ using (8.65); user u updates $\mathbf{v}_u^{(k+1)}$ using (8.66)
8. BS b updates $\boldsymbol{\rho}_b^{(k+1)}$ using (8.68); user u updates $\boldsymbol{\xi}_u^{(k+1)}$ using (8.67)
9. **end for**
10. $k := k + 1.$
11. **until** ADMM convergence, i.e. $||\tilde{\boldsymbol{\theta}}_b - \tilde{\boldsymbol{\phi}}_b|| \leq \epsilon^{\mathrm{ADMM}}$
12. Obtain $(\{\mathbf{w}_u^*\}, \{t_u^*\}, \{\mu_u^*\}, \{\pi_b^*\}; \boldsymbol{\xi}_u^*, \boldsymbol{\rho}_b^*)$, the solution from the ADMM
13. $l := l + 1; k := 0$
14. $(\{\mathbf{w}_u^{(l)}\}, \{t_u^{(l)}\}, \{\mu_u^{(l)}\}, \{\pi_b^{(l)}\}) := (\{\mathbf{w}_u^*\}, \{t_u^*\}, \{\mu_u^*\}, \{\pi_b^*\});(\tilde{\mathbf{v}}_b^{(0)}, \mathbf{v}_u^{(0)}; \boldsymbol{\rho}_b^{(0)}, \boldsymbol{\xi}_u^{(0)}) := (\tilde{\mathbf{v}}_b^*, \mathbf{v}_u^*; \boldsymbol{\rho}_b^*, \boldsymbol{\xi}_u^*)$
15. **until** SCA convergence, i.e. $|| \sum_{u\in\mathcal{U}}(A_u^{(l)}t_u^{(l)} - A_u^{(l+1)}t_u^{(l+1)}) - \sum_{b\in\overline{\mathcal{B}}}(B_b^{(l)}\pi_b^{(l)} - B_b^{(l+1)}\pi_b^{(l+1)}||_2 \leq \epsilon^{\mathrm{SCA}}$

Finally, the Lagrangian multipliers are updated as follows:

$$\boldsymbol{\xi}_u^{(k+1)} = \boldsymbol{\xi}_u^{(k)} + d(\boldsymbol{\theta}_u^{(k+1)} - \boldsymbol{\phi}_u^{(k+1)}), \tag{8.67}$$

$$\boldsymbol{\rho}_b^{(k+1)} = \boldsymbol{\rho}_b^{(k)} + d(\tilde{\boldsymbol{\theta}}_b^{(k+1)} - \tilde{\boldsymbol{\phi}}_b^{(k+1)}). \tag{8.68}$$

Since $\boldsymbol{\phi}_u^{(k+1)}$ and $\tilde{\boldsymbol{\phi}}_b^{(k+1)}$ are already available at user u and BS b, respectively, updating $\boldsymbol{\xi}_u^{(k+1)}$ and $\boldsymbol{\rho}_b^{(k+1)}$ does not require additional exchanged information. In summary, the main steps of the distributed procedure are outlined in Algorithm 8.5.

8.3.2.4 Numerical Examples

We consider an area with radius 500 m centered at the origin where the macro-cell BS is placed. There are four small cell BSs ($B = 4$) placed at (300 m, 0), (0, 300 m), (−300 m, 0) and (0, −300 m). The number of antennas equipped at the BSs are $K_0 = 8$ and $K_b = 6$ for all $b \in \mathcal{B}$. There are $U = 6$ users randomly generated inside the area. We use the modified Okumura-Hata urban, i.e. (8.42), for pathloss and the standard deviation of the shadowing is set as 8 dB. The noise power density is $N_0 = -174$ dBm/Hz and the operation bandwidth is 1 MHz. The maximum transmission power at BSs are $P_0 = 40$ dBm and $P_b = 35$ dBm for all $b \in \mathcal{B}$; we take $p_{ub}^{\mathrm{cirCo}} = p^{\mathrm{cirCo}}$ for all u, b; the power amplifier's efficiency at the BSs are $\lambda_0 = 0.96$ and $\lambda_b = 0.93$ for all $b \in \mathcal{B}$; we take the QoS thresholds as $\overline{\mathbf{r}} \triangleq \{\overline{R}_u\} = [1.1;\ 0.91;\ 0.75;\ 1.3;\ 1.22;\ 1.15]$ Mbits/s.

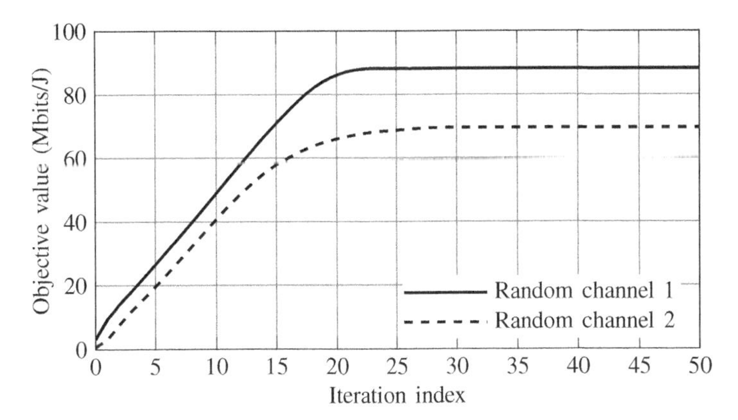

Figure 8.6 Convergence behavior of the SCA procedure over two random channel realizations. The total circuit power P^{cir} is 30 dBm.

Figure 8.6 plots the convergence performance of the SCA procedure solving (8.51) over two random channel realizations. We can observe from the figure that the algorithm converges within a few iterations (less than 30 iterations) in all cases of the considered channels.

Figure 8.7 depicts the average energy efficiency performance of the noncoherent transmission and the coordinated transmission as the functions of p^{cirCo}. Here, for the coordinated transmission, each of the users only receives information from the nearest BS; the circuit power for this scheme is $P^{\text{cir}} = \sum_{b\in\overline{\mathcal{B}}} p_b^{\text{cirTx}} + \sum_{u\in\mathcal{U}} p_u^{\text{cirRx}} + \sum_{u\in\mathcal{U}} p_{ub_u}^{\text{cirCo}}$, where b_u is the BS serving user u. The main observation is that when p^{cirCo} is small the non-coherent transmission outperforms the coordinated transmission in terms of EE due to the gain of the joint processing transmission. As p^{cirCo} increases, the EE performance of both schemes decrease; however, the reduction speed of the

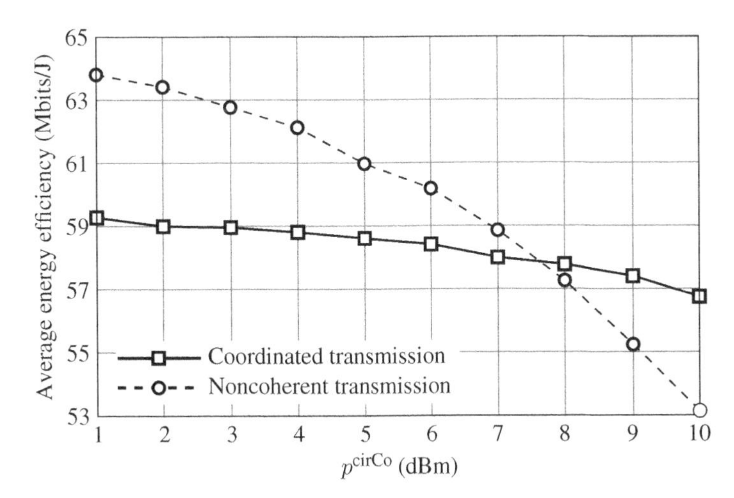

Figure 8.7 Average energy efficiency performance of the non-coherent transmission scheme and coordinated transmission scheme with different p^{cirCo}; $p_u^{\text{cirRx}} = 10$ dBm for all u, $p_0^{\text{cirTx}} = 30$ dBm, and $p_b^{\text{cirTx}} = 15$ dBm for all $b \in \mathcal{B}$.

non-coherent transmission is faster since its circuit consumption power increases faster compared to the coordinated transmission scheme. As a result, when p^{cirCo} is large enough, coordinated transmission outperforms non-coherent transmission.

The results in Figure 8.7 suggest an approach to further improving the energy efficiency performance for the networks that is appropriately controlling the operation modes of the nodes. The approach involves dealing with Boolean (or discrete) variables. The interested reader is referred to [22, Section IV] and [9, Section IV] for examples.

8.4 Conclusion

Energy efficiency optimization for wireless communications usually involves dealing with fractional programs due to the EE definition. We have introduced the optimization tools for overcoming various classes of fractional programs including concave fractional programs, max–min fractional programs, and generalized non-convex fractional programs. While the efficient optimal solutions for the first two classes are available, that for the last is still an open problem. We have also introduced ADMM as an efficient tool for developing distributed implementations.

We have presented the applications of the provided optimization tools in optimizing energy efficiency for dense networks including spectrum sharing networks and dense small cell networks. In each scenario, centralized and distributed solutions have been provided.

The provided tools mainly rely on the framework of convex optimizations where there are many efficient algorithms, e.g. first-order methods, solving a large-scale problem. In addition, ADMM does not only build distributed algorithms but also separates a large problem into small subproblems, and in some cases, a subproblem can be solved via a simple analytical formula. Therefore, these tools are well suited to ultra-dense networks.

Bibliography

1 Ericsson White Paper, 5G energy performance, Apr. 2015.

2 M. Kamel, W. Hamouda, and A. Youssef, "Ultra-dense networks: a survey," *IEEE Communications Surveys Tutorials*, vol. 18, no. 4, pp. 2522–2545, Fourth quarter 2016.

3 J. Hoydis, M. Kobayashi, and M. Debbah, "Green small-cell networks," *IEEE Vehicular Technology Magazine*, vol. 6, no. 1, pp. 37–43, Mar. 2011.

4 EARTH, "Most suitable efficiency metrics and utility functions," Technical report, URL https://www.ict-earth.eu.

5 C. Isheden, Z. Chong, E. Jorswieck, and G. Fettweis, "Framework for link-level energy efficiency optimization with informed transmitter," *IEEE Transactions on Wireless Communications*, vol. 11, no. 8, pp. 2946–2957, Aug. 2012.

6 S. Schaible and J. Shi, "Recent developments in fractional programming: single ratio and max-min case," *Nonlinear Analysis and Convex Analysis*, pp. 493–505, 2004.

7 W. Dinkelbach, "On nonlinear fractional programming," *Management Science*, vol. 13, no. 7, pp. 492–498, Oct. 1967.

8 S. Schaible and Köln, "Parameter-free convex equivalent and dual programs of fractional programming problem," *Zeitschrift für Operations Research*, pp. 187–196, Apr. 1973.
9 Q. D. Vu, L. N. Tran, R. Farrell, and E. K. Hong, "Energy-efficient zero-forcing precoding design for small cell networks," *IEEE Transactions on Communications*, vol. 64, no. 2, pp. 790–804, Feb 2016.
10 O. Tervo, A. Tölli, M. Juntti, and L. N. Tran, "Energy-efficient beam coordination strategies with rate-dependent processing power," *IEEE Transactions on Signal Processing*, vol. 65, no. 22, pp. 6097–6112, Nov. 2017.
11 K.-G. Nguyen, L.-N. Tran, O. Tervo, Q.-D. Vu, and M. Juntti, "Achieving energy efficiency fairness in multicell MISO downlink," *IEEE Communications Letters*, vol. 19, no. 8, pp. 1426–1429, Aug. 2015.
12 A. Charnes and W.W. Cooper, "Programming with linear fractional functionals," *Naval Research Logistics Quarterly*, vol. 9, no. 3-4, pp. 181–186, 1962.
13 J.P. Crouzeix, J.A. Ferland, and S. Schaible, "An algorithm for generalized fractional programs," *Journal of Optimization Theory and Applications*, vol. 47. no. 1, pp. 35–49, Sep. 1985.
14 J.-P. Crouzeix and J.A. Ferland, "Algorithms for generalized fractional programming," *Mathematical Programming*, vol. 52, no. 2, pp. 191–207, May 1991.
15 B.R. Marks and G.P. Wright, "A general inner approximation algorithm for nonconvex mathematical programs," *Operations Research*, vol. 26, no.4, pp. 681–683, Aug. 1978.
16 A. Beck, A. Ben-Tal, and L. Tetruashvili, "A sequential parametric convex approximation method with applications to nonconvex truss topology design problems," *Journal of Global Optimization*, vol. 47, no. 1, pp. 29–51, 2010.
17 T. Dinh Quoc and M. Diehl, "Sequential convex programming methods for solving nonlinear optimization problems with DC constraints," *ArXiv e-prints*, Jul. 2011.
18 S. Boyd, N. Parikh, E. Chu, B. Peleato, and J. Eckstein, "Distributed optimization and statistical learning via the alternating direction method of multipliers," *Foundations and Trends in Machine Learning*, vol. 3, no.1, pp. 1–122, 2011.
19 S. Bubeck, "Convex optimization: algorithms and complexity," *Found. Trends Mach. Learn.*, vol. 8, 3–4, pp. 231–357, Nov. 2015.
20 D.E. Charilas and A.D. Panagopoulous, "Multiaccess radio network enviroments," *IEEE Vehicular Technology Magazine*, vol. 5, no. 4, pp. 40–49, Apr. 2010.
21 W. Zhuang and M. Ismail, "Cooperation in wireless communication networks," *IEEE Wireless Communications*, vol. 19, no.2, pp. 10–20, 2012.
22 Q. D. Vu, L. N. Tran, M. Juntti, and E. K. Hong, "Energy-efficient bandwidth and power allocation for multi-homing networks," *IEEE Transactions on Signal Processing*, vol. 63, no.7, pp. 1684–1699, Apr. 2015.
23 Y. Choi, H. Kim, S. wook Han, and Y. Han, "Joint resource allocation for parallel multi-radio access in heterogeneous wireless networks," *IEEE Transactions on Wireless Communications*, vol. 9, no. 11, pp. 3324–3329, Nov. 2010.
24 J. Miao, Z. Hu, K. Yang, C. Wang, and H. Tian, "Joint power and bandwidth allocation algorithm with QoS support in heterogeneous wireless networks," *IEEE Communications Letters*, vol. 16, no. 4, pp. 479–481, Apr. 2012.

25 P. Xue, P. Gong, J.H. Park, D. Park, and D.K. Kim, "Radio resource management with proportional rate constraint in the heterogeneous networks," *IEEE Transactions on Wireless Communications*, vol. 11, no. 3, pp. 1066–1075, Mar. 2012.
26 M. Ismail and W. Zhuang, "Decentralized radio resource allocation for single-network and multi-homing services in cooperative heterogeneous wireless access medium." *IEEE Transactions on Wireless Communications*, vol. 11, no. 11, pp. 4085–4095, Nov. 2012a.
27 M. Ismail and W. Zhuang, "A distributed multi-service resource allocation algorithm in heterogeneous wireless access medium," *IEEE Journal on Selected Areas in Communications*, vol. 30, no. 2, pp. 425–432, Feb. 2012b.
28 D. Niyato and E. Hossain, "Dynamics of network selection in heterogeneous wireless networks: an evolutionary game approach," *IEEE Transactions on Vehicular Technology*, vol. 58, no.4, pp. 2008–2017, May 2009.
29 S. Boyd and L. Vandenberghe, *Convex Optimization*, 1st edn. Cambridge, 2004.
30 L. Xiao and S. Boyd, "Fast linear iterations for distributed averaging," *Syst. Control Lett.*, vol. 53, pp. 65–78, 2004.
31 E. Björnson, M. Kountouris, and M. Debbah, "Massive MIMO and small cells: improving energy efficiency by optimal soft-cell coordination," in *20th International Conference on Telecommunications (ICT)*, pp. 1–5, May 2013.

Part III

Applications of Ultra-dense Networks

9

Big Data Methods for Ultra-dense Network Deployment

Weisi Guo[1], Maria Liakata[2], Guillem Mosquera[3], Weijie Qi[4], Jie Deng[5] and Jie Zhang[4]

[1] *School of Engineering, University of Warwick, UK*
[2] *Department of Computer Science, University of Warwick, UK*
[3] *Mathematics Institute, University of Warwick, UK*
[4] *Department of Electronic and Electrical Engineering, University of Sheffield, UK*
[5] *School of Electronic Engineering and Computer Science, Queen Mary University of London, UK*

9.1 Introduction

As we accelerate into the twenty-first century, we are seeing increased human digital activity through the compounded effects of urbanization and the proliferation of smart devices (see Figure 9.1). Data demand has risen at a super-linear rate, with a 10× growth from 2011 to 2017. Beyond these statistical trends, networks are also experiencing increased data demand complexity from people and machines. Enabling understanding and exploitation of these complexities is important. This is especially so for the deployment of ultra-dense network (UDN) nodes, which include small cells, relays, distributed antennas/reflectors, and other heterogeneous network elements deployed in a spatially dense formation [1]. Besides the wireless notions of cell planning, economically efficient UDN deployment requires precise knowledge of traffic patterns and usage contexts, as well as the shortfalls in existing deployment to reduce outage and customer complaints. For the first time, the analysis of big data sets gives us an opportunity to do this.

This chapter outlines big data methods for UDN deployment from the perspective of two data science approaches: (1) identify spatial-temporal traffic and service challenge patterns to aid UDN targeted deployment and (2) identify social community patterns to assist ultra-dense peer-to-peer and device-to-device networking. Having access to consumer data, and being able to analyse it in the wireless network context, allows us to improve deployment and operations of UDNs.

9.1.1 The Economic Case for Big Data in UDNs

Over the past decade, wireless and social networks have increasingly made available meta-data with spatial components of varying resolutions. Connected network devices under the Internet-of-Everything (IoE) paradigm can potentially leverage on a combination of real-time data streams and other databases (i.e. geographical information system data, commercial data, census data, engineering data) to derive social context and

Ultra-dense Networks for 5G and Beyond: Modelling, Analysis, and Applications,
First Edition. Edited by Trung Q. Duong, Xiaoli Chu and Himal A. Suraweera.

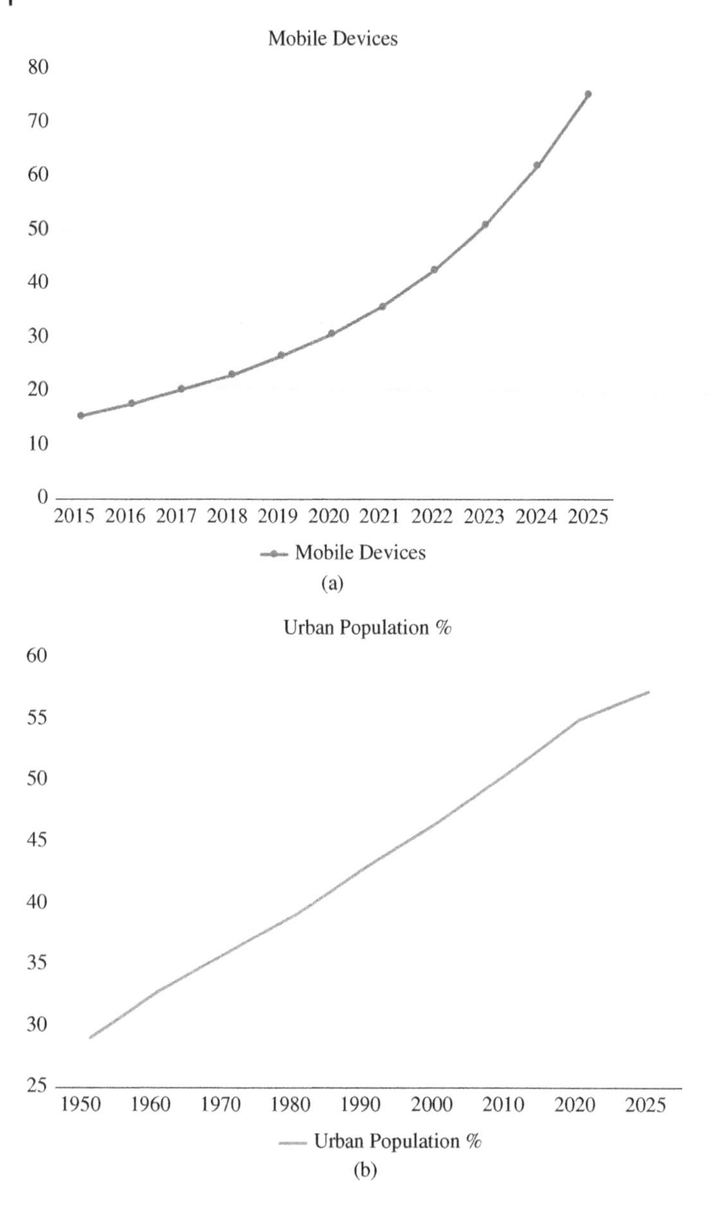

Figure 9.1 Consistent growth in (a) number of new wireless devices (millions) [2] and (b) percentage of world population living in urban areas [3].

improve the performance of underlying 4G/5G wireless network performance. Wireless networks are a fundamental cornerstone of the global digital economy and a key component in our daily lives. Each year, the mobile cellular network industry contributes 3.6% to the global GDP ($2.4 trillion), supports 10.5 million jobs, and has contributed to $336 billion of public funding [4].

Over the past two decades, most of the capacity growth has been as the result of spectrum reuse from cell densification. Although dense heterogeneous networks have been widely considered for future mobile networks, it has been shown that the network capacity and packet delay cannot be further improved when the density of small cells goes beyond a certain value due to the excessive inter-cell interference and the limited capacity of backhaul links. Moreover, mobile users and their data demands are not uniformly distributed in space or over time. In dense urban areas, large crowds can form, disperse, migrate, and demand wireless services in spontaneous and difficult to predict ways. The patterns of data demand and mobility are changing more rapidly than at any previous point in history. Whilst traffic demand is increasing exponentially, network operator revenue remains flat, and there is an urgent need to reduce both the capital expenditure (CAPEX) and operational expenditure (OPEX) for network infrastructure. The importance of big data analysis lies in its scalable potential to extract consumer patterns for proactive network optimization. Thus, data analytics for network optimization is becoming a significant part of data analytics in the telecom market. For example, according to Ericsson, configuring the network to deliver more reliable services and monitoring quality-of-experience (QoE) to proactively correct any potential problems are two of five areas that can transform telecommunications.

Machine learning has seen rapid advances, both in the areas of supervised and unsupervised learning, Bayesian and neural network approaches. In this chapter, we focus primarily on recent work on using machine learning to characterize mobile data demand from heterogeneous data sources and contextualize meta information. As such, this is not a chapter that focuses on optimal machine learning algorithms, but rather it aims to inspire the reader to think how machine learning and data/complexity science techniques can improve the service provisioning in wireless networks.

9.1.2 Chapter Organization

In this chapter, we review appropriate research on heterogeneous big data analytics to optimize the deployment of wireless networks. In the first half of the chapter, we will examine how social media data can quantify traffic patterns across various temporal and spatial scales. The methodological aspects of this work includes structured data analytics, such as clustering and social community detection. In the second half of the chapter, we will examine how social media data can improve our understanding of sentiment toward different service topics, and reinforce our understanding of poor QoE reasons and locations for targeted improvement. As the data are often text based and not numerical, the methodological aspects of this work includes unstructured data analytics, such as natural language processing (NLP) and surveys.

9.2 Structured Data Analytics for Traffic Hotspot Characterization

9.2.1 Social Media Mapping of Hotspots

1. *Background to Social Media Data* Over the past few years, large volumes of data have been transforming businesses to deliver higher precision and more personalized services. Increasing availability of real-time online social network (OSN) data can provide network operators with an opportunity to analyze and combine existing cell planning and operational practices. The advantage of using social media data over operators

own data are as follows: (1) it can discover the overall traffic demand across all wireless networks (RATs) and operators (i.e. the whole market, as opposed to its own), (2) it can uncover textual data about how people feel about various aspects of service, and (3) it provides updated data demand: current base stations were deployed by the operator many years ago and new data trends will have emerged due to changes in the city. Until recently, mapping traffic patterns using OSN data have not been studied extensively in academia and the closest example is using OSN data to infer channel occupancy in cognitive radio [5] and customer complaints [6, 7]. In this section, we demonstrate that structured OSN data can act as a superior predictor for long-term and short-term data demand.

2. *Traditional Methods* Traditionally, the population data of residents and businesses is widely used in traditional cell planning and spectrum purchases to both gauge the number of potential subscribers and estimate the long-term traffic demand. The data are an important input to deciding where BSs are deployed. Recent research in [8] show that there is weak spatial correlation (adjusted $R^2 = 0.29$) between the density of BSs and the 3G traffic demand,[1] and an adequate spatial correlation (adjusted $R^2 = 0.57$) between the census population data and the 3G traffic demand. The paper compared outlier wards to gain a better understanding on why the traffic data does nt correlate well with the BS density and population density. They went on to construct a rank of the densities and compare the rank difference. It was found that one particular ward (among the outlier wards) has the largest rank difference between traffic and both BS density and population density: Marylebone High Street. The area is host to a number of major commercial streets (bounded by Oxford Street in the south) and tourist attractions (Sherlock Holmes and Madame Tussaud to the north). The traffic demand is ranked one of the highest, and yet both the working and residential population (fifth lowest) and BS density are of the lowest. This indicates that existing macro BSs are not well deployed to meet the **current** traffic patterns, and static population census data alone cannot act as an accurate predictor for both the long-term traffic demand or the short-term traffic temporal patterns. Therefore, there is motivation to examine how Twitter data can be used to provide a better estimate for long- and short-term traffic demand.

3. *Long-term Spatial Traffic Demand Estimation* One of the key metrics that drive small cell deployment is the expected traffic load (demand) in its small coverage area. For deployment of small cells, the short-term temporal variation of the traffic demand is of less interest than the long-term spatial variation. In order to obtain high spatial resolution data, the researchers in [8] plot the spatial traffic demand pattern against the Twitter activity intensity over the Greater London area. Figure 9.2 shows an example of geo-tagged social media data over th Greater London area with aggregated densities at the ward level. The **hypothesis** is that the Twitter activity level can be used as a proxy for estimating the real wireless traffic demand in both the UL and DL channels. There has already been studies that show that the number of Tweets is highly correlated with the number of people in confined spaces (i.e. a stadium or an airport) [9]. Therefore, the authors in [8] go a step further and infer the traffic demand directly. The data available for analysis from Twitter record only UL Tweets, which consume negligible bandwidth.

1 Here, 3G traffic was taken in 2012, which had already seen a significant rise in social media and multimedia data.

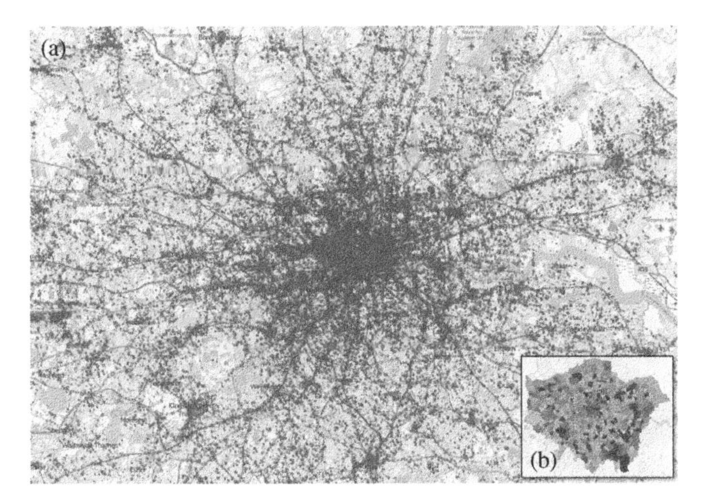

Figure 9.2 Social media data: (a) geo-tagged data over the Greater London area and (b) aggregated densities at the ward level.

Therefore, it is not immediately obvious why small volumes of UL data should be representative of overall UL and DL data demand, especially given the variety of multimedia and social media applications. Yet the authors hypothesize that Twitter activity is closely related to other multi-media activities, simply because average human behaviour associates Twitter uplink with all other mobile activities. In fact, in this section the authors show that the Twitter activity level is highly correlated with both the UL and DL traffic demands.

In Figure 9.3, the ward average 3G data load (UL and DL) is plotted against the number of Tweets per second. Each data point represents a ward, which approximately equals the coverage area of a BS. The results show that a loglinear relationship exists between estimated traffic load $\hat{r}$ (kbps) and the Twitter activity level n (Tweets/s):

$$\log_{10}(\hat{r}_i) = a_i \log_{10}(n) + b_i, \tag{9.1}$$

where [$a_{UL} = 0.86$ kb/Tweet $b_{UL} = 1.97$ kbps], and [$a_{DL} = 0.88$ kb/Tweet, $b_{DL} = 2.37$ kbps]. Alternatively, this can be expressed as a **power law**: $\hat{r}_i = 10^{b_i}(n)^{a_i}$. A polynomial least-squares regression is used with the minimum number of parameters P that maximizes the adjusted R^2 value, as increasing the parameters P will trade off improved accuracy versus a decreased adjusted R^2 value. The correlation achieved is high: $R^2 = 0.71$ for UL and $R^2 = 0.79$ for DL. In other words, the regression model on Twitter data explains 71–79% of the variations in the 3G traffic data. This supports our initial hypothesis that OSN data can be used as a more reliable predictor for traffic demand than the census data. Furthermore, OSN data is operator neutral and radio-access technology (RAT) neutral, potentially giving insight on all mobile customers. In general, it is also worth noting that the data rate in UL and DL is fairly low, showing that whilst the aggregate demand is increasing rapidly, the average demand per second remains below the capacity of current BSs. The outlier wards are ones with high average traffic loads and correspond to tourism hotspots in the wards situated in the Westminster and City of London boroughs.

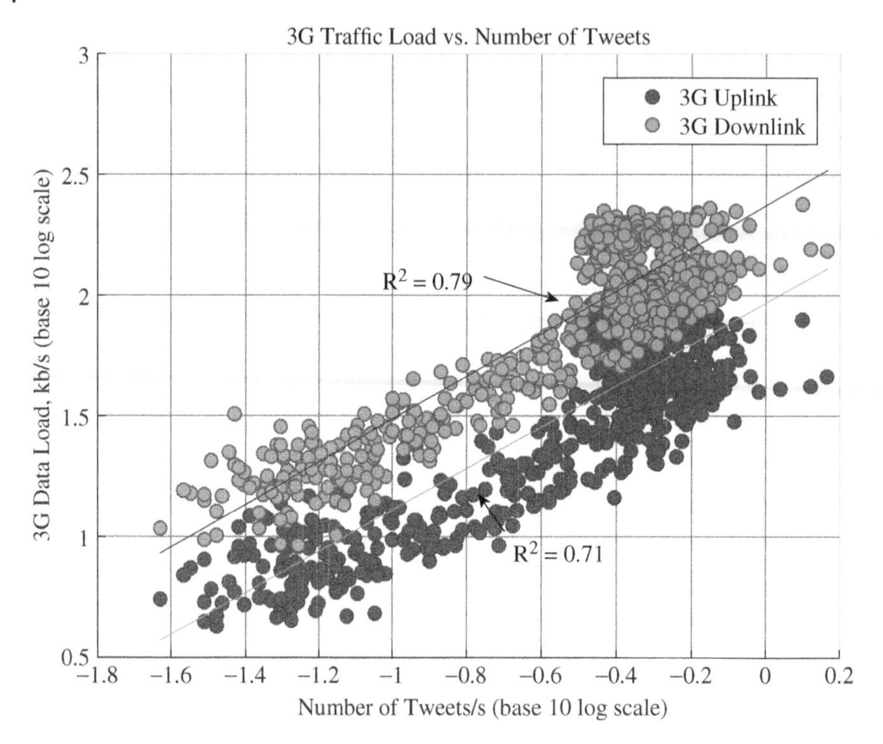

Figure 9.3 Temporal correlation of average 3G traffic load (demand) versus number of tweets over 7 days.

In terms of caveats, the analysis in [8] has only been utilized on geo-tagged Tweets. That is to say, we do not understand how the total number of Tweets correlate with the traffic demand, since most Tweets do not have an accurate location (approximately 1% of Tweets are geo-tagged in the London area). Whilst an increasing number of Tweets and other OSN data are becoming geo-tagged, most data cannot be used to identify the traffic pattern. For long-term traffic prediction, historical Twitter and OSN data are more than sufficient. Over several weeks and a large area, this paper was able to show with 0.6 million Tweets that a power law relationship between existing geo-tagged Tweets and data demand. However, how accurately can these data be used to predict short-term traffic demand asks for further investigation.

4. *Short-term Temporal Traffic Demand Estimation* As mentioned previously, another challenge is whether the current Twitter data can be used to accurately predict upcoming short-term traffic demand (i.e. either in a few hours' time or the same time a few days later). This has strong applications in self-organizing-network (SON) operations [10], such as load balancing. In all of these cases, the upcoming short-term temporal patterns in the traffic demand are of interest. By using cross-correlation between Twitter intensity $n[k]$ (complex conjugate $n * [k]$), and the 3G traffic load $r[k]$ defined as $\sum_{m=-N}^{+N} n * [m]r[m+k]$ with lag m (minutes), the authors in [8] examined the normalized correlation value. Figure 9.4 shows the normalized cross-correlation value against the lag value. The results show that for a lag of approximately 120 minutes, the correlation is strong (> 0.9), meaning that the traffic can be accurately predicted for the next

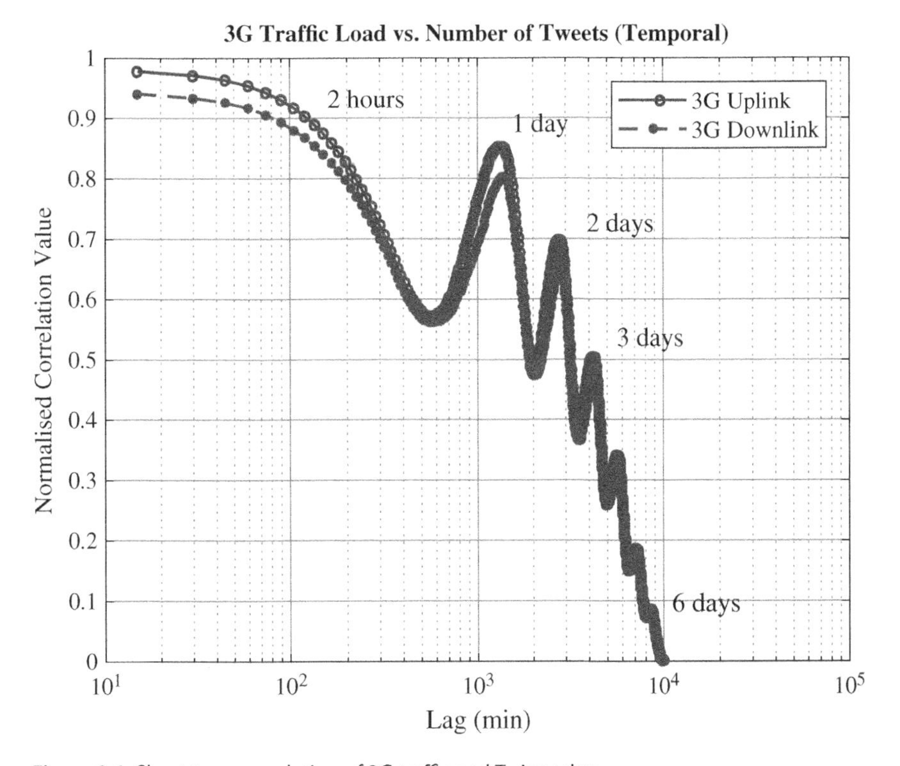

Figure 9.4 Short-term correlation of 3G traffic and Twitter data.

2 hours using the current Twitter activity level. At the same time the next day, the correlation remains strong (> 0.85), but this value falls on the third day to 0.7 and on the fourth day to 0.5. The correlation trend is very similar for UL and DL traffic. That is to say, current Twitter data can be used to predict the UL and DL traffic for the next 2 hours and for the same time on the next day. However, future traffic prediction will need continuous monitoring of geo-tagged Twitter data, which can be expensive to obtain.

9.2.2 Community and Cluster Detection

Clustering of user data to identify areas of demand significance and subsequent cell deployment should consider the irregular radio propagation environment in urban areas and contextual information such as community structures. Current geometric cluster filters (e.g. Gaussian) are not suitable for urban small cells, where obstacles highly impact cell coverage areas. From a mobile traffic demand perspective, the inference of hidden structures that characterize the spatial temporal fluctuations of subscribers' mobile service consumption is an open problem. In this section, we will review how big data can allow us to understand social structures that can in turn reinforce clustering methods. Our traditional understanding of radio-wave propagation in urban areas is that of a highly spatially embedded energy loss, leading to hotspots that are likely to have small and highly irregular geometries. Yet, what is far less understood is the role of social structures in defining hotspots. Social community structures arise when a group of interactions have a stronger mutual coupling than with other interactions. Increasingly available data, such as mobile call detail records (CDR) can reveal and quantify the strength of community interactions (e.g. detecting community boundaries from CDRs in Belgium [11]). Table 9.1 summarizes the community detection methods.

1. *Social Community Detection Methods* There are a number of ways to detect and define community structures from the underlying data of interactions between different wireless entities. However, given the ill-defined nature of network communities, selecting a suitable detection method is still discretionary to the researcher's needs and intuition, both in terms of computing complexity and data characteristics. Here we present the main general classes of community detection methods currently in use across the literature, referring to their strengths and weaknesses.

(a) *Spectral Graph Clustering* A long known result in graph theory is that clusters of highly inter-connected vertices can be recovered through the study of the *eigenvalue spectrum* of the adjacency or Laplacian matrices of the graph [12]. It can be seen that, for most non-sparse networks, one or more eigenvalues of these matrices appear as outliers in the complex plane with respect to the rest. The aim of spectral graph clustering then is to use the eigenvectors corresponding to such outlying eigenvalues to project the vertices to a metric space where they can be easily clustered using unsupervised techniques such as *k-means*. Unfortunately, there's no guarantee of finding convergence in the clustering step for sparse networks, which is typically the case of real-world networks.

(b) *Modularity* Many community detection methods are based on the optimization of some cluster quality function Q over the space of possible partitions of the network. The most popular quality function was introduced by Girvan and Newman, so-called *modularity* computes, for every cluster in a partition – the difference between the empirical intra-cluster edge density and the corresponding density expected in a chosen null-model [12]. In this sense, modularity is a quality function of the form $Q = \sum_{ij}(A_{ij} - P_{ij})\delta(c_i, c_j)$, where A_{ij} is the empirical adjacency matrix, P_{ij} the expected edge weight in the null model, and $\delta(c_i, c_j) = 1$ when nodes i and j belong to the same community but vanish otherwise. The configuration model, i.e. a degree-preserving randomized version of the network, is usually chosen as a null model. The rationale is that a complete randomization of a network destroys any existent community structure. Therefore, a *modularity-maximizing partition* ensures the presence of meaningful communities. However, such maximization over the partition space has been shown to be **NP-hard**, and consequently a number of more or less successful heuristics have been developed in order to find approximated solutions, for instance, greedy optimization

Table 9.1 Summary of the main methods for detecting communities in complex networks

Detection method	Community indicator
Spectral clustering	Eigenspace closeness
Modularity optimization	Higher link density
Statistical inference	Higher link likelihood
Spin–spin interactions	Low energy domains
Coupled oscillators	Phase synchronization
Markov processes	Random walk confinement

of modularity in the street network of Greater London using the Louvain method [13]. Furthermore, modularity optimization suffers from a systematic bias called **resolution limit**: this effect prevents the method from finding small communities in large networks, effectively making it unreliable for some multi-scale real-world applications.

(c) *Statistical Inference* Statistical inference is a logical approach towards community detection when both the generative network model of the data and the number of communities present in it are supposed to be known. The empirical network is treated as a sample from the generative model with a given set of parameters, and the aim is to infer such parameters using likelihood maximization techniques. The most popular generative model is the stochastic block model (SBM), where the main parameters are the probabilities of finding edges between nodes inside every community and between nodes across every different pair of communities [12]. Despite being exact and conceptually simple, the method necessarily needs to sample across the space of possible partitions of the network to find a guaranteed maximum of likelihood, making it **NP-complete**. Multiple approximations to the problem exist that guarantee high-probability recovery of the community structure. As in the case of modularity, this method suffers from the **resolution limit**.

(d) *Network Dynamics* A more recent approach to the study of networks consist in analyzing the properties of dynamical processes running on the graph [14]. The assumption here is that the interplay between dynamics and structure can yield a better understanding of the mesoscopic organization of the network. In the case of community detection, three main classes of dynamical processes running on the nodes of the network are usually considered: spin–spin interaction models, synchronization of phase oscillators, and diffusion processes. Interestingly, the first two processes can be shown to be dual to the latter, i.e. equivalent to Markov chain modeling the diffusion of random walks across the network. Furthermore, using different conditions one can see that clustering nodes based on diffusion flow characteristics is equivalent to optimizing modularity in one extreme and to performing spectral clustering in the other: studying Markov processes on the graph is therefore one of the most versatile ways of unveiling the community structure of a network. These diffusion processes can be easily interpreted in the context of OSN data, insomuch as they represent the flow of information, interests, and influence across the underlying social interaction network. Finally, random walk diffusion can be used to study the community structure of multiplex networks, a generalization that the aforementioned non-dynamical methods cannot naturally accommodate.

9.2.3 Machine Learning for Clustering in Heterogeneous UDNs

1. *Application to HetNets* Heterogeneous Networks (HetNets) is becoming a popular topic to meet rapidly growing user capacity requirements since it was brought up in 3GPP standards [15]. HetNets apply an innovative topology to replace traditional macrocell network with multi-tier cell networks, including different types of small cells, such as a picocell, femtocell, and so on [16]. As a result, macrocells still play the role for long-range coverage while lower-tier cells provide a better quality of service (QoS) in hotspot areas [17]. Although small cells may have a lower transmission power, their high mobility and small scale ensure that they can be easily installed in hotspot areas. HetNets are also designed to dramatically increase user equipment (UE) capacity because of their

ability of enhancing spectrum reuse efficiency [18]. In general, HetNets innovate the mobile network into a multi-layer, high-mobile, large-scale, and hence complicated system. It is required to merge a wide range of wireless devices, controlling protocols, and features into one network [19]. However, the unique design of HetNets will also bring issues that may limit the development of it. First, the transmission power for various tier networks may have huge differences, thus UE will prefer to stay in a higher-tier network due to its better receiving signal. As a result, the efficiency of small cell networks will be relatively low due to load unbalancing, and the design of HetNets will be of no use [20]. Second, HetNets require a small cell network to have high mobility so that the hotspot can be quickly detected and covered. Also, the allocation of new UEs and their handover among different tiers will demand coordination of both networks [21]. These features will increase the computing complexity of the whole system and may lead to an exponential growth of signal overheads for a large-scale network. Therefore, an efficient UE classification for HetNets is in urgent need, which cannot only re-balance the load but also reduce computing complexity to adapt huge-scale networks; the feature of machine learning (ML) may be capable to fulfill this requirement of addressing increasing dimensional complexity in urban environments.

2. *Self-Organization Network* In order to reduce the rising computing complexity and high maintenance cost due to the large scale of network, introducing AI-based techniques will gradually replace manual operating in both the installing and maintenance phases, and also adapt various environments to enhance system total throughput and cover the range and QoS of users, which can be realized because of the self-organizing network (SON) feature of AI techniques [22]. An SON mainly contains the following three characteristics. First is self-configuration: given a current operating network, any newly allocated node should be able to set the configuration automatically so that it can adapt the network without further manual involvement. Second is self-healing: this feature requires the network to detect system failure and apply a compensation method to fix the issue according to current network situation. Third is self-optimization: the network will automatically find the optimal nodes for deployment and coordination so that the system performance is secured. For wireless network, the coordination among all cells will be optimized to achieve better QoS, coverage, interference mitigation and so on. The optimization algorithm may be different if the optimization objective is different [23].

For HetNets specifically, the allocation of BS, backhaul issues, and the coordination of various tiers of network should be designed before installing the system. However, the small cell network is supposed to be highly mobile to adapt the fast-changing data traffic map, so that the configuration and parameters of a small cell network will continuously and rapidly change[24]. Therefore, self-configuration in real time with minimized manual work is essential in HetNets [25]. Furthermore, the high mobility of a small cell network requires the small cell to be close to UE, and small cells may be allocated in a lamp pole or bus station, which leave them lacking protection [26]. The feature of SON enables HetNets to track mobile devices, monitor the condition of the small cell, provide a short-term healing measurement, and even a long-term network redesign to maintain the system to operate normally without aggressive human intervention. Finally, the continuous changing small cell network will make the default parameters or configuration unsuitable, which demands that HetNets have the ability to optimize the parameters in real time so that the performance of the system can be maintained. This is the function of the third feature of SON self-optimization [27].

3. *Machine Learning Algorithms* Machine learning (ML) is a concept that enables the system to self-organize and self-build by analyzing data. It applies algorithms that allow the system to modify itself without being explicitly programmed [28]. In general, ML can be separated into two categories: unsupervised learning and supervised learning.

(a) *Unsupervised Learning* Generally speaking, the main objective of supervised learning is to establish a model from the training data with labels. Unlike supervised learning, there is no implication of the pattern or correct answer for the input; also the output is not provided in unsupervised learning. The objective of this learning is to find the pattern of the input data set, in which certain data sets will follow more often than other sets [29].

The most common algorithm of unsupervised learning is clustering. Its objective is to cluster the input data sets into several subsets, in which elements may generally follow the same pattern [30]. According to the parameters given by the data set, the algorithm will automatically distribute data elements into groups, and each group has similar elements (this similarity is basically defined by their parameters). Once clustering is finished, we can study the pattern of each group and make decisions according to the result, such as making measures to mitigate negative parameters or finding outliers that are not suitable for this group [31].

The application of clustering is distributed among various fields. Short-range weather predictions can be realized through collecting and clustering daily weather conditions as a data base [32]. For a financial field, clustering can be applied to analyze stock prices and find a potential manipulation factor [33]. Image compression use clustering to group image pixels according to their RGB value (the parameters) so that the pixels with similar colors or similar patterns will be assembled for a better compression [34]. The clustering algorithm is also widely applied in geography. Through clustering referent vectors of the self-organizing map, the model can be used to analyse and measure the colour of the ocean [35].

(b) *Supervised Learning* In order to implement a supervised learning algorithm, the following two requirements should be satisfied: (1) the target parameters or variables and predictor variables are clarified and listed and (2) enough samples implying the correct values for target parameters or variables should be given. The algorithm will learn from these given data to analyse the pattern between input variables and output results. Therefore, a common supervised learning model will follow similar methodology to implement and analyse the algorithm [36]. The initial step is to collect the training data set. This set should contain predefined values of the parameters and output result variables. For example, a list of patients with the name of their illness will be the result variables; meanwhile, each patient is attached with their gender, age, and occupation as predefined values of the parameters. Normally, this training set is an incomplete set because we cannot collect all of the patients' data and, most importantly, we cannot collect the data for new patients because it has not happened at this time point. The algorithm can only generate the model to find the pattern between input and output for given data.

As a result, the following procedure will be the evaluation phase of the generated model. For this phase, we need a test data set. The characteristic of this data set is the same as the training data one. The result variables, however, should be held first for later evaluation. The model generated from the training data set according to the machine learning algorithm is then applied to the test data, and the predicted result variables will be achieved. After that, the predicted result variables will be compared with provided ones of the test data set and the performance of this model can be evaluated. In the end, the model will be modified to mitigate the error rate for the given test data set.

Nevertheless, this modified model may not be satisfied enough to predict unseen data. We need another validation data set and apply the same modified model to it as we did for the test data set. Further modification will be added to this model till the error rate for the validation data set is also mitigated to a minimum, and the final version of this model can be applied to predict unseen data.

(c) *Overfitting versus Underfitting* When we try to evaluate the performance of a supervised machine learning model, we may introduce a terminology overfitting. It is normally used to test the adaptation of a newly built model in statistics [37]. Both overfitting and underfitting will lead to a poor performance of the proposed model.

For the training data set, the error rate of the model will continue to drop as the model keeps learning and modifying itself. However, if the training phase is too long, the complexity of the model will increase and start to take less relevant and even irrelevant information (noise) into account, such as Name in the patient's list when making a diagnosis. In such an occasion, overfitting happens. Although the generality and accuracy of the model increases as more variables are taking effect, the complexity will also inevitably increase, especially for unnecessary computations.

Later for the test data set and validation data set, the error rate of the model will also drop from the beginning of learning as the model is still in the underfitting phases. However, the error rate will start to increase as the complexity of the model increases. The problem is due to the long training phase of the model. The model has taken account of too much irrelevant information from the training data set and the importance of useful information is surpassed during computations. Furthermore, the computing time will also increase because of the high complexity caused by unnecessary information. The model has become the personalized version for the training data set, and is not reliable in predicting the result for the test and validation set [38].

(d) *K-means Clustering Algorithm and K-Nearest Neighbour Algorithm* The K-means clustering algorithm (KCA) is one of the unsupervised learning algorithms and is specifically used to solve the clustering problem. By analyzing the characteristics and comparing the dissimilarity of a group of data, K-means is able to divide them into k groups and iteratively make new decisions as the group grows without further supervision [39]. In other words, this unsupervised algorithm will efficiently reduce computing complexity because it can adapt the fast-changing user traffic map through a self-organizing network (SON), and is thus suitable to solve our first issue.

However, the K-means algorithm will only do clustering according to data characteristics and faithfully reflects the real-time situation. It can be foreseen that most of UEs will still be allocated to a macrocell group due to its high transmission power advantage. As a result, we need to further modify the algorithm to fit our HetNets scenario, and cell range expansion (CRE) may be the key to solve this issue. CRE is a technique of enhanced inter-cell interference coordination (eICIC) and was first mentioned in 3GPP release 10 [40]. It affects the UE association decision by adding a virtual bias to the receiving power from a small cell. Therefore, the situation of load unbalancing can be mitigated by applying CRE.

After clustering the existing UEs through the K-means algorithm and realizing UE offloading in HetNets, the next step will be the prediction of a new UE's association when they enter the system. Classification is a category of the supervised machine learning algorithm, which is a popular research field and has been widely applied in pattern recognition, statistics, medical science, and so on [41]. Like the traditional supervised learning algorithm, classification requires a list of target result variables combined with predefined parameters. The machine learning model will analyse this training data set along with their parameters or predictor variables and manage to find the pattern. If the training data set is provided in clustering type, such as categorized files, this model will be able to classify a new element into its belonged clustering according to its parameters, so that the prediction can be realized. K-nearest neighbour (K-NN) is a popular classification algorithm and can also be considered as an instance-based learning [42]. Before classification, the model should have a set of training data as a reference. During the classification phase, the newly entered element will be compared with the stored reference training data and will be classified into the cluster with the most similar parameter pattern.

After understanding the function of KCA and K-NN, we may combine these two machine learning algorithms to implement an SON system for HetNets UE partition. First, we can apply KCA to existing UEs and obtain the UE clustering pattern and store it. Second, we can apply K-NN and use this stored UE clustering pattern as the reference training set. As a result, any unclassified UE entering this HetNets system will be rapidly assigned to its suitable tier network without further human intervention and the computing complexity will also be reduced. However, K-NN still has some issues before being applyied to the HetNets system. First, K-NN normally considers all the neighbouring elements to be of the same importance when countering the number of votes from the stored training data set [43]. Therefore, neighboring elements from different clusters may contribute the same vote weight. This is clearly not the situation for HetNets, especially for classification between the macrocell tier and the small cell tier, due to unbalanced transmission power. Second, once the new element is assigned to the cluster according to its parameter pattern, no further definition of this element is added. As a result, the effect of this new element with respect to the cluster pattern may be neglected. Nevertheless, if the number of new elements is large enough, the effects may accumulate and change the stored cluster pattern [41]. This issue is even more severe for our scenario, because we have discussed that the network is designed to adapt a fast-changing traffic map. As a result, these two issues should be addressed if we want to apply it to UE classification in HetNets.

(e) *K-means Clustering Algorithm (KCA)* The major contribution of current research is to design modified K-means clustering algorithms to solve the offloading problem for HetNets, which can be called a user-based K-means clustering algorithm (UBKCA). This class of algorithm involves the HetNets background knowledge and applies the eICIC technique. In particular, a center user group set is established to reduce computing complexity and CRE bias is introduced to enhance the performance of the algorithm in the offloading factor. After that, in order to realize classification and prediction for new elements entering HetNets, two methods have been applied for comparison. The first method is to obtain the decision boundary for current clusters through curve-fitting. The second method is to apply the K-NN classification algorithm for obtained clusters. The protocol for future self-optimization is also established so that the system can realize self-organization in UE partition and adapt the change of traffic map automatically.

The K-means clustering algorithm is already widely used to solve partition problems in many fields [24]. It is considered to be capable in clustering big data because of its fast speed, automation, and high adaption. Therefore, it is prudent to first apply k-means clustering to decide user association in HetNets, which will solve the first issue. In general, KCA can be formulated as a mathematical computing problem P [44]:

$$\begin{aligned} \text{minimize } P(M,O) = & \sum_{i=1}^{k}\sum_{j=1}^{n} M_{i,j} d(R_j, O_i) \\ \text{subject to } & \sum_{i=1}^{k} M_{i,j} = 1, \quad 1 \le i \le n, \\ & M_{i,j} \in \{0,1\}, \quad 1 \le j \le k. \end{aligned} \tag{9.2}$$

We need to divide a data set $R = \{R_1, R_2, ...R_n\}$ into k clusters, and $O = \{O_1, O_2, ...O_k\}$ is the set of all cluster centers. Here, M is an $n \times k$ matrix which presents all element decisions and $M_{i,j}$ is either 0 or 1. The KCA will then be applied so that the partition will minimize the aforementioned equation. For implementing the k-means clustering algorithm, we need to analyze data using the following steps (see Figure 9.5):

- Find the suitable partition number by analyzing data, which means the data will be divided into k groups.
- Randomly pick k elements and assume them to be the center of each group.
- Allocate all the remaining $n - k$ elements to their nearest center to form clusters and then calculate the new center for each cluster.
- Repeat the steps till the system converges.

In order to reduce computing complexity for the algorithm, we use the Euclidean distance method to calculate the total distance between the elements and centers in Eq. (9.1). The equation is as follows, where $p_1, p_2, ...p_n$ represents all the elements within the data set and $q_1, q_2, ..., q_n$ represents the corresponding clustering centers [45]:

$$d(p,q) = \sqrt{(p_1 - q_1)^2 + (p_2 - q_2)^2 + \cdots + (p_n - q_n)^2}. \tag{9.3}$$

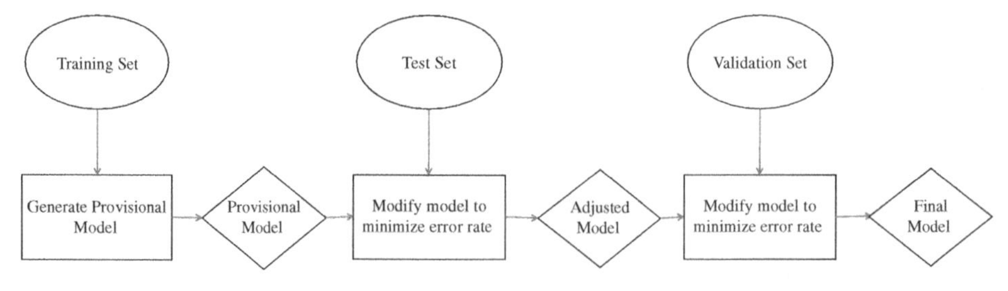

Figure 9.5 Lexicon and machine learning based NLP algorithm with filtering, tokenization, and opinion score classification.

9.3 Unstructured Data Analytics for Quality-of-Experience Mapping

An important component in modern business is the mining and analysis of data obtained from the customer base. In the past 5 years, the proliferation of online social networks (OSN) platforms (i.e. Twitter, Facebook, Foursquare, Wechat) has led to the development of new unstructured data analytic tools suited to social media data (i.e. natural language processing). In the context of mobile network operators, their core future business is to accommodate the rapid growth in mobile data demand. However, any overall network capacity growth has to take into account blackspots. In the United Kingdom alone, 17 million customers suffer from blackspots at home and up to 30% of customers experience service delivery problems in certain economic areas (UK British Infrastructure Group Report 2016). Investment in both small cell deployments and adaptive self-organizing networks (SONs) has been widely recognized as a cornerstone of current 4G LTE-Advanced and future 5G cellular network architecture. However, current small cells are often deployed without high-resolution traffic and QoE knowledge, leading to potentially poor profit returns. Existing practices of using long-term statistical traffic data from macro base stations (BSs) have poor spatial resolution and lack context (i.e. cannot differentiate between indoor and outdoor traffic).

Large volumes of real-time OSN data provide network operators with an opportunity to analyze and combine existing cell planning and operation practices. In recent years, a number of enterprises have already been using OSN data to act as a proxy for traffic demand and guide small cell planning. Recent research by ourselves used geo-tagged Tweets to show that Twitter data, despite its low data size (kB per Tweet), is an accurate proxy for both uplink and downlink 3G data demand (which explains the over 71–79% of variations in traffic) [8]. This supports other social science research, which shows that human activity and density is highly correlated with social media data intensity [9]. Other related research has examined the opportunity to use geo-tagged OSN data to add spatial analysis on top of existing temporal and semantic studies of QoE complaints [6] and to detect core network problems [7]. Yet, most current research in wireless network engineering have not exploited the unstructured information in Tweets, which have the potential to convert text language to uncover the sentiment toward specific topic areas (i.e. signal quality) and understand the perceived quality-of-experience (QoE). This section will discuss the potential to leverage a large-scale analysis with natural language processing (NLP) techniques to address the usefulness of OSN data.

Current identification of QoE blackspots requires customer service to receive complaints through voice calls and other social media platforms. These are typically human processed and fed to the radio and network engineering teams. However, this only captures direct complaints to the operator and delays remedies. Automated data mining of customer sentiment (i.e. complaints) is increasingly used to obtain real-time understanding of QoE for different service issues. The simplest customer sentiment analysis uses sentiment lexicon and dictionary approaches. These typically use n-gram keywords to estimate the sentiment of a message toward a target specific topic. We will review both lexicon-based and more sophisticated machine learning (ML) methods (i.e. Naive Bayes, a support vector machine with different Gaussian and non-Gaussian filters) to understand QoE.

9.3.1 Topic Identification

1. *Probabilistic Topic Modeling* The most widely used technique for detecting topics of interest is probabilistic topic modeling. This is an unsupervised method which assumes that a collection of data are permeated by hidden underlying topics, each of which is uncovered as a set of *word distributions* that define the topics. There are numerous variants on probabilistic topic modeling to suit different data lengths, but the core of most is *latent Dirichlet allocation (LDA)* for larger data sets, where a prior Dirichlet distribution of topics and an iterative generative process is used to discover the conditional distribution of topics given the data and the conditional distribution of words to topics. The iterative step involves computing the above two probabilities given the previous assignment to topics and then reassign each word to a topic using the probabilities, assuming that each word is being generated by its topic. Recent work has shown that LDA can reveal similar topics representing a set of documents as traditional social science survey methodologies [46].

The goal in opinion mining, including sentiment analysis and stance detection, is to discover opinions about topics or issues mentioned in documents or conversations as a two- or three-way supervised classification problem. While originally sentiment analysis was developed for product reviews and involved classification at the document level (e.g. one sentiment per review or Tweet), in recent years it has moved toward aspect based sentiment and target based sentiment analysis where the objective is to obtain the sentiment toward specific entities (e.g. politicians, topics) or aspects (e.g. the safety of a neighborhood), resulting in multiple types of sentiment per document (e.g. per Tweet) [47, 48].

2. *Classification Techniques* Most systems obtain features from the text either as sequences of words (e.g. n-gram probabilistic $(n-1)$th order Markov model) or more recently as combinations of word embeddings, where each word encountered in the vocabulary is represented by a *word vector* obtained through iterative methods operating on large word co-occurrence matrices. While for some time support vector machines (SVMs) have been used as the classifier of choice for sentiment analysis, in recent years these have been replaced by neural network approaches and in particular convolutional neural networks (CNN) and recurrent neural networks, depending on the particular problem. CNNs require no domain expertise and automatically create increasing levels of abstraction to distort the input data space into linearly separable data (see Figure 9.1b). CNNs are particularly suited to the synthesis of sentence representations from individual words while recurrent neural networks and most specifically long short-term memory networks (LSTMs) are usually preferred for sequential classification tasks. In the classification process, the strength of topic interdependencies can also be uncovered. For example, a data set may be classified to belong to a number of topic layers, with a vector L indicating the classification *strength* or the *accuracy* of classification for each topic layer. A threshold is then used to decide which layers are selected as being inter-related.

Recent work has sought to incorporate a social network structure as embeddings, which are used to help create notions of linguistic homophily that can guide the focus of neural network approaches in sentiment analysis [49]. Another line of relevant work in NLP involves inferring characteristics of users from their posts on social media. Such characteristics include mood, socioeconomic status [50], gender, and age [51]. The problem is usually tackled as a supervised classification task with authors in this area using linear and logistic regression models as well as non-linear regression and kernel methods such as Gaussian processes. Challenges involve ethics concerns on appropriateness of data labeling and biases in data sampling.

9.3.2 Sentiment

A general flow model for the NLP algorithm and a lexicon based example is given in Figure 9.6. We first filter all Tweets and identify ones that are relevant to the topic domain (QoE for cellular networks). After which, we classify each Tweet complaint with sentiment labels to infer polarity of the filtered Tweets. To assign each Tweet with a sentiment score we first apply tokenization prefiltering to remove language noise and transform all text to a common lower case format with no punctuation. We then extract single word unigram features independently to determine the orientation of the Tweet.

1. *Opinion Lexicon* First, a dictionary of words is obtained through a bootstrapping process using WordNet, which offers a semantic relations hip among words [52]. It is built utilizing the adjectives synonym set and the antonym set available in WordNet to predict the semantic orientations of text used in a Tweet sentence. This method enables us to find the average semantic orientation of Tweets in simplistic ways that are independent of the context of the text under analysis. This technique was successfully implemented in previous research to analyse sentiment [53], but has not been applied to urban contexts to understand the underlying sources of happiness. The algorithm then calculates the score of each Tweet by simply subtracting the number of occurrences of negative words from the number of positive occurrences for each Tweet. A threshold is

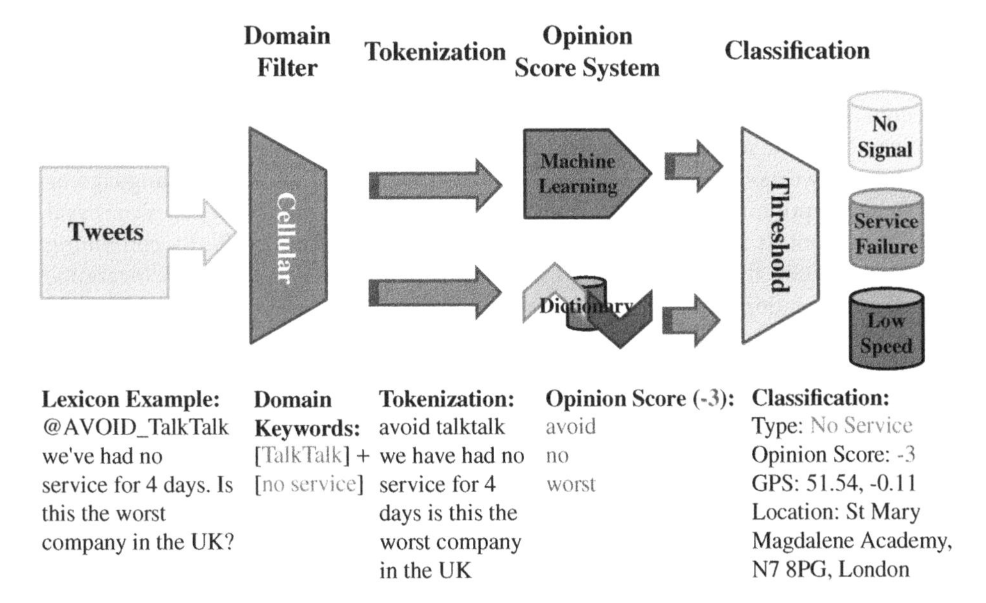

Figure 9.6 Lexicon and machine learning based NLP algorithm with filtering, tokenization, and opinion score classification.

usually implemented before classification in order to reduce the number of false positive opinion results.

2. *Machine Learning* We consider two techniques: SVM with a Gaussian kernel and Naive Bayes (NB). Both techniques depend on the volume of training data, which is manually annotated by the authors, of which 80 representative example texts were available from existing complaint data on Twitter. In considering the accuracy performance of machine learning approaches, the $F1$ score is often used, which considers both the precision and the recall statistics. Precision p is the number of correct positive results divided by the number of all positive results, and r is the number of correct positive results divided by the number of positive results that should have been returned. The $F1$ score is the harmonic mean of precision and recall:

$$F1 = 2\frac{pr}{p+r}. \tag{9.4}$$

The accuracy results in Figure 9.7a show that SVM with a Gaussian kernel has a similar $F1$ score performance to Naive Bayes, superseding it for low sample sizes. It is worth noting that the training data size for more complex complaints is usually orders of magnitude higher, but for simpler Tweets with a limited number of keywords, the sample required is quite small.

Commonly occurring **false positives** are removed from the data (i.e. poor train signal). It is worth noting that filtering by operator names or the operator's official Hashtag is not useful in most cases, as most complaints do not use the associated name or Hashtag of the operator. The results in Figure 9.7b compare the threshold size and the false positive Tweet numbers for Lexicon and ML techniques using the full training data set. The results clearly demonstrate the superior performance of ML techniques over Lexicon based techniques, especially for SVM.

Location Context and Validation Location context is useful for understanding how people use data in different social and physical environments. Location context attributes can be collected by combining several data sets, but one innovative way pioneered in recent years is to use existing knowledge to train machine learning classifiers that mine social media and mobile network data. Existing studies have shown that NLP of Tweets can reveal broad geographical areas from either the language or the specific mentioning of locations. Further studies have shown that it can also reveal the location context and social value of co-located customers [37–39]. By combining social media/OSN data with national databases on business registration locations, and business location premise shape files [13], we can now cluster Tweets to specific businesses categories.

9.3.3 Data-Aware Wireless Network (DAWN)

Deployment optimization traditionally considers several traffic-independent factors, such as propagation loss and interference. Optimization, therefore, naturally assumes either a uniform user demand distribution or uses long-term statistics (i.e., census data) to spatially weight the optimization. Moving beyond stationary radio planning is important as urban areas become more dynamic, more complex (tourists, commuters, changing urban landscape), and demand becomes more stochastic. As such, we propose a new concept of data-aware wireless network (DAWN) deployment to adapt the

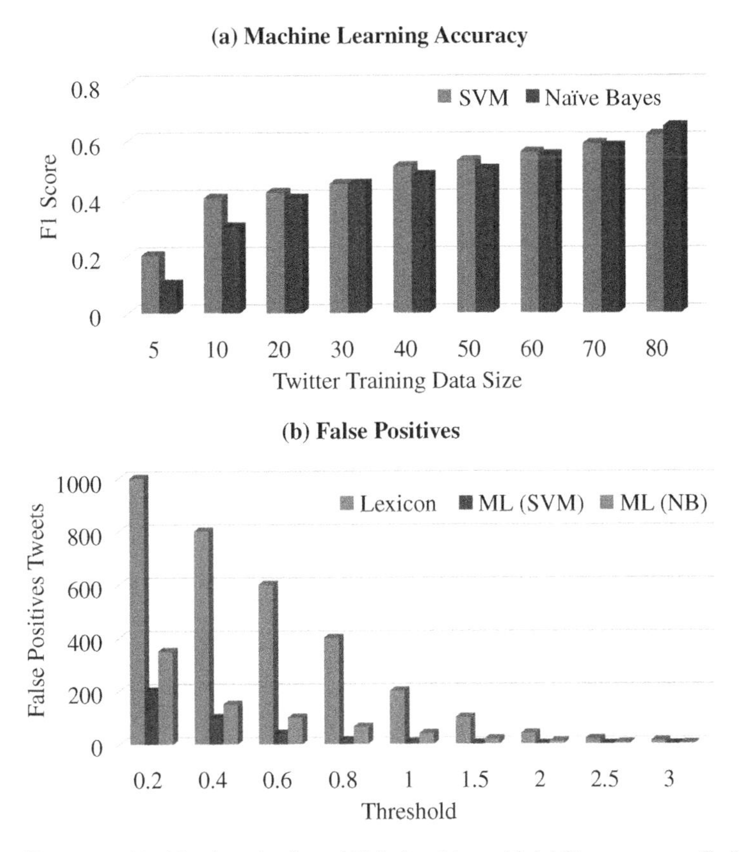

Figure 9.7 Machine learning based NLP algorithm with (a) $F1$ score versus Twitter training data size and (b) false positive versus threshold.

process to match long-term demand patterns accurately by integrating understanding of both traffic hotspots and service blackspots. Validation of the accuracy of these combined techniques can be done using existing signal databases from Apps, such as OpenSignal, to validate the blackspot findings. More sophisticated ground based signal testing and survey of residents can be used to further validate the suspected areas of high traffic and poor QoE and calibrate the hotspot and QoE mappings for different geographic regions. Such triple verification using combined big data, ground surveys, and radio testing is important when introducing new methodologies into existing practice. This iterative cycle of product design, prototyping, and testing is essential to the reliability and refinement of the outputs.

1. *Case Study* By using the previously mentioned NLP techniques, we were able to show that the majority of customer complaints were a result of poor signal (86%) before 4G rollout in 2012 (see Figure 9.8a). By checking against keywords related to indoor areas, it was found that many (36%) of these cases are persistent. After 4G rollout in 2016 (see Figure 9.8b), the complaints were mainly (66%) of service failures that span from a few hours to several days. This may indicate a problem related to operating the

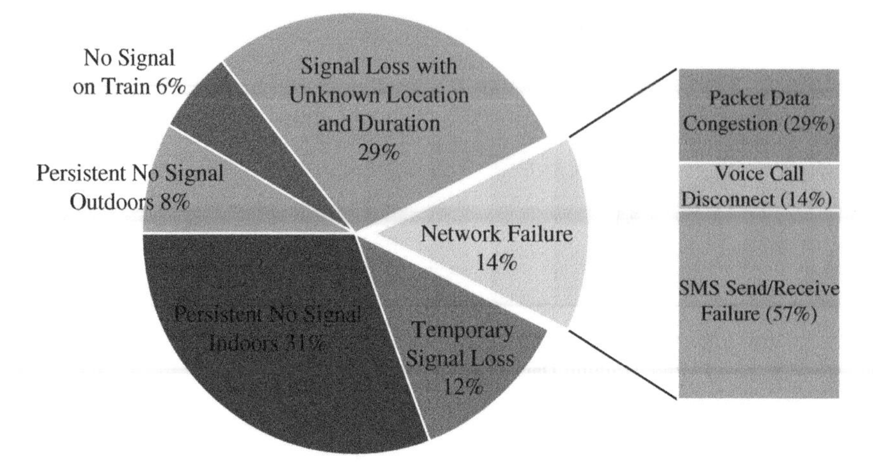

(a) 2012 (Pre-4G) QoE Problems Classification in Greater London

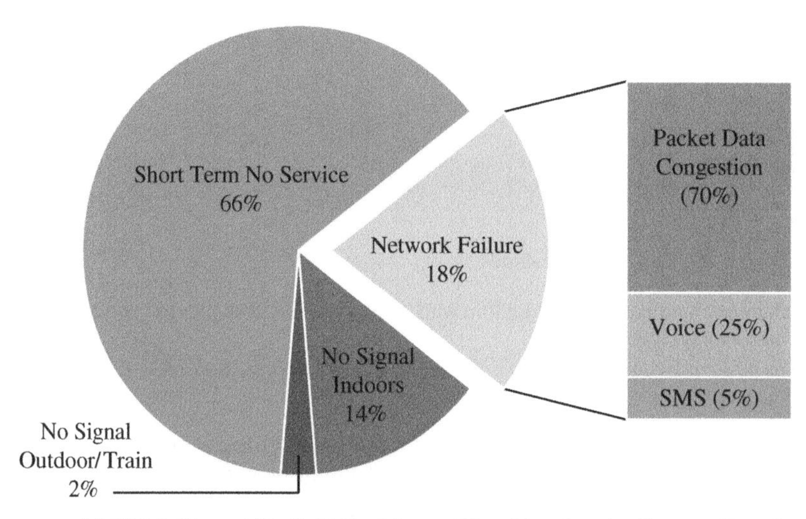

(b) 2016 (Post-4G) QoE Problems Classification in Greater London

Figure 9.8 Text analysis of Tweets: results show the customer complaints for pre-4G rollout in 2012 and post-4G rollout in 2016.

4G network. Signal complaints (especially in indoor and on trains) were lower both proportionally and in absolute numbers.

Several QoE complaint areas were identified. These are plotted as star symbols in Figure 9.9a and b. Figure 9.9c shows an example area centred on London Bridge Station. There are numerous QoE complaints inside the station (black star), despite several macro-BSs nearby, yielding an excellent outdoor signal. Therefore, this presents small cell operators with an opportunity to both target the persistent poor QoE blackspot inside the station (black circular zone), as well as several Twitter hotspots near the station (red circular zones). The authors also use OpenSignal to reveal poor

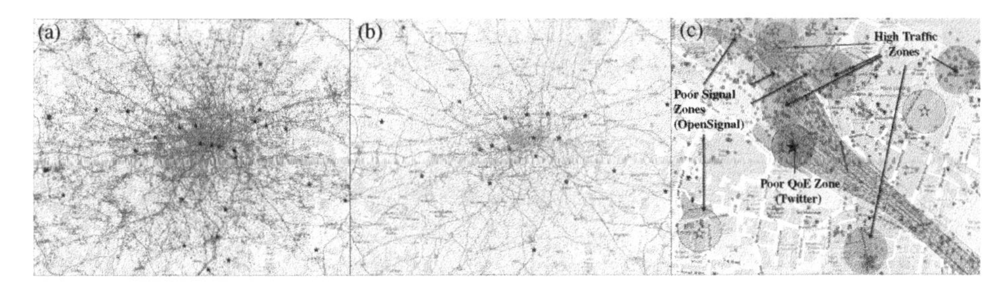

Figure 9.9 (a) Identified poor QoE blackspots (pre-4G rollout in 2012), (b) identified poor QoE blackspots (post-4G rollout in 2016), (c) case study of London Bridge showing both high traffic hotspots and poor signal (OpenSignal data) and poor QoE (Twitter data) blackspots.

signal measurement reports in this area with a spatial resolution of 60 m. These are represented by blue stars with a blue circular zone. The combined picture in Figure 9.9c shows that overlapping circular zones represent an opportunity to address one or more of the following: (i) high traffic demand (inferred from Twitter intensity); (ii) poor signal strength (measurement reports from OpenSignal); and (iii) poor QoE (data mined from Twitter).

2. *Real-time OSN Data Analysis and Integration* In terms of opportunities, OSN data can not only guide small cell deployment but can also be integrated into software-defined network (SDN) platforms to enable SON capabilities and make time- and cost-critical traffic decisions [10]. Network operators can perform large-scale data analytics on both their internal low-level CDR data, as well as high-level OSN data to feedback decisions to the control plane functions that manage the network. In order to integrate the structured (see [8]) and unstructured data analysis into real-time SON algorithms on SDN platforms, we present a real-time processing model.

The real-time model is built based on the common distributed processing framework Spark [54] to leverage dynamics of a Tweets load. As the design shows in Figure 9.10, the online model first opens a network port to absorb Tweet streams. Due to the fact that the cellular network performance various from BS to BS, tweets are then handled separately based on their location area. Thus, a location distributor is designed to separate Tweets. Though time based window size is more commonly used in stream mining, the authors use the number of tweets captured as the measure of window size, due to the fact that NLP based QoE mining requires a relatively high number of Tweets to guarantee the accuracy and the Tweets frequency varies over different areas. Then a proprietary data persistence layer after the NLP processing is implemented, so the NLP mining thread is called on demand only after sufficient OSN data are collected. Finally, the result QoE found is updated for the network operators' reference.

In detail, the parameters considered in the online model are BS coverage range, windows size, and sliding interval. The BS coverage range can be altered based on a range of optimization parameters such as minimum service requirement, energy saving, and load balance. The windows size is the number of Tweets processed in one batch: i.e. the authors found in their earlier work [8] that the number of Tweets is no longer correlated with the traffic after approximately 16 200 Tweets, so this should determine the window size. The sliding interval is how often the evaluation process performs. From our calculation of existing Twitter data, we found that the QoE event detected usually lasts for

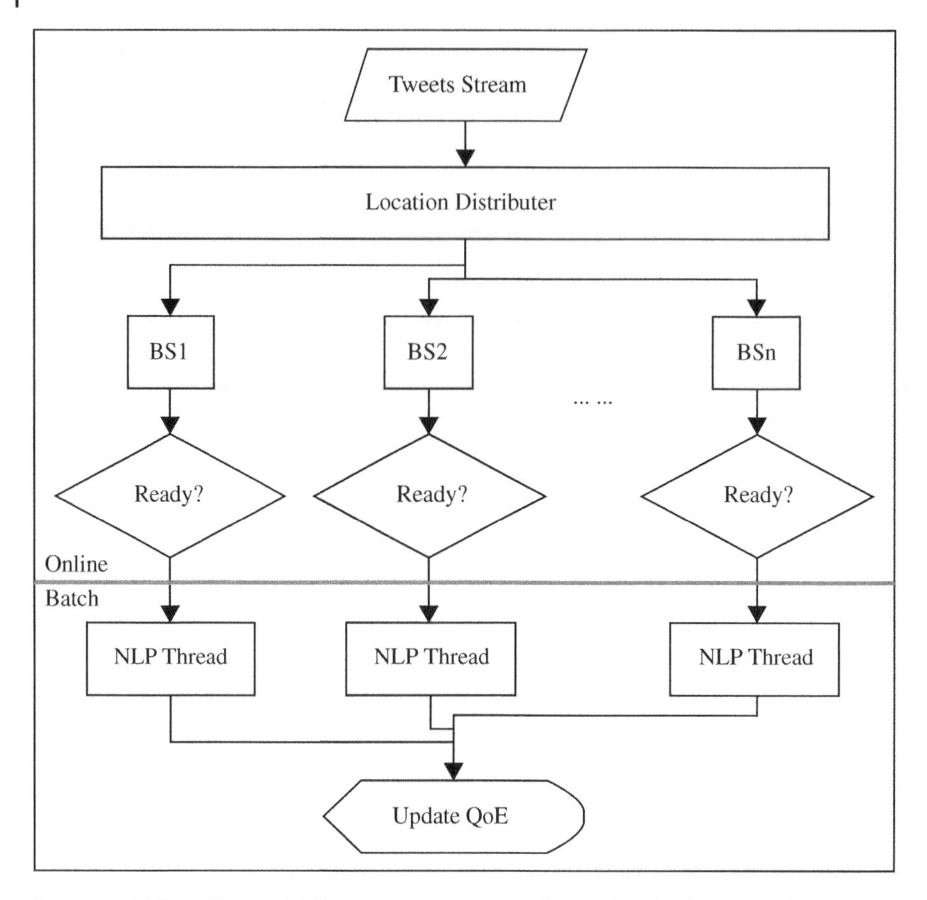

Figure 9.10 The online model that processes Tweets and gives QoE feedback in real time.

18 000 Tweets. Thus the authors choose half the period, every 9000 Tweets to update the QoE. However, in practice, the value should be set according to available computation resources and network operators' preferences.

9.4 Conclusion

As we accelerate into the twenty-first century, we are seeing increased human digital activity through the compounded effects of urbanization and the proliferation of smart devices (see Figure 9.1). Data demand has risen at a super-linear rate, with a 10× growth from 2011 to 2017. Beyond these statistical trends, networks are also experiencing increased data demand complexity from people and machines. Enabling understanding and exploitation of these complexities is important. This is especially so for the deployment of ultra-dense network (UDN) nodes, which include small cells, relays, distributed antennas/reflectors, and other heterogeneous network elements deployed in a spatially dense formation. Besides the wireless notions of cell planning, economically efficient UDN deployment requires precise knowledge of traffic patterns and usage contexts, as well as the shortfalls in existing deployment to reduce outage and customer complaints. For the first time, the analysis of big data sets gives us an opportunity to do this.

In this chapter, we review appropriate research on heterogeneous big data analytics to optimize the deployment of wireless networks. In the first half of the chapter, we examined how social media data can quantify traffic patterns across various temporal and spatial scales. The methodological aspects of this work includes structured data analytics, such as clustering and social community detection. In the second half of the chapter, we examined how social media data can improve our understanding of sentiment toward different service topics and reinforce our understanding of poor QoE reasons and locations for targeted improvement. The methodological aspects of this work includes unstructured data analytics, such as natural language processing (NLP) and surveys.

Bibliography

1 "Ultra dense network (UDN) white paper," Nokia White Paper, 2016.
2 "Cisco visual networking index: global mobile data traffic forecast update, 20162021 white paper," Cisco, 2017.
3 "World urbanization prospects – the united nations," UN Habitat, 2014.
4 "The mobile economy," GSMA, 2014.
5 K. Kotobi, P. Mainwaring, C. Tucker, and S. Bilen, "Data-throughput enhancement ssing data mining-informed cognitive radio," *Electronics*, vol. 4, no. 2, pp. 221–238, 2015.
6 T. Qiu, J. Feng, Z. Ge, J. Wang, J. Xu, and J. Yates, "Listen to me if you can: tracking user experience of mobile network on social media," in *ACM Internet Measurement Conference (IMC)*, 2010.
7 K. Takeshita, M. Yokota, and K. Nishimatsu, "Early network failure detection system by analyzing Twitter data," in *IEEE International Symposium on Integrated Network Management (IM)*, Ottawa, ON, 2015, pp. 279–286.
8 B. Yang, W. Guo, B. Chen, G. Yang, and J. Zhang, "Estimating mobile traffic demand using twitter," *IEEE Wireless Communications Letters*, vol. 4, no. 4, pp. 380–383, Aug. 2016.
9 F. Botta, H.S. Moat, and T. Preis, "Quantifying crowd size with mobile phone and Twitter data," *Royal Society Open Science*, vol. 2, May 2015.
10 A. Sathiaseelan, M.S. Seddiki, S. Stoyanov, and D. Trossen, "Social SDN: online social networks integration in wireless network provisioning," in *ACM Proceedings of SIGCOMM*, Illinois, USA , 2014, pp. 375–376.
11 R. Lambiotte, V. Blondel, C. de Kerchove, E. Huens, C. Prieur, Z. Smoreda, and P.V. Dooren, "Geographical dispersal of mobile communication networks," *Physica A*, vol. 387, no. 21, pp. 5317–5325, 2008.
12 M.E.J. Newman, "Spectral methods for community detection and graph partitioning," *Phys. Rev. E*, vol. 88, Oct. 2013.
13 S. Law, "Defining Street-based Local Area and measuring its effect on house price using a hedonic price approach: the case study of metropolitan London," *Cities*, vol. 60, pp. 166–179, 2017.

14 J.-C. Delvenne, S.N. Yaliraki, and M. Barahona, "Stability of graph communities across time scales," in *Proceedings of the National Academy of Sciences (PNAS)*, vol. 107, no. 29, pp. 12755–12760, 2010.
15 S. Yeh, S. Talwar, G. Wu, N. Himayat, and K. Johnsson, "Capacity and coverage enhancement in heterogeneous networks," *IEEE Wireless Communications*, vol. 18, no. 3, pp. 32–38, June 2011.
16 H. Elsawy, E. Hossain, and D. Kim, "HetNets with cognitive small cells: user offloading and distributed channel access techniques," *IEEE Communications Magazine*, vol. 51, no. 6, pp. 28–36, June 2013.
17 K.I. Pedersen, Y. Wang, B. Soret, and F. Frederiksen, "eICIC functionality and performance for LTE HetNet co-channel deployments," in *IEEE Vehicular Technology Conference (VTC Fall)*, Quebec City, QC, 2012, pp. 1–5.
18 D. Lopez-Perez, I. Guvenc, G. de la Roche, M. Kountouris, T. Quek, and J. Zhang, "Enhanced intercell interference coordination challenges in heterogeneous networks," *IEEE Wireless Communications*, vol. 18, no. 3, pp.22–30, June 2011.
19 X. Wang, X. Li, and V.C.M. Leung, "Artificial intelligence-based techniques for emerging heterogeneous network: state of the arts, opportunities, and challenges," *IEEE Access*, vol. 3, pp. 1379–1391, 2015.
20 J. Oh and Y. Han, "Cell selection for range expansion with almost blank subframe in heterogeneous networks," in *Proc. IEEE Intl Symp. Personal Indoor and Mobile Radio Communications (PIMRC12)*, Sydney, NSW, 2012, pp. 653–657.
21 X. Gelabert, G. Zhou, and P. Legg, "Mobility performance and suitability of macro cell power-off in LTE dense small cell HetNets," in *IEEE 18th International Workshop on Computer Aided Modeling and Design of Communication Links and Networks (CAMAD)*,Berlin, 2013, pp. 99–103.
22 O.G. Aliu, A. Imran, M.A. Imran, and B. Evans, "A survey of self organisation in future cellular network," *IEEE Commun. Surv. Tuts.*, vol. 15, no. 1, pp. 336–361, 2013.
23 D. Lopez-Perez, I. Guvenc, G.D.L. Roche, M. Kountouris, T.Q. Quek, and J. Zhang, "Enhanced intercell interference coordination challenges in heterogeneous networks," *IEEE Wireless Communications*, vol. 18, no. 3, pp. 22–30, June 2011.
24 C. Lin, L.J. Greenstein, and R.D. Gitlin, "A microcell/macrocell cellular architecture for low- and high-mobility wireless users," *IEEE Journal on Selected Areas of Communications (JSAC)*, vol. 11, no. 6, pp. 885–891, Aug. 1993.
25 M. Peng, D. Liang, Y. Wei, J. Li, and H.-H. Chen, "Self-conguration and self-optimization in LTE-advanced heterogeneous networks," *IEEE Communications Magazine*, vol. 51, no. 5, pp. 36–45, May 2013.
26 J. Hoydis, M. Kobayashi, and M. Debbah, "Green small-cell networks," *IEEE Veh. Technol. Mag.*, vol. 6, no. 1, pp. 37–43, Mar. 2011.
27 H. Hu, J. Zhang, X. Zheng, Y. Yang, and P. Wu, "Self-configuration and self-optimization for LTE networks," *IEEE Communications Magazine*, vol. 48, no. 2, pp. 94–100, Feb . 2010.
28 F. Sebastiani, "Machine learning in automated text categorization," *ACM Computing Surveys*, vol. 34, no. 1, Mar. 2002.
29 S. Goldman and Y. Zhou, "Enhancing supervised learning with unlabeled data," *International Machine Learning Workshop*, 2000.

30 E. Chandra and V.P. Anuradha, “A survey on clustering algorithms for data in spatial database management systems,” *International Journal of Computer Applications*, vol. 24, 2011.

31 V.J. Hodge and J. Austin, “A survey of outlier detection methodologies,” *Artificial Intelligence Review*, vol. 22, no. 2, pp. 85–126, Oct. 2004.

32 J.M. Gutirrez, R. Cano, A.S. Cofio, and M.A. Rodrguez, “Clustering methods for statistical down-scaling in short-range weather forecast,” *Monthly Weather Review*, vol. 132, 2004.

33 S. Ni, N. Pearson, and D. Poteshman, “Stock price clustering on option expiration dates,” *Journal of Financial Economics*, vol. 78, 2005.

34 T. Kanungo, D.M. Mount, N.S. Netanyahu, C. Piatko, R. Silverman, and A.Y. Wu, “An efficient$k$-means clustering algorithm: analysis and implementation,” *IEEE Trans. Patt. Anal. Mach. Intell.*, vol. 24, no. 7, pp. 881–892, July 2002.

35 M. Yacoub, F. Badran, and S. Thiria, “A topological hierarchical clustering: application to ocean color classification,” *Artificial Neural Networks—ICANN*, 2001, pp. 492–499, 2001.

36 D.T. Larose, “k-Nearest neighbor algorithm,” *Discovering Knowledge in Data: An Introduction to Data Mining*, 2005.

37 W. van der Aalst, V. Rubin, H. Verbeek, B. van Dongen, E. Kindler, and C. Gunther, “Process mining: a two-step approach to balance between underfitting and overfitting,” *Software and Systems Modeling*, vol. 9, 2010.

38 A.A. Freitas, “Understanding the crucial differences between classification and discovery of association rules. *A position paper,” ACM SIGKDD Explorations*, vol. 2, no. 1, pp. 65–69, 2000.

39 K. Wagstaff, C. Cardie, S. Rogers, and S. Schroedl, “Constrained k−means clustering with background knowledge,” in *Proc. 18th International Conference on Machine Learning*, 2001, pp. 577–584.

40 S. Deb, P. Monogioudis, J. Miernik, and J. Seymour, “Algorithms for enhanced inter-cell interference coordination (eICIC) in LTE HetNets,” *IEEE/ACM Transactions on Networking*, vol. 22, no. 1, pp. 137–150, Feb. 2014.

41 J.M. Keller, M.R. Gray, and J.A. Givens, “A fuzzy k-nearest neighbor algorithm,” *IEEE Trans. Syst., Man, Cybern.*, vol. 15, no. 4, pp. 580–585, July–Aug. 1985.

42 W. Cheng and E. Hullermeier, “Combining instance-based learning and logistic regression for multilabel classification,” *Proc. ECML/PKDD*, vol. 76, no. 2-3, pp. 211–225, 2009.

43 I. Rahal and W. Perrizo, “An optimized approach for KNN text categorization using P-trees,” in *ACM Symposium on Applied Computing*, 2004, pp. 613–617.

44 Z. Huang, “Extensions to the k−means algorithm for clustering large data sets with categorical values,” *Data Mining Knowledge Discovery*, vol. 2, no. 3, pp. 283–304, 1998.

45 K.A.A. Nazeer and M.P. Sebastian, “Improving the accuracy and efficiency of the k-means clustering algorithm,” in *International Conference on Data Mining and Knowledge Engineering (ICDMKE), Proceedings of the World Congress on Engineering (WCE-2009)*, vol. 1, July 2009.

46 E. Baumer, D. Mimno, S. Guha, E. Quan, and G. Gay, “Comparing grounded theory and topic modeling: extreme divergence or unlikely convergence?” *Journal of the*

Association for Information Science and Technology, vol. 68, no. 6, pp. 1397–1410, 2017.

47 B. Wang, M. Liakata, A. Zubiaga, and R. Procter, "Tdparse: multi-target-specific sentiment recognition on twitter," in *Proceedings of the 15th Conference of the European Chapter of the Association for Computational Linguistics*, Valencia, Spain: Association for Computational Linguistics, April 2017, pp. 483–493.

48 S. Marzieh, G. Bouchard, M. Liakata, and S. Riedel, "Sentihood: targeted aspect based sentiment analysis dataset for urban neighbourhoods," in *COLING 2016, 26th International Conference on Computational Linguistics, Proceedings of the Conference*, 2016, pp. 1546–1556.

49 Y. Yang and J. Eisenstein, "Overcoming language variation in sentiment analysis with social attention," *Transactions of the Association for Computational Linguistics (TACL)*, vol. 5, 2017.

50 V. Lampos, N. Aletras, J. Geyti, B. Zou, and I. Cox, *Inferring the Socioeconomic Status of Social Media Users Based on Behaviour and Language*, 2016.

51 D. Nguyen, R. Gravel, R. Trieschnigg, and T. Meder, "How old do you think I am? A study of language and age in Twitter," 2013.

52 M. Hu, B. Liu, and S.M. Street, "Mining and summarizing customer reviews," in *ACM Conference on Knowledge Discovery and Data Mining (SIGKDD)*, 2004, pp. 168–177.

53 J. Fiaidhi, O. Mohammed, S. Mohammed, S. Fong, and T.H. Kim, "Opinion mining over Twitterspace: classifying tweets programmatically using the R approach," in *ACM Int. Conf. Digit. Inf. Manag. (ICDIM)*, Macau, 2012, pp. 313–319.

54 A. Spark, "Lightning-fast cluster computing," 2013.

10

Physical Layer Security for Ultra-dense Networks under Unreliable Backhaul Connection

Huy T. Nguyen[1], *Nam-Phong Nguyen*[2], *Trung Q. Duong*[2] *and Won-Joo Hwang*[1]

[1] *Department of Information and Communication Engineering, Inje University, South Korea*
[2] *School of Electronics, Electrical Engineering and Computer Science, Queen's University Belfast, UK*

In the explosion of handsets and associated services, the demand for transferring information and communication tends to grow in an uncontrollable manner. To meet the requirements of reliable communications and high data rate systems, ultra-dense networks (UDNs) are considered as a solution by deploying multiple small cell base stations (SBSs) in the traditional macrocell base station (MBS) [1], the coverage and throughput of the whole system can be improved significantly. In such network infrastructures, there are important links, namely backhaul, which connect the core network with multiple SBSs [2]. In the traditional cellular systems, wired backhaul is mostly used since it has high reliability communication. However, deploying wired backhaul in the large-scale systems would lead to an increase in the cost of maintenance services. As a result, wireless backhaul has been focused as an alternative solution to offer cost-efficiency and flexibility [3]. Even though wireless backhaul has shown many advantages compared to wired backhaul, its reliability could not fully be achieved due to the nature of broadcasting and propagation effects [4]. The investigation of backhaul reliability impacts has been widely discussed [5–7]. In particular, the authors in [5] analyzed the performance of the coordinated multi-point system under the existing multiple wireless backhauls. The asymptotic performance of the cooperative system with two distinct diversity combining schemes, i.e. selection combining and maximum ratio transmission, has been derived to show the importance of backhaul reliability impacts [6]. For the purpose of extending the investigation of unreliable backhaul links, the authors in [7] have proposed the performance analytical framework in the context of cognitive heterogeneous networks.

Since those studies have shown the important impacts of backhaul reliability level on the performance limitation, the theoretical analysis in various environments could bring fundamental knowledge on the wireless system design. In recent years, the transmission of confidential information among nodes has gained security issues in the presence of malicious eavesdroppers [8]. One of the security methods is encryption, which encodes the message at the transmitter and then performs the decoding process at the receiver. However, those schemes usually perform high-complexity algorithms and could not directly handle the eavesdropping attacks. Therefore, physical layer security (PLS) has emerged as an information-theoretical approach in order to make the confidential

Ultra-dense Networks for 5G and Beyond: Modelling, Analysis, and Applications,
First Edition. Edited by Trung Q. Duong, Xiaoli Chu and Himal A. Suraweera.

message securely acquired at the receiver [9, 10]. PLS methods can either rely on the cooperative transmission or generate a jamming signals target on the eavesdroppers by exploiting the benefits of physical characteristics [11, 12]. By exploiting the advantages of both cooperative transmission and jamming noise, this chapter contributes the analytical information in terms of secrecy performance under the impacts of unreliable backhaul links. In addition, the two-phase transmitter/relay selection scheme is provided to reflect the dense network impacts on security enhancement.

10.1 Backhaul Reliability Level and Performance Limitation

Since the end-to-end connections mainly depend on wireless backhaul links, we first present the effects of the backhaul reliability level as well as its impacts on the performance limitation. Let us consider a spectrum sharing system consisting of an MBS, denoted by S, K SBSs denoted by T_k ($k = 1, 2, ..., K$) connected to S via independent and identical wireless backhaul links, as shown in Figure 10.1. A decode-and-forward (DF) relay R connects to a secondary user D. In the network, a primary user P shares the same spectrum with the K SBSs and R.

Since the K transmitters and R reuse the same spectrum with the primary network, their transmit powers are constrained by the maximal transmit power $\mathcal{P}_{\max}$ and the peak interference power $\mathcal{Q}$. Mathematically, the transmit powers at each SBS and relay can be expressed as

$$\tilde{P}_{\mathrm{k}} = \min(\mathcal{P}_{\max}, \mathcal{Q}/|h_{k,p}|^2) \text{ and } \tilde{P}_{\mathrm{r}} = \min(\mathcal{P}_{\max}, \mathcal{Q}/|g_r|^2), \tag{10.1}$$

where $h_{k,p}$ and g_r are primary channel coefficients of the links $T_k \rightarrow$ P and R$\rightarrow$ P, respectively.

The instantaneous SNR at R in the first time slot and D in the second time slot are respectively given as

$$\gamma_1^{k,r} = \min\left(\overline{\gamma}_{\mathcal{P}}|h_{k,r}|^2, \frac{\overline{\gamma}_{\mathcal{Q}}}{|h_{k,p}|^2}|h_{k,r}|^2\right) \text{ and } \gamma_2^r = \min\left(\overline{\gamma}_{\mathcal{P}}|f_r|^2, \frac{\overline{\gamma}_{\mathcal{Q}}}{|g_r|^2}|f_r|^2\right), \tag{10.2}$$

where $\overline{\gamma}_{\mathcal{P}} = \mathcal{P}_{\max}/\sigma_n^2, \overline{\gamma}_{\mathcal{Q}} = \mathcal{Q}/\sigma_n^2$, with σ_n^2 being the noise variance; $h_{k,r}$ and f_r are secondary channel coefficients of the links $T_k \rightarrow$ R and R$\rightarrow$ D, respectively.

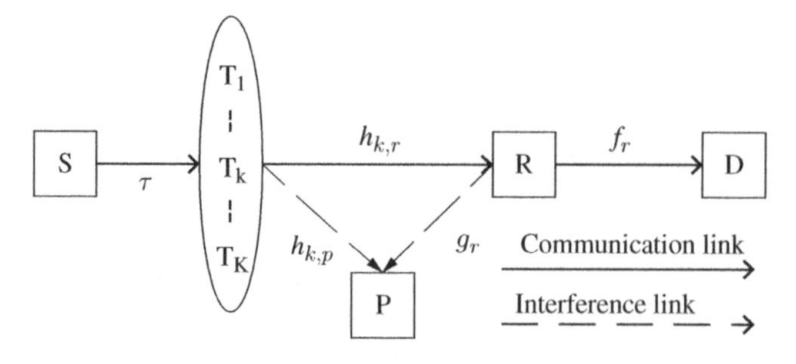

Figure 10.1 The spectrum sharing system in the presence of unreliable backhaul links. Reprinted with permission from [13].

In UDN infrastructures, the message delivered from S to T_k must go through wireless backhaul links and is affected by propagation delay or network congestion; the receiving status at each SBS T_k is either success or failure [5]. Let us model the reliability of those links between S and the K transmitters $\mathbb{I}_k$ as a Bernoulli process with successful transmission probability τ, i.e. $\Pr(\mathbb{I}_k - 1) - \tau$ and $\Pr(\mathbb{I}_k - 0) = 1 - \tau$. Assuming that the best user selection is applied, the transmitter T_{k^*} is selected among the K transmitters such that the instantaneous SNR of the first time slot is largest, which then yields

$$k^* = \arg\max_{[k\in K]}(\gamma_1^{k,r}\mathbb{I}_k). \tag{10.3}$$

Hence, the instantaneous end-to-end SNR is given by

$$\gamma_{DF} = \min(\gamma_1^{k^*,r}\mathbb{I}_{k^*}, \gamma_2^r). \tag{10.4}$$

To provide a more practical channel model, we assume that all channel coefficients undergo identical Nakagami-m fading, i.e. the primary and secondary channels are distributed according to gamma distribution. Let us denote the corresponding severity parameters and mean powers for primary channels as $\{\mu_p^i, \Omega_p^i\}$ and secondary channels as $\{\mu_s^i, \Omega_s^i\}$, where $i \in \{1, 2\}$ denotes the ith time slot, respectively.

10.1.1 Outage Probability Analysis under Backhaul Reliability Impacts

In order to derive analytical framework for the considered system, we obtain the closed-form expression of outage probability by deriving the cumulative distribution function (CDF) of the SNRs for the first and second hop, which are given by [7]

$$\begin{aligned} F_{\gamma_1^{k,r}\mathbb{I}_k}(x) &= 1 - \tau[\Phi_1^1(x) + \Phi_2^1(x)], \\ F_{\gamma_2^r}(x) &= 1 - [\Phi_1^2(x) + \Phi_2^2(x)], \end{aligned} \tag{10.5}$$

where

$$\begin{aligned} \Phi_1^i(x) &= \sum_{m=0}^{\mu_s^i-1} \frac{\Lambda_i}{m!}\left(\frac{\mu_s^i}{\overline{\gamma}_P\Omega_s^i}\right)^m x^m e^{-(\mu_s^i x/\overline{\gamma}_P\Omega_s^i)}, \\ \Phi_2^i(x) &= \sum_{n=0}^{\mu_s^i-1}\sum_{g=0}^{\mu_p^i+n-1}\sum_{l=0}^{g} \binom{\mu_p^i+n-1}{\mu_p^i-1}\binom{g}{l}\frac{1}{g!} \\ &\left(\frac{\mu_s^i}{\overline{\gamma}_P\Omega_s^i}\right)^g \frac{e^{-(\overline{\gamma}_Q\mu_p^i/\overline{\gamma}_P\Omega_p^i)}}{\varepsilon_i^{g-n-l}} \frac{x^{n+l}e^{-(\mu_s^i x/\overline{\gamma}_P\Omega_s^i)}}{(1+\varepsilon_i x)^{\mu_p^i+n}}, \end{aligned} \tag{10.6}$$

and

$$\Lambda_i = \frac{\Upsilon(\mu_p^i, \overline{\gamma}_Q\mu_p^i/\overline{\gamma}_P\Omega_p^i)}{\Gamma(\mu_p^i)}, \quad \varepsilon_i = \frac{\mu_s^i\Omega_p^i}{\overline{\gamma}_Q\Omega_s^i\mu_p^i}.$$

In Eq. (10.6), $\Upsilon(\alpha, x)$ and $\Gamma(.)$ are the lower incomplete gamma function and gamma function, respectively. Applying the best user selection amongst transmitters, the end-to-end SNRs at the relay can be obtained by

$$\gamma_1^{k^*,r}\mathbb{I}_{k^*} = \max_{k=1,\ldots,K}\gamma_1^{k,r}\mathbb{I}_k, \tag{10.7}$$

which yields the CDF of random variable (RV) $\gamma_1^{k^*,r}\mathbb{I}_{k^*}$ as

$$F_{\gamma_1^{k^*,r}\mathbb{I}_{k^*}}(x) = 1 + \sum_{k=1}^{K} \binom{K}{k} (-1)^k \tau^k \widetilde{\sum} \frac{x^{\psi_1^1} e^{-\beta x}}{(1+\varepsilon_1 x)^{\psi_2^1}}, \tag{10.8}$$

where

$$\widetilde{\Psi}_{a_n} \triangleq \sum_{b_n=0}^{\mu_p^1+n-2} b_n a_{b_n+1}, \quad \beta \triangleq k\mu_s^1/\overline{\gamma}_P \Omega_s^1, \quad \psi_1^1 \triangleq \sum_{\vartheta=0}^{\mu_s^1-1} \vartheta u_{\vartheta+1}$$
$$+ \sum_{t=0}^{\mu_s^1-1} t w_{t+1} + c_1 + c_2 + \cdots + c_{\mu_s^1}, \quad \psi_2^1 \triangleq \sum_{t=0}^{\mu_s^1-1} (\mu_p^1 + t) w_{t+1},$$

and $\widetilde{\sum}$ is a shorthand notation, which can be extracted from the binomial and multinomial theorem as in [7].

The outage probability, defined as the probability that the SNR falls below the threshold θ, is calculated as

$$\mathcal{O} \triangleq \Pr(\gamma_{DF} \leq \theta) = F_{\gamma_{DF}}(\theta), \tag{10.9}$$

where the CDF of the instantaneous end-to-end SNR $F_{\gamma_{DF}}(x)$ is given by

$$F_{\gamma_{DF}}(x) = 1 - [1 - F_{\gamma_1^{k^*,r}\mathbb{I}_{k^*}}(x)][1 - F_{\gamma_2^r}(x)]. \tag{10.10}$$

10.1.2 Performance Limitation

To give full insights into the impacts of unreliable backhaul links, the asymptotic expression of outage probability is thus derived. From the definition of RVs $\gamma_1^{k^*,r}\mathbb{I}_{k^*}$ and γ_2^r, it can easily be observed that $\sum_{j=0}^{\mu_s^1-1}(.)$ and $\sum_{g=0}^{\mu_s^2-1}(.)$ are dominated by $j = 0$ and $g = 0$, respectively. On the other hand,

$$\lim_{\delta\to\infty} \Gamma(\alpha, \gamma/\delta)/\Gamma(\alpha) \approx 1. \tag{10.11}$$

Thus, the asymptotic outage in the high-SNR regime can be obtained as

$$\mathcal{O}^{\infty} \overset{\overline{\gamma}_P\to\infty}{=} \prod_{k=1}^{K} (1-\tau). \tag{10.12}$$

From the above asymptotic derivation, it can be said that the backhaul reliability level is the main factor that determines the limitation of system performance. In other words, such observations again verify that unreliable backhaul links have a strong impact on the end-to-end performance.

10.1.3 Numerical Results

In this section, we conduct the link-level simulation to show the backhaul reliability impacts. The curves obtained via link-level simulations are denoted by Ex, whereas analytically derived curves are denoted by An. We use a fixed $\theta = 3$ dB for the computation of outage probability. Without any loss of generality, the channel mean powers are set as $\Omega_s^1 = \Omega_p^1 = \Omega_s^2 = 16, \Omega_p^2 = 4$. We also assume that the increase of the peak

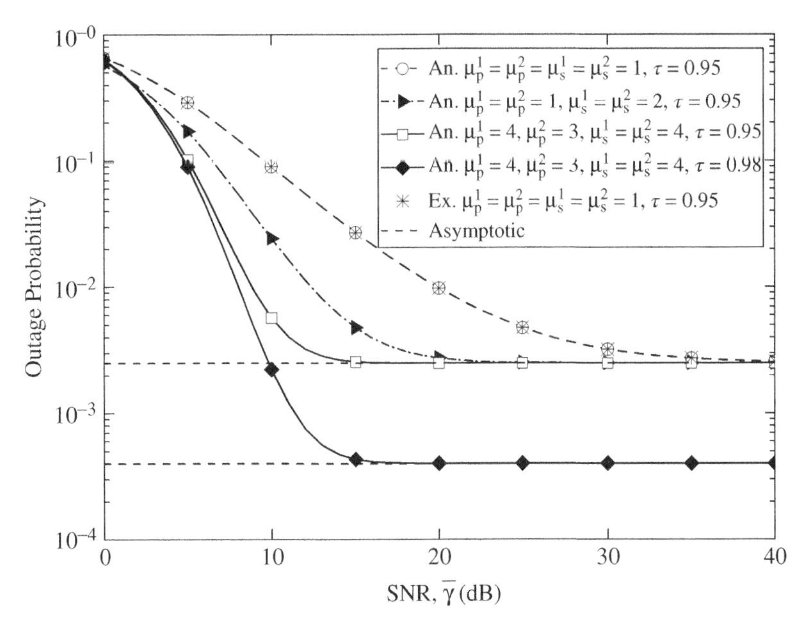

Figure 10.2 The impacts of fading parameters and backhaul reliability on the outage probability for $K = 2$.

interference power $\overline{\gamma}_Q$ is proportional to the increase of the maximum transmit power $\overline{\gamma}_P$, i.e. $\overline{\gamma}_Q/\overline{\gamma}_P = \kappa$, where κ is a constant. Let us denote the average SNR as $\overline{\gamma} = \overline{\gamma}_P$.

From Figure 10.2, it can be observed that the increase in both backhaul reliability and fading parameters gives an impact on the outage probability. In particular, the system performance is improved with large fading factors. However, it remains unchanged in a high-SNR regime. On the other hand, the increase in backhaul reliability level dramatically reduces the outage probability, which proves that the system performances are mainly determined by backhaul reliability.

10.2 Unreliable Backhaul Impacts with Physical Layer Security

Several works have been proposed to improve the PLS method [14–16]. However, the ideal wireless backhaul is almost assumed in those works, which incorrectly reflect the practical scenarios. Very recently, the authors in [17] have analyzed the effects of unreliable backhaul links on various secrecy performance metrics but lacked consideration of the impacts of dense networks. Inspired by the PLS in the UDN scenario [18], the two-phase transmitter/relay selection scheme was developed to maximize SNR at the relays while minimizing the signal-to-interference-plus-noise ratio (SINR) of the eavesdroppers with the aid of a friendly jammer. In addition, the theoretical framework was introduced to analyze the backhaul reliability impacts on secrecy performance.

Let us consider the dense system as illustrated in Figure 10.3. In this network, there is an MBS (Macro-BS) connected to the core network and K SBSs T_k, $k \in \{1, 2, \cdots, K\}$,

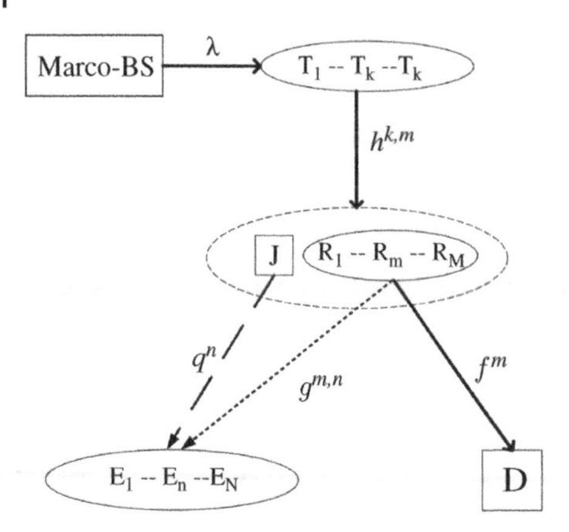

Figure 10.3 The scenario of PLS in UDN, where dense SBSs are deployed to communicate with a group of relays in the presence of malicious eavesdroppers and connect to the MBS via unreliable backhaul links. Reprinted with permission from [18].

are connected to the Macro-BS via unreliable backhaul links. These K SBSs communicate with a user D via M DF relaying nodes R_m, $m \in \{1, 2, \dots, M\}$. Furthermore, there are a jamming node J and N eavesdroppers E_n, $n \in \{1, 2, \dots, N\}$, in the network. Let us assume that the location of K SBSs is far away from the location of D and group of eavesdroppers, so we do not consider the direct link of $\mathrm{T}_k \rightarrow \mathrm{D}$ and $\mathrm{T}_k \rightarrow \mathrm{E}_n, \forall k \in K, \forall n \in N$. We also assume that all the transmitters and receivers are equipped with a single antenna and operated in the half-duplex mode. We suppose that J and D cooperate with each other. Thus, the complete nulling of the jamming signal at D can be achieved.

The channel state information (CSI) is assumed to be available at all active nodes, since it is a common assumption for PLS literature [17]. As the confidential messages are transmitted from the Macro-BS to SBSs, they must pass through the dedicated wireless backhaul links. However, due to the nature of broadcasting signals, the SBSs can either successfully receive or drop. Thus, the backhaul reliability metrics can be modeled by the Bernoulli process with a successful probability of λ_k [5, 7]. To make a practical model of the small-scale fading, we assume that all channels are distributed according to the gamma distribution, where the CDF and probability density function (PDF) of RV $\chi \sim \mathrm{Ga}(\mu_\chi, \tilde{\eta}_\chi)$, where $\chi \in \{h^{k,m}, g^{m,n}, f^m, q^n\}$ are given as

$$f_\chi(x) = \frac{1}{(\mu_\chi - 1)!(\tilde{\eta}_\chi)^{\mu_\chi}} x^{\mu_\chi - 1} e^{(-x/\tilde{\eta}_\chi)},$$

$$F_\chi(x) = 1 - e^{(-x/\tilde{\eta}_\chi)} \sum_{i=0}^{\mu_\chi - 1} \frac{1}{i!} (x/\tilde{\eta}_\chi)^i, \tag{10.13}$$

where $\mu_\chi \in \{\mu_k^m, \mu_m^n, \mu_j^n, \mu_d^m, \}$ denote the positive fading severity parameters and $\tilde{\eta}_\chi \in \{\tilde{\eta}_k^m, \tilde{\eta}_m^n, \tilde{\eta}_j^n, \tilde{\eta}_d^m\}$ are the scaling factors.

The received signal at R_m from a particular SBS T_k is given by

$$\mathrm{y}_R^{k,m} = \sqrt{P_t \alpha_T^{k,m}} \mathrm{h}^{k,m} \mathbb{I}_k \mathrm{x} + \mathrm{n}_R^{k,m}, \tag{10.14}$$

where P_t is the transmit power and $n_R^{k,m} \sim \mathcal{CN}(0, \sigma_n^2)$ is an additive noise vector at R_m. $\mathbb{I}_k$ recalls the backhaul reliability, which is modeled as a Bernoulli process. From Eq. (10.14), the instantaneous SNR between T_k and R_m in the first time slot can be expressed as

$$\gamma_R^{k,m} = \frac{P_t\alpha_T^{k,m}||\mathrm{h}^{k,m}||^2}{\sigma_n^2}\mathbb{I}_k = \tilde{\alpha}_T^{k,m}||\mathrm{h}^{k,m}||^2\mathbb{I}_k = \lambda^{k,m}\mathbb{I}_k, \tag{10.15}$$

where

$$\tilde{\alpha}_T^{k,m} \triangleq \frac{P_t\alpha_T^{k,m}}{\sigma_n^2}, \quad \lambda^{k,m} \sim \mathrm{Ga}(\mu_k^m, \eta_k^m) \quad \text{with} \quad \eta_k^m = \frac{\tilde{\alpha}_T^{k,m}\mathbb{E}\{||\mathrm{h}^{k,m}||^2\}}{\mu_k^m}.$$

The received signals at E_n and D from R_m are, respectively, given by

$$\begin{aligned} y_E^{m,n} &= \sqrt{P_r\alpha_E^{m,n}}g^{m,n}x + \sqrt{P_j\alpha_J^n}q^n v + n_E^{m,n}, \\ y_D^m &= \sqrt{P_r\alpha_D^m}f^m x + n_D^m, \end{aligned} \tag{10.16}$$

where P_r and P_j are the transmit powers at the relays and J, respectively, $n_E^{m,n} \sim \mathcal{CN}(0,\sigma_n^2)$ and $n_D^m \sim \mathcal{CN}(0,\sigma_n^2)$ denote the noise vectors at E_n and D, respectively.

Thus, the instantaneous SINR between R_m and E_n can be written as

$$\gamma_E^{m,n} = \frac{P_r\alpha_E^{m,n}||\mathrm{g}^{m,n}||^2}{\sigma_n^2 + P_j\alpha_J^n||\mathrm{q}^n||^2} = \frac{\tilde{\alpha}_E^{m,n}||\mathrm{g}^{m,n}||^2}{1+\tilde{\alpha}_J^n||\mathrm{q}^n||^2} = \frac{\lambda^{m,n}}{1+\lambda_J^n}, \tag{10.17}$$

where

$$\tilde{\alpha}_E^{m,n} \triangleq \frac{P_r\alpha_E^{m,n}}{\sigma_n^2}, \quad \tilde{\alpha}_J^n \triangleq \frac{P_j\alpha_J^n}{\sigma_n^2}, \quad \lambda^{m,n} \sim \mathrm{Ga}(\mu_m^n, \eta_m^n), \quad \lambda_J^n \sim \mathrm{Ga}(\mu_j^n, \eta_j^n) \quad \text{with}$$

$$\eta_m^n = \frac{\tilde{\alpha}_E^{m,n}\mathbb{E}\{||\mathrm{g}^{m,n}||^2\}}{\mu_m^n}, \quad \eta_j^n = \frac{\tilde{\alpha}_J^n\mathbb{E}\{||\mathrm{q}^n||^2\}}{\mu_j^n}.$$

In the second time slot, the instantaneous SNR between R_m and D can be expressed as

$$\gamma_D^m = \frac{P_r\alpha_D^m||\mathrm{f}^m||^2}{\sigma_n^2} = \tilde{\alpha}_D^m||\mathrm{f}^m||^2 = \lambda_D^m, \tag{10.18}$$

where

$$\tilde{\alpha}_D^m \triangleq \frac{P_r\alpha_D^m}{\sigma_n^2}, \quad \lambda_D^m \sim \mathrm{Ga}(\mu_d^m, \eta_d^m) \quad \text{with} \quad \eta_d^m = \frac{\tilde{\alpha}_D^m\mathbb{E}\{||\mathrm{f}^m||^2\}}{\mu_d^m}.$$

We assume that the unreliable backhaul links are independent from the indices of the K transmitters, i.e., $\lambda_k = \lambda, \forall k$. We also assume that the positive fading severity parameter μ_χ and the scaling factor η_χ are identically varied among the K transmitters, M relays, and N eavesdroppers; i.e. it can be rewritten as $\mu_T = \mu_k^m$, $\mu_E = \mu_m^n$, $\mu_J = \mu_j^n$, $\mu_D = \mu_d^m$, and $\eta_T = \eta_k^m$, $\eta_E = \eta_m^n$, $\eta_J = \eta_j^n$, $\eta_D = \eta_d^m$, $\forall k, m, n$.

10.2.1 The Two-Phase Transmitter/Relay Selection Scheme

In the purpose of enhancing the security level, the selection scheme is proposed with two different phases in order to maximize the achievable performance while reducing the eavesdropping attacks from the eavesdroppers. In the first phase, each relay chooses the best transmitter among the K SBSs to maximize their achievable SNR. The selected transmitter can be mathematically expressed as

$$\text{Phase 1:} \quad k^* = \arg \max_{k=1,\ldots,K} \gamma_R^{k,m}, \tag{10.19}$$

where $\gamma_R^{k,m}$ recalls the instantaneous SNR at R_m via T_k. From Eq. (10.19), the statistical property of the instantaneous SNR via the best transmitter T_{k^*} is given in the following theorem.

Theorem 10.1 Given K independent and identical unreliable backhaul connections, the CDF of the received SNR at R_m via the best transmitter is given as

$$\begin{aligned} F_{\gamma_R^{k^*,m}}(x) &= [F_{\gamma_R^{k,m}}(x)]^K = \sum_{k=0}^{K} \binom{K}{k} (-1)^k \left(\lambda e^{-x/\eta_T} \sum_{l=0}^{\mu_R-1} \frac{1}{l!} \left(\frac{x}{\eta_T} \right)^l \right)^k \\ &= 1 + \sum_{k=1}^{K} \sum_{\omega_1,\ldots,\omega_{\mu_R}}^{k} \binom{K}{k} \left(\frac{k!}{\omega_1! \ldots \omega_{\mu_R}!} \right) \frac{(-1)^k \lambda^k}{\prod_{t=0}^{\mu_R-1} (t!(\eta_T)^t)^{\omega_{t+1}}} x^{\sum_{t=0}^{\mu_R-1} t\omega_{t+1}} e^{-kx/\eta_T}. \end{aligned} \tag{10.20}$$

In the second phase, the relay is selected such that the SINR between the particular relay and N eavesdroppers is minimized. It can be formulated as

$$\text{Phase 2:} \quad m^* = \arg \min_{m=1,\ldots,M} \gamma_E^{m,n^*}, \tag{10.21}$$

where $\gamma_E^{m,n^*} = \max(\gamma_E^{m,1}, \ldots, \gamma_E^{m,N})$ is the maximum instantaneous SINR between R_m and N eavesdroppers. The CDF of γ_E^{m,n^*} is given in the following lemma.

Lemma 10.1 For the independent and identically distributed (i.i.d.) Nakagami-m fading channels, the CDF of the instantaneous SINR between R_m and N eavesdroppers is given as

$$F_{\gamma_E^{m,n^*}}(x) = \widehat{\sum_{N,n,\mu_E}} e^{-\varphi_1^N x} x^{\varphi_2^N} \left(\frac{1}{\eta_J} + \frac{x}{\eta_E} \right)^{-\varphi_3^N}, \tag{10.22}$$

where

$$\varphi_1^N \triangleq n/\eta_E, \quad \varphi_2^N \triangleq \sum_{t=0}^{\mu_E-1} t\vartheta_{t+1}, \quad \varphi_3^N \triangleq \sum_{\eta_1=0}^{0} (\mu_J + \eta_1)\mu_{1,\eta_1+1}$$
$$+ \sum_{\eta_2=0}^{1} (\mu_J + \eta_2)\mu_{2,\eta_2+1} + \cdots + \sum_{\eta_{\mu_E}=0}^{\mu_E-1} (\mu_J + \eta_{\mu_E})\mu_{\mu_E,\eta_{\mu_E}+1}$$

and $\widehat{\sum_{N,n,\mu_E}}$ is the shorthand notation, which can be extracted as in [18].

Proof: The proof is given in Appendix A.

Given M independent relays, the statistical property of the achievable SINR at the eavesdroppers via R_m is given in the following theorem.

Theorem 10.2 For the i.i.d. Nakagami-m fading channels, the PDF of the instantaneous SINR between R_m and the eavesdroppers, denoted by $\gamma_E^{m^*,n^*}$, is given later by.

$$f_{\gamma_E^{m^*,n^*}}(x) = \mathcal{Q}\widetilde{\sum_E} e^{-\tilde{\varphi}_1 x}(\mathcal{B}_1 x^{\tilde{\varphi}_2} - \mathcal{B}_2 x^{\tilde{\varphi}_2-1} + \mathcal{B}_3 x^{\tilde{\varphi}_2+1})\left(\frac{1}{\eta_J} + \frac{x}{\eta_E}\right)^{-\tilde{\varphi}_3}, \tag{10.23}$$

where $\mathcal{Q} \triangleq \frac{MN}{(\eta_J)^{\mu_J}(\mu_J - 1)!}$, $\mathcal{B}_1 \triangleq \frac{1/\eta_J + \mu_J + j - i}{\eta_E}$, $\mathcal{B}_2 \triangleq \frac{i}{\eta_J}$, $\mathcal{B}_3 \triangleq \frac{1}{(\eta_E)^2}$,

$\tilde{\varphi}_1 \triangleq 1/\eta_E + \varphi_1^{N-1} + \varphi_1^{mN}$, $\tilde{\varphi}_2 \triangleq \varphi_2^{N-1} + \varphi_2^{mN} + i$, $\tilde{\varphi}_3 \triangleq \varphi_3^{N-1} + \varphi_3^{mN} + \mu_J + j + 1$

and $\widetilde{\sum_E}$ is the shorthand notation of

$$\widetilde{\sum_E} \triangleq \widehat{\sum_{N-1,l,\mu_E}} \widehat{\sum_{mN,r,\mu_E}} \sum_{m=0}^{M-1} \sum_{i=0}^{\mu_E-1} \sum_{j=0}^{i} \binom{M-1}{m}\binom{i}{j}(-1)^m \frac{1}{i!(\eta_E)^i}\Gamma(\mu_J + j). \tag{10.24}$$

Proof: The proof is given in Appendix B.

Due to DF relaying protocol at the relays, the selected relay processes the information after receiving the signals and directly forwards it to the receiver. The instantaneous end-to-end SNR at D from the m^*th relay, denoted by $\tilde{\gamma}_{DF}^{m^*}$, is mathematically given by

$$\tilde{\gamma}_{DF}^{m^*} = \min(\gamma_R^{k^*,m^*}, \gamma_D^{m^*}), \tag{10.25}$$

where $\gamma_D^{m^*}$ recalls the instantaneous SNR between the m^*th relay and D in the second time slot. The CDF of the RV $\gamma_D^{m^*}$ is given as

$$F_{\gamma_D^{m^*}}(x) = 1 - e^{-x/\eta_D} \sum_{q=0}^{\mu_D-1} \frac{1}{q!}\left(\frac{x}{\eta_D}\right)^q. \tag{10.26}$$

According to Eq. (10.25), the statistical property of the instantaneous end-to-end SNR $\tilde{\gamma}_{DF}^{m^*}$ can be obtained by the following theorem.

Theorem 10.3 For the proposed dense networks with unreliable backhaul links, the CDF of the instantaneous SNR at D via the m^*th relay is given by

$$F_{\tilde{\gamma}_{DF}^{m^*}}(x) = 1 + \widetilde{\sum_D} x^{\beta} e^{-\Phi x}, \tag{10.27}$$

where $\beta = \sum_{t=0}^{\mu_R-1} t\omega_{t+1} + q$, $\Phi = k/\eta_T + 1/\eta_D$, and $\widetilde{\sum_D}$ is the shorthand notation of

$$\widetilde{\sum_D} = \sum_{k=1}^{K} \sum_{q=0}^{\mu_D-1} \sum_{\omega_1,\ldots,\omega_{\mu_R}}^{k} \binom{K}{k}\left(\frac{k!}{\omega_1!\ldots\omega_{\mu_R}!}\right)(-1)^k \lambda^k \frac{1}{q!\prod_{t=0}^{\mu_R-1}(t!(\eta_T)^t)^{\omega_{t+1}}}\left(\frac{1}{\eta_D}\right)^q. \tag{10.28}$$

Proof: According to the definition of RV $\tilde{\gamma}_{DF}^{m^*}$, which is given in Eq. (10.25), the CDF of $\tilde{\gamma}_{DF}^{m^*}$ can be expressed as

$$F_{\tilde{\gamma}_{DF}^{m^*}}(x) = 1 - [1 - F_{\gamma_R^{k^*,m^*}}(x)][1 - F_{\gamma_D^{m^*}}(x)]. \tag{10.29}$$

By substituting Eqs. (10.20) and (10.26) into Eq. (10.29) and after some simple manipulations, the CDF of instantaneous end-to-end SNR at D is obtained as in Eq. (10.27).

10.2.2 Secrecy Outage Probability with Backhaul Reliability Impact

In order to observe the backhaul reliability impact on the security level, the closed-form expression of the secrecy outage probability is provided. We assume that the eavesdroppers's CSI is not available in the considered network. In this case, the transmitters encode and send the confidential message with the constant secrecy rate of θ. If the instantaneous secrecy capacity, denoted by C_S in bits/s/Hz, is greater than θ, the secrecy gain is guaranteed. Otherwise, information-theoretic security is compromised.

The secrecy capacity C_S can be expressed as [14]

$$C_S = \frac{1}{2}\left[\log_2(1+\tilde{\gamma}_{DF}^{m^*}) - \log_2(1+\gamma_E^{m^*,n^*})\right]^+, \tag{10.30}$$

where $\log_2(1+\tilde{\gamma}_{DF}^{m^*})$ is the instantaneous capacity at D with respect to the m^*th relay and $\log_2(1+\gamma_E^{m^*,n^*})$ is the instantaneous capacity of the wiretap channel between the m^*th relay and n^*th eavesdropper.

The secrecy outage probability, which is defined as the probability that the secrecy capacity falls below the given rate threshold, can be expressed as [17]

$$\begin{aligned} \mathcal{P}_{out}(\theta) &= Pr(C_S < \theta) \\ &= \int_0^\infty F_{\tilde{\gamma}_{DF}^{m^*}}(2^{2\theta}(1+x)-1) f_{\gamma_E^{m^*,n^*}}(x)dx. \end{aligned} \tag{10.31}$$

From Eq. (10.31), the closed-from expression for the secrecy outage probability is given in the following theorem.

Theorem 10.4 For the dense networks with unreliable backhaul links, the secrecy outage probability with a two-phase transmitter/relay selection scheme is given as in.

$$\mathcal{P}_{out}(\theta) = 1 + \mathcal{Q}\widetilde{\sum_D}\widetilde{\sum_E}\sum_{\alpha=0}^{\beta}\binom{\beta}{\alpha}(\Upsilon-1)^{\beta-\alpha}(\Upsilon)^{\alpha}\eta_E^{\tilde{\varphi}_3}e^{-\Phi(\Upsilon-1)}(\mathcal{O}_1 - \mathcal{O}_2 + \mathcal{O}_3), \tag{10.32}$$

where $\Upsilon \triangleq 2^{2\theta}$, $\epsilon \triangleq \eta_E/\eta_J$, and

$$\begin{aligned} \mathcal{O}_1 &= \mathcal{B}_1\Gamma(\tilde{\varphi}_2+\alpha+1)\epsilon^{\tilde{\varphi}_2+\alpha+1-\tilde{\varphi}_3}\Psi(\tilde{\varphi}_2+\alpha+1, \tilde{\varphi}_2+\alpha+2-\tilde{\varphi}_3, \epsilon(\Phi\Upsilon+\tilde{\varphi}_1)), \\ \mathcal{O}_2 &= \mathcal{B}_2\Gamma(\tilde{\varphi}_2+\alpha)\epsilon^{\tilde{\varphi}_2+\alpha-\tilde{\varphi}_3}\Psi(\tilde{\varphi}_2+\alpha, \tilde{\varphi}_2+\alpha+1-\tilde{\varphi}_3, \epsilon(\Phi\Upsilon+\tilde{\varphi}_1)), \\ \mathcal{O}_3 &= \mathcal{B}_3\Gamma(\tilde{\varphi}_2+\alpha+2)\epsilon^{\tilde{\varphi}_2+\alpha+2-\tilde{\varphi}_3}\Psi(\tilde{\varphi}_2+\alpha+2, \tilde{\varphi}_2+\alpha+3-\tilde{\varphi}_3, \epsilon(\Phi\Upsilon+\tilde{\varphi}_1)). \end{aligned}$$

Proof: The proof is given in Appendix C.

10.2.3 Secrecy Performance Limitation under Backhaul Reliability Impact

To provide full insights into the impacts of unreliable backhaul connections, the asymptotic expression for the secrecy outage probability is given in the following theorem.

Theorem 10.5 Given the fixed set $\{\tilde{\alpha}_T, \tilde{\alpha}_E, \tilde{\alpha}_J\}$, the asymptotic expression for secrecy outage probability is given as

$$\mathcal{P}_{out}^{\infty}(\theta) \overset{\tilde{\alpha}_D \to \infty}{=} 1 + \mathcal{Q} \widetilde{\sum_{D^\infty}} \widetilde{\sum_{E}} \sum_{\alpha=0}^{\tilde{\beta}} \binom{\tilde{\beta}}{\alpha} (\Upsilon - 1)^{\tilde{\beta}-\alpha} (\Upsilon)^{\alpha} \eta_E^{\tilde{\varphi}_3} e^{-\tilde{\Phi}(\Upsilon-1)} (\widehat{\mathcal{O}_1} - \widehat{\mathcal{O}_2} + \widehat{\mathcal{O}_3}), \quad (10.33)$$

where

$$\tilde{\beta} = \sum_{t=0}^{\mu_R-1} t\omega_{t+1}, \quad \tilde{\Phi} = \frac{k}{\eta_T}, \quad \widetilde{\sum_{D^\infty}} = \sum_{k=1}^{K} \sum_{\omega_1,\ldots,\omega_{\mu_R}}^{k} \binom{K}{k} \left(\frac{k!}{\omega_1! \ldots \omega_{\mu_R}!} \right) \frac{(-1)^{k-1} \lambda^k}{\prod_{t=0}^{\mu_R-1} (t!(\eta_T)^t)^{\omega_{t+1}}},$$

and

$$\widehat{\mathcal{O}_1} = \mathcal{B}_1 \Gamma(\tilde{\varphi}_2 + \alpha + 1) \epsilon^{\tilde{\varphi}_2+\alpha+1-\tilde{\varphi}_3} \Psi(\tilde{\varphi}_2 + \alpha + 1, \tilde{\varphi}_2 + \alpha + 2 - \tilde{\varphi}_3, \epsilon(\tilde{\Phi}\Upsilon + \tilde{\varphi}_1)),$$
$$\widehat{\mathcal{O}_2} = \mathcal{B}_2 \Gamma(\tilde{\varphi}_2 + \alpha) \epsilon^{\tilde{\varphi}_2+\alpha-\tilde{\varphi}_3} \Psi(\tilde{\varphi}_2 + \alpha, \tilde{\varphi}_2 + \alpha + 1 - \tilde{\varphi}_3, \epsilon(\tilde{\Phi}\Upsilon + \tilde{\varphi}_1)),$$
$$\widehat{\mathcal{O}_3} = \mathcal{B}_3 \Gamma(\tilde{\varphi}_2 + \alpha + 2) \epsilon^{\tilde{\varphi}_2+\alpha+2-\tilde{\varphi}_3} \Psi(\tilde{\varphi}_2 + \alpha + 2, \tilde{\varphi}_2 + \alpha + 3 - \tilde{\varphi}_3, \epsilon(\tilde{\Phi}\Upsilon + \tilde{\varphi}_1)).$$

Proof: Observing the CDF of RV $\gamma_D^{m^*}$ in (10.26), we find that $e^{-x/\eta_D} \overset{\tilde{\alpha}_D \to \infty}{\approx} 1$ and $\sum_{q=0}^{\mu_D-1}$ is dominated by $q = 0$ when $\tilde{\alpha}_D \to \infty$. Thus,

$$\lim_{\eta_D \to \infty} F_{\gamma_D^{m^*}}(x) \approx 0, \quad (10.34)$$

which yields the CDF of the instantaneous end-to-end SNR of D as

$$F_{\tilde{\gamma}_{DF}^{m^*}}(x) = 1 + \widetilde{\sum_{D^\infty}} x^{\tilde{\beta}} e^{-\tilde{\Phi}x}. \quad (10.35)$$

By substituting Eqs. (10.23) and (10.35) into Eq. (10.31), the asymptotic outage probability is obtained as in Eq. (10.33).

From Eq. (10.33), we observe that since the backhaul links are unreliable, the secrecy diversity gain is not achievable. Thus, the limitation on secrecy outage probability is determined as a constant.

10.2.4 Numerical Results

The link-level simulations are conducted to verify the impacts of backhaul reliability and fading parameters on the secrecy outage probability. The curves obtained via link-level simulations are denoted by Ex, whereas the curves for analytical results are denoted by An. The rate threshold is normalized as 1 in the computation of secrecy outage probability.

In Figure 10.4, the plot illustrates the secrecy outage probability for various M and N. The network parameters are set as $K = 3$, $\lambda = 0.995$, $\{\mu_R, \mu_E, \mu_J, \mu_D\} = \{2, 2, 2, 3\}$, and $\{\tilde{\alpha}_T, \tilde{\alpha}_E, \tilde{\alpha}_J\} = \{10, 10, 10\}$ dB. From the figure, the good matches between simulation curves and analytical curves are clearly presented. On the other hand, all curves converg to the asymptotic limitation, which proves the correctness of our asymptotic analysis.

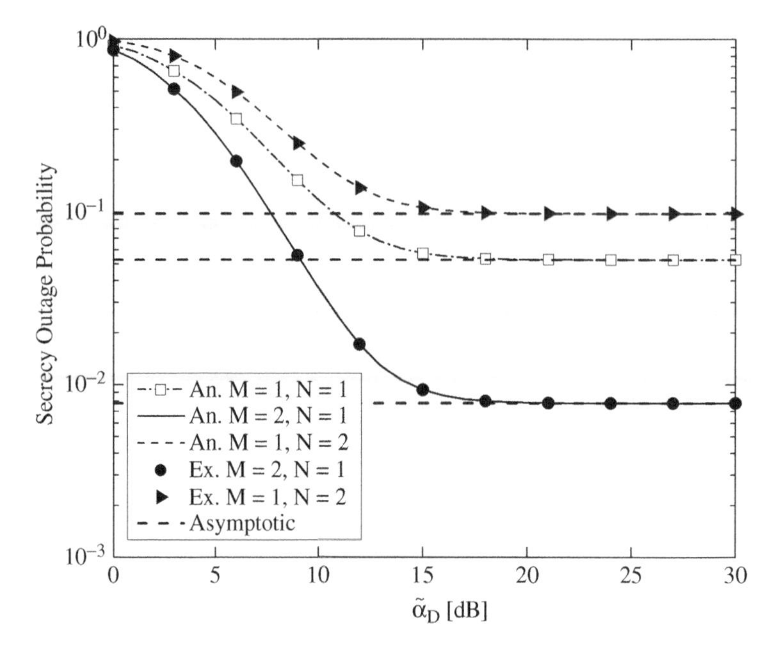

Figure 10.4 Secrecy outage probability with dense networks impact.

We also observe that dense networks have a strong impact on the secrecy outage probability. In particular, the secrecy outage probability becomes lower when more relays join into the network and get higher when the number of eavesdroppers increases. This is due to the fact that when the number of relays increases, the secrecy rate becomes higher as a result of the reduction in the wiretap channel capacity.

Figure 10.5 plots the secrecy outage probability with various effects of backhaul reliability level λ and fading parameters. We set $K = 3, M = 1, N = 1$ and $\{\tilde{\alpha}_T, \tilde{\alpha}_E, \tilde{\alpha}_J\} = \{10, 10, 10\}$ dB. It can be observed that the secrecy performance is significantly improved when the backhaul links tend to be more reliable, i.e. $\mathcal{P}_{out}(\theta)$ when $\lambda = 0.95$ is lower than $\mathcal{P}_{out}(\theta)$ when $\lambda = 0.85$. Furthermore, the increase in fading severity parameters in main channels and the decrease in fading severity parameters in wiretap channels lead to the enhancement of secrecy performance. Otherwise, the secrecy performance tends to get worse. This plot also shows the good match between the simulation results and analytical results, where all curves converge on the asymptotic line.

Appendix A

Proof of Lemma 10.1

According to Eq. (10.16), the distribution of the RV $\gamma_E^{m,n}$ is analytically calculated as

$$F_{\gamma_E^{m,n}}(x) = \mathbb{E}\{F_{\tilde{\alpha}_E^{m,n}||g^{m,n}||^2}((1+y)x) \mid \tilde{\alpha}_J^n||q^n||^2 = y\}. \tag{10.36}$$

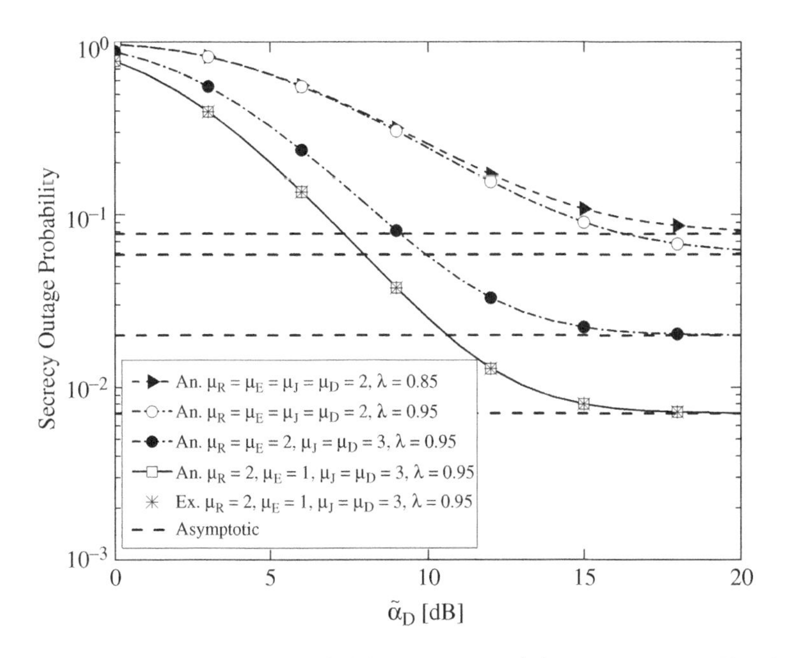

Figure 10.5 Secrecy outage probability with various fading parameters and backhaul reliability.

Since the Nakagami-m fading channels between the particular relay and the eavesdroppers are i.i.d. RVs, the CDF of the RV γ_E^{m,n^*} is given as

$$F_{\gamma_E^{m,n^*}}(x) = [F_{\gamma_E^{m,n}}(x)]^N = \sum_{n=0}^{N} \binom{N}{n} (-1)^n \left(\frac{e^{-x/\eta_E}}{(\eta_J)^{\mu_J}(\mu_J - 1)!} \right)^n \times \underbrace{\left(\sum_{l=0}^{\mu_E-1} \sum_{r=0}^{l} \binom{l}{r} \frac{\Gamma(\mu_J + r)}{l!(\eta_E)^l} \left(\frac{1}{\eta_J} + \frac{x}{\eta_E} \right)^{-(\mu_J+r)} x^l \right)^n}_{\mathcal{J}_1}, \tag{10.37}$$

By applying binomial and multinomial theorems, $\mathcal{J}_1$ can be extracted as in [18].

Appendix B

Proof of Theorem 10.2

According to the definition of the RV $\gamma_E^{m^*,n^*}$ in Eq. (10.21), which is given by $\gamma_E^{m^*,n^*} = \min(\gamma_E^{1,n^*}, \ldots, \gamma_E^{M,n^*})$, the PDF of $\gamma_E^{m^*,n^*}$ can be mathematically expressed based on the order statistics as

$$f_{\gamma_E^{m^*,n^*}}(x) = M f_{\gamma_E^{m,n^*}}(x)[1 - F_{\gamma_E^{m,n^*}}(x)]^{M-1}. \tag{10.38}$$

From Eq. (10.38), $f_{\gamma_E^{m,n^*}}(x)$ can be calculated by taking the first derivative of the CDF of the RV γ_E^{m,n^*}, which was derived in Eq. (10.22). The expression $f_{\gamma_E^{m,n^*}}(x)$ is thus obtained as

$$\begin{aligned}
f_{\gamma_E^{m,n^*}}(x) &= \frac{\partial F_{\gamma_E^{m,n^*}}(x)}{\partial x} \\
&= \frac{N}{(\eta_J)^{\mu_J}(\mu_J-1)!}\sum_{i=0}^{\mu_E-1}\sum_{j=0}^{i}\binom{i}{j}\frac{\Gamma(\mu_J+j)}{i!(\eta_E)^i}\left(\frac{1/\eta_J+\mu_J+j-i}{\eta_E}x^i\right. \\
&\quad \left.-\frac{i}{\eta_J}x^{i-1}+\frac{1}{(\eta_E)^2}x^{i+1}\right)e^{-x/\eta_E} \\
&\quad \times\left(\frac{1}{\eta_J}+\frac{x}{\eta_E}\right)^{-(\mu_J+j+1)}\widehat{\sum_{N-1,l,\mu_E}} e^{-\varphi_1^{N-1}x}x^{\varphi_2^{N-1}}\left(\frac{1}{\eta_J}+\frac{x}{\eta_E}\right)^{-\varphi_3^{N-1}}. \qquad (10.39)
\end{aligned}$$

where $\{\varphi_1^{N-1}, \varphi_2^{N-1}, \varphi_3^{N-1}\}$ is the set of parameters corresponding to $\widehat{\sum_{N-1,l,\mu_E}}$.

Again binomial and multinomial theorems for $[1-F_{\gamma_E^{m,n^*}}(x)]^{M-1}$ yield

$$[1-F_{\gamma_E^{m,n^*}}(x)]^{M-1} = \sum_{m=0}^{M-1}\binom{M-1}{m}(-1)^m\widehat{\sum_{mN,r,\mu_E}} e^{-\varphi_1^{mN}x}x^{\varphi_2^{mN}}\left(\frac{1}{\eta_J}+\frac{x}{\eta_E}\right)^{-\varphi_3^{mN}}, \qquad (10.40)$$

where $\{\varphi_1^{mN}, \varphi_2^{mN}, \varphi_3^{mN}\}$ is the set of parameters corresponding to $\widehat{\sum_{mN,r,\mu_E}}$.

By substituting Eqs. (10.39) and (10.40) into (10.38), and after some manipulations, the PDF of the RV $\gamma_E^{m^*,n^*}$ is thus obtained as in Eq. (10.23).

Appendix C

Proof of Theorem 10.4

From the definition of the secrecy outage probability, by substituting (10.27) and (10.23) into Eq. (10.31), the secrecy outage probability expression is thus obtained as

$$\begin{aligned}
\mathcal{P}_{out}(\theta) &= \int_0^\infty \left[1+\widetilde{\sum_D}(\Upsilon-1+\Upsilon x)^\beta e^{-\Phi(\Upsilon-1+\Upsilon x)}\right] \\
&\quad \times \mathcal{Q}\widetilde{\sum_E} e^{-\tilde{\varphi}_1 x}(\mathcal{B}_1 x^{\tilde{\varphi}_2}-\mathcal{B}_2 x^{\tilde{\varphi}_2-1}+\mathcal{B}_3 x^{\tilde{\varphi}_2+1})\left(\frac{1}{\eta_J}+\frac{x}{\eta_E}\right)^{-\tilde{\varphi}_3} dx \\
&= 1+\mathcal{Q}\widetilde{\sum_D}\widetilde{\sum_E}\sum_{\alpha=0}^{\beta}\binom{\beta}{\alpha}(\Upsilon-1)^{\beta-\alpha}(\theta)^\alpha e^{-\Phi(\Upsilon-1)} \\
&\quad \times \underbrace{\int_0^\infty e^{-(\Phi\Upsilon+\tilde{\varphi}_1)x}(\mathcal{B}_1 x^{\tilde{\varphi}_2+\alpha}-\mathcal{B}_2 x^{\tilde{\varphi}_2+\alpha-1}+\mathcal{B}_3 x^{\tilde{\varphi}_2+\alpha+1})\left(\frac{1}{\eta_J}+\frac{x}{\eta_E}\right)^{-\tilde{\varphi}_3} dx,}_{\mathcal{J}_2}
\end{aligned} \qquad (10.41)$$

where $\mathcal{J}_2$ can be evaluated with the help of [[19], Eq. (2.3.6.9)], where $\Psi(a,b,c) = 1/\Gamma(a)\int_0^\infty e^{-ct}t^{b-a}(1+t)^{b-a-1}dt$ denotes the confluent hypergeometric function [[20], Eq. (9.211.4)]. After some manipulations, we arrive at Eq. (10.32).

Bibliography

1 X. Ge, H. Cheng, M. Guizani, and T. Han, "5G wireless backhaul networks: challenges and research advances," *IEEE Network*, vol. 28, no. 6, pp. 6–11, Nov. 2014.
2 X. Ge, S. Tu, G. Mao, C.-X. Wang, and T. Han, "5G ultra-dense cellular networks," *IEEE Wirel. Commun.*, vol. 23, no. 1, pp. 72–79, Feb. 2016.
3 U. Siddique, H. Tabassum, E. Hossain, and D.I. Kim, "Wireless backhauling of 5G small cells: challenges and solution approaches," *IEEE Wirel. Commun.*, vol. 22, no. 5, pp. 22–31, Oct. 2015.
4 M. Coldrey, H. Koorapaty, J.-E. Berg, Z. Ghebretensae, J. Hansryd, A. Derneryd, and S. Falahati, "Small-cell wireless backhauling: a non-line-of-sight approach for point-to-point microwave links," in *Proc. IEEE Veh. Technol. Conf.*, Quebec City, Canada, Sep. 2012, pp. 1–5.
5 T.A. Khan, P. Orlik, K.J. Kim, and R.W. Heath, "Performance analysis of cooperative wireless networks with unreliable backhaul links," *IEEE Commun. Lett.*, vol. 19, no. 8, pp. 1386–1389, Aug. 2015.
6 K.J. Kim, T. Khan, and P. Orlik, "Performance analysis of cooperative systems with unreliable backhauls and selection combining," *IEEE Trans. Veh. Technol.*, vol. 66, no. 3, pp. 2448–2461, Mar. 2017.
7 H.T. Nguyen, D.-B. Ha, S.Q. Nguyen, and W.-J. Hwang, "Cognitive heterogeneous networks with unreliable backhaul connections," *J. Mobile Netw. Appl.*, pp. 1–14, 2017.
8 L. Dong, Z. Han, A.P. Petropulu, and H.V. Poor, "Improving wireless physical layer security via cooperating relays," *IEEE Trans. Signal Process.*, vol. 58, no. 3, pp. 1875–1888, Mar. 2010.
9 N. Yang, L. Wang, G. Geraci, M. Elkashlan, J. Yuan, and M. Di Renzo, "Safeguarding 5G wireless communication networks using physical layer security," *IEEE Commun. Mag.*, vol. 53, no. 4, pp. 20–27, Apr. 2015.
10 H. Hui, A.L. Swindlehurst, G. Li, and J. Liang, "Secure relay and jammer selection for physical layer security," *IEEE Signal Process. Lett.*, vol. 22, no. 8, pp. 1147–1151, Aug. 2015.
11 J. Yang, I.-M. Kim, and D.I. Kim, "Optimal cooperative jamming for multiuser broadcast channel with multiple eavesdroppers," *IEEE Trans. Wireless Commun.*, vol. 12, no. 6, pp. 2840–2852, Jun. 2013.
12 T.M. Hoang, T.Q. Duong, N.-S. Vo, and C. Kundu, "Physical layer security in cooperative energy harvesting networks with a friendly jammer," *IEEE Wireless Commun. Lett.*, vol. 6, no. 2, pp. 174–177, Jan. 2017.
13 H.T. Nguyen, T.Q. Duong, and W.-J. Hwang, "Multiuser relay networks over unreliable backhaul links under spectrum sharing environment," *IEEE Commun. Lett.*, vol. 21, no. 10, pp. 2314–2317, Oct. 2017.

14 N. Yang, H.A. Suraweera, I.B. Collings, and C. Yuen, "Physical layer security of TAS/MRC with antenna correlation," *IEEE Trans. Inf. Forensics Security*, vol. 8, no. 1, pp. 254–259, Jan. 2013.

15 L. Wang, M. Elkashlan, J. Huang, R. Schober, and R.K. Mallik, "Secure transmission with antenna selection in MIMO Nakagami-*m* fading channels," *IEEE Wireless Commun.*, vol. 13, no. 11, pp. 6054–6067, Nov. 2014.

16 T.M. Hoang, T.Q. Duong, H.A. Suraweera, C. Tellambura, and H.V. Poor, "Cooperative beamforming and user selection for improving the security of relay-aided systems," *IEEE Trans. Commun.*, vol. 63, pp. 5039–5051, Dec. 2015.

17 K.J. Kim, P.L. Yeoh, P.V. Orlik, and H.V. Poor, "Secrecy performance of finite-sized cooperative single carrier systems with unreliable backhaul connections," *IEEE Trans. Signal Process.*, vol. 64, no. 17, pp. 4403–4416, Sep. 2016.

18 H.T. Nguyen, J. Zhang, N. Yang, T.Q. Duong, and W.-J. Hwang, "Secure cooperative single carrier systems under unreliable backhaul and dense networks impact," *IEEE Access*, vol. 5, pp. 18310–18324, Sep. 2017.

19 A. Prudnikov, Y.A. Brychkov, and O. Marichev, *Integrals and Series, Vol. 1: Elementary Functions*, 4th edn. London: Gordon and Breach, 1998.

20 I.S. Gradshteyn and I.M. Ryzhik, *Table of Integrals, Series, and Products*. New York: Academic Press, 2007.

11

Simultaneous Wireless Information and Power Transfer in UDNs with Caching Architecture

Sumit Gautam, Thang X. Vu, Symeon Chatzinotas and Björn Ottersten

Interdisciplinary Centre for Security, Reliability and Trust, University of Luxembourg, Luxembourg

11.1 Introduction

The exponential increase in the usage of wireless devices like smart-phones, wearable gadgets, or connected vehicles has not only posed substantial challenges to meet the performance and capacity demands [1], but also revealed some serious environmental concerns with alarming energy consumption and CO_2 emissions [2]. These concerns become more significant as the forecast number of devices will exceed 50 billions by the end of 2020 [3]. Recent developments in the upcoming paradigm of Internet-of-Things (IoT) emphasize the interconnection between equipments, commodities, functionalities, and customers, with or without human mediation. Since most of these connecting operations involve wireless sensor nodes or equivalent battery-limited devices that may not be continuously powered, energy becomes a sparse and pivotal resource. These challenges require the future communication networks to have not only efficient energy management but also the capability of being self-powered from redundant energy sources, which is known as energy harvesting (EH).

Among potential EH techniques, simultaneous wireless information and power transfer (SWIPT) has received much attention as the key enabling technique for future IoT networks. The basic premise behind SWIPT is to allow concurrent data reception and EH from the same radio frequency (RF) input signal. Considering rapid drainage of battery sources in wireless devices, it has almost become essential to take up such techniques in order to compensate for this issue. Since the conventional receiver architectures are capable of performing information decoding with a focus on increasing the data rate only and are unable to harvest energy, this calls for alternative receiver architectures to support SWIPT [4–9]. Two notable architectures have been proposed based on time switching (TS) and power splitting (PS) schemes [10, 11]. In the former, the received signal is switched between the information decoder and energy harvester. It is noted in the TS scheme that full received power is used for either information decoding or energy harvesting. In the latter, both the information decoder and energy harvester are active simultaneously, each of them receives part of the signal power. A comprehensive review on SWIPT is presented in [12].

Another major problem the future networks have to face is network congestion, which usually occurs during peak hours when the network resource is scarce. The cause of this

Ultra-dense Networks for 5G and Beyond: Modelling, Analysis, and Applications,
First Edition. Edited by Trung Q. Duong, Xiaoli Chu and Himal A. Suraweera.

congestion is mainly due to the fact that replicas of a common content may be demanded by various mobile users. A promising solution to overcome network congestion is to shift the network traffic from peak hours to off-peak times via content placement or caching [13]. In (off-line) caching, there is usually a placement phase and a delivery phase. In the placement phase, which usually occurs during off-peak times when the network resources are abundant, popular content is prefetched in distributed caches close to end users. The latter usually occurs during peak hours when the actual users' requests are revealed. If the requested content is available in the user's local storage, it can be served without being requested from the core network. Various advantages brought by caching have been observed in terms of backhaul's load reduction [13, 14] and system energy efficiency improvement [15].

Furthermore, densification of these networks is also crucial for the future of wireless networks in order to tackle heavy congestion scenarios. However, ultra-dense networks (UDNs) inflict significant issues in terms of dynamic system foundation and implementation of algorithmic models [16]. Recent studies reveal that UDNs within macrocells are expected to provide not only an extended coverage with enhanced system capacity but also an appreciable quality of experience (QoE) to the end-user [17]. In addition, targeting the enhancement of system performance at the end-user, which is assisted by a chosen relay from a pool of cooperative subnetwork of relays within the UDNs, is an interesting problem. In this vein, by leveraging the concept of cooperative systems within UDNs, where relays are superficially equipped with energy harvesting and caching capabilities, we propose a framework for relay selection to serve the end-user.

A generic concept outlining joint EH and caching has recently been proposed for 5G networks to exploit the benefits of both techniques. A so-called framework GreenDelivery is proposed in [2], which enables efficient content delivery in small cells based on energy harvesting small base stations. In [18], an online energy-efficient power control scheme for EH with a caching capability is developed. In particular, by adopting the Poisson distribution for the energy sources, a dynamic programming problem is formulated and solved iteratively by using numerical methods. The authors in [19] propose a caching mechanism at the gateway for the energy harvesting based IoT sensing service to maximize the hit rate. In [20], the authors investigate the performance of heterogeneous vehicular networks with a renewable energy source. A network planning problem is formulated to optimize cache size and energy harvesting rate subject to backhaul capacity limits. We note that these works either address an abstract EH with general external energy sources or consider EH separated from caching.

In this chapter, we investigate the performance of EH based cooperative networks within the UDNs, in which the relay nodes are equipped with both SWIPT and caching capabilities. In particular, we aim at developing a framework to realize the integration of SWIPT with caching architectures. The considered system is assumed to operate in the TS based mode, since the PS counterpart imposes complex hardware design challenges of the power splitter [11]. The main contributions of this chapter are threefold, listed as follows.

1. First, we introduce a novel architecture for cache-assisted SWIPT relaying systems under the TS based and decode-and-forward (DF) relaying protocol and study the interaction between caching capacity and SWIPT in the considered system.
2. Second, an optimization problem is formulated to maximize the throughput of the (serving) link between the relay and destination, taking into account the caching

capacity, minimum harvested energy, and quality-of-service (QoS) constraints. By using the Karush-Kuhn-Tucker (KKT) conditions with the aid of the Lambert function, a closed-form solution of the formulated problem is obtained. Based on this result, the best relay will be selected for cooperation.

3. Third, we formulate an optimization problem to maximize the energy stored at the relay subjected to the QoS constraint. Similar to the previous problem, a closed-form solution is obtained by using the KKT conditions and the Lambert function. The effectiveness of the proposed schemes are demonstrated via intensive numerical results, through which the impacts of key system parameters are observed.

The remainder of this chapter is organized as follows. Section 11.2 describes the system model and relevant variables. Section 11.3 presents the problem for maximization of the link throughput between the relay and destination. Section 11.4 maximizes the stored energy at the relay. In Section 11.5, numerical results are presented to demonstrate the effectiveness of the proposed architectures. Finally, Section 11.6 concludes the chapter.

11.2 System Model

We consider a generic TS based SWIPT system, which consists of one source, K relays, and one destination, as depicted in Figure 11.1. Due to limited coverage, e.g., transmit power limit or blockage, there is no direct connection between the source and the destination. The considered model can find application on the downlink where the base station plays the source's role and sends information to a far user via a small- or femto-cell base station. The relays operate in DF mode and are equipped with a single antenna. We consider a general cache-aided SWIPT model, in which each relay contains

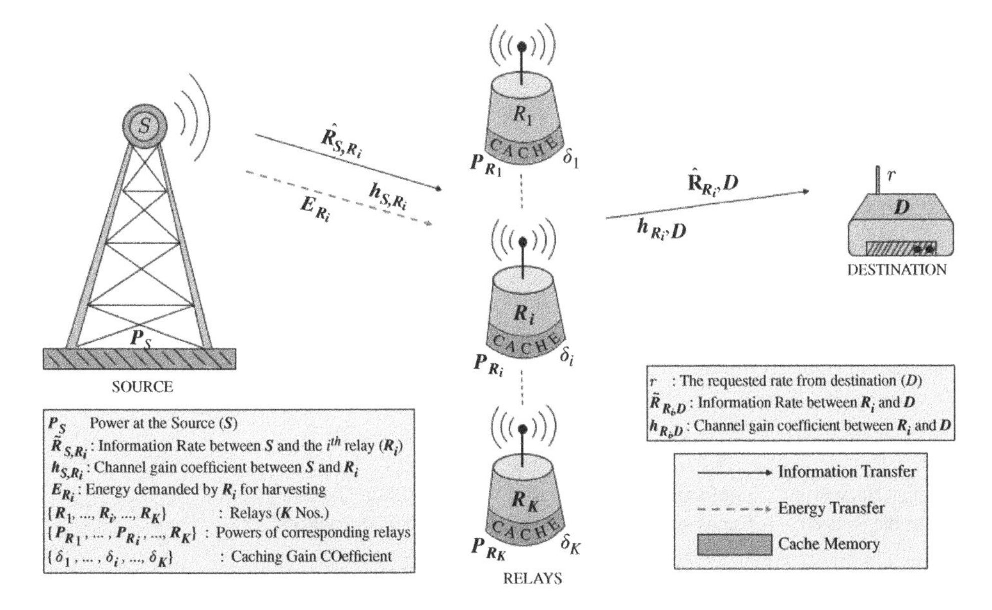

Figure 11.1 System model for SWIPT with caching.

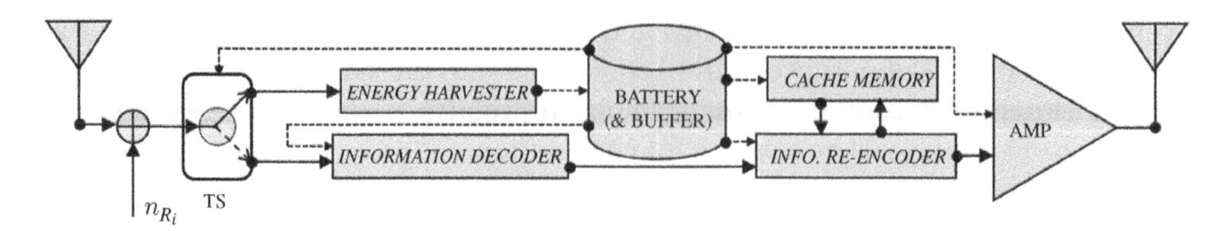

Figure 11.2 Proposed DF relay transceiver design for hybrid SWIPT and caching with a time switching (TS) architecture.

an information decoder, an energy harvester, and a cache in order to store or exchange information. The block diagram of a typical relay architecture is shown in Figure 11.2.

We consider block Rayleigh fading channels in which the channel coefficients remain constant within a block (or coherence time) and independently change from block to block. One communication session (broadcasting and relaying) takes place in T seconds (T does not exceed the coherence time). In the broadcasting phase, the source–relay links are active for information decoding (at the relay) and energy harvesting. The relays employ the TS scheme so that the received signal is first provided to the the energy harvester for some fraction of the time allocated for the transmitter-relay communication link, and then to the information decoder for the remaining fraction. This mechanism is also known as the harvest-then-forward protocol [21]. In the relaying phase, the selected relay forwards the information to the destination. Full channel state information (CSI) is assumed to be available at a centralized base station, which performs all the computations and informs the relevant devices via adequate signaling. The relay with the best reward will be selected for sending information to the destination. Details on relay selection will be presented in the next sections.

Figure 11.3 presents a convention for allocation of time fractions in the TS scheme: (i) energy harvesting at the relay, (ii) information processing at the relay, and (iii) information forwarding to the destination. The link between the transmitter and the ith relay with a cache is active for a fraction of $\beta_i T$ seconds, while the link between the ith relay and the destination is active for the remaining $(1 - \beta_i)T$, where $0 \leq \beta_i \leq 1$. Furthermore, we assume that the energy harvesting at the relays takes place for a fraction of $\alpha_i \beta_i T$ seconds and the information decoding at the relay takes place for a fraction of $(1 - \alpha_i)\beta_i T$ seconds, where $0 \leq \alpha_i \leq 1$. For ease of representation, we assume normalized time to use energy and power interchangeably without loss in generality.

11.2.1 Signal Model

We define the information transfer rates from the source to the ith relay as $\hat{R}_{S,R_i}$ and from the ith relay to the destination as $\hat{R}_{R_i,D}$, with $1 \leq i \leq K$, where $i \in \mathbb{Z}$.

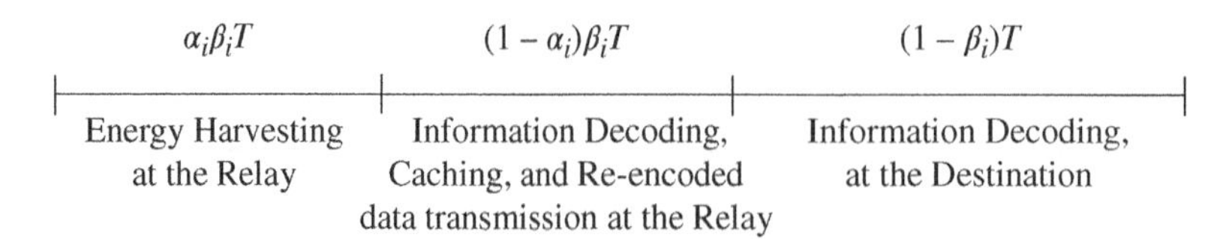

Figure 11.3 Convention assumed for distribution of time to investigate the rate maximization problem.

Let h_{S,R_i} and $h_{R_i,D}$ denote the channel coefficients between the source and the ith relay and between the ith relay and the destination, respectively, and d_{S,R_i} and $d_{R_i,D}$ denote the distance between the source and ith relay, and the distance between the ith relay and destination, respectively. Furthermore, P_S and P_{R_i} denote the transmit power at the source and the ith relay, respectively.

Let $x \in \mathbb{C}$ be the transmitted symbol by the source satisfying $E\{|x|^2\} = 1$. The signal received at the ith relay is given by

$$y_{R_i} = \sqrt{P_S}\; d_{S,R_i}^{-\vartheta/2}\; h_{S,R_i}\; x + n_{R_i}, \tag{11.1}$$

where ϑ is the path loss exponent and n_{R_i} is the additive white Gaussian noise (AWGN) at the relay, which is an independent and identically distributed (i.i.d.) complex Gaussian random variable with zero mean and variance $\sigma^2_{n_{R_i}}$.

Upon receiving the desired signal from the source, the relay decodes to obtain the estimate of the original signal. Then the (selected) relay re-encodes and forwards it to the destination. The signal received at the destination as transmitted by the ith relay is given by

$$y_D = \sqrt{P_{R_i}}\; d_{R_i,D}^{-\vartheta/2}\; h_{R_i,D}\; \tilde{x} + n_D, \tag{11.2}$$

where $\tilde{x}$ is the transmit symbol from the relay and n_D is the AWGN at the destination node, which is an i.i.d. complex Gaussian random variable with zero mean and variance $\sigma^2_{n_D}$. We note that in the proposed cached-aid architecture, the relayed symbol can be either a decoded source symbol or from the relay's cache.

The effective signal-to-noise ratio (SNR) of the source-relay and relay-destination links are given as

$$\gamma_{S,R_i} = \frac{P_S\; d_{S,R_i}^{-\vartheta}\; |h_{S,R_i}|^2}{\sigma^2_{n_{R_i}}}; \quad \gamma_{R_i,D} = \frac{P_{R_i}\; d_{R_i,D}^{-\vartheta}\; |h_{R_i,D}|^2}{\sigma^2_{n_D}}. \tag{11.3}$$

By assuming a Gaussian codebook, the achievable information rate on the source-relays link is

$$\hat{R}_{S,R_i} = B\log_2(1 + \gamma_{S,R_i}), \qquad \forall i, \tag{11.4}$$

and the achievable information rate at the destination is

$$\hat{R}_{R_i,D} = B\log_2(1 + \gamma_{R_i,D}), \qquad \forall i, \tag{11.5}$$

where B is the channel bandwidth.

11.2.2 Caching Model

A general caching model is considered at the relays. In particular, in addition to the information sent from the source, the relays have access to the information stored in their individual caches to serve the destination. For robustness, we assume that for $i = 1, \ldots, K$, the ith relay does not have information about the content popularity. Therefore, it will store $0 \le \delta_i \le 1$ parts of every file in its cache [13, 15].[1] For convenience, we call δ_i as the caching coefficient throughout the chapter. This caching

1 This caching method is also known as probabilistic caching.

scheme will serve as the lower bound benchmark compared to the case where priori information of content popularity is available. When the destination requests a file from the library, δ_i parts of that file are already available at the ith relay's cache. Therefore, the source needs to send only the remainder of that file to the relay.

11.2.3 Power Assumption at the Relay

For robustness, we assume that the relays are powered by an external source, E_{ext}, on top of the harvested energy. This general model allows the impact of various practical scenarios to be analyzed. As a use case, the purely SWIPT relay is obtained by setting E_{ext} to zero. The harvested energy at the ith relay is given by

$$E_{R_i} = \zeta \alpha_i \beta_i (P_S d_{S,R_i}^{-\vartheta} |h_{S,R_i}|^2 + \sigma_{n_{R_i}}^2), \tag{11.6}$$

where ζ is the energy conversion efficiency of the receiver.

In the following couple of sections, we address the problems for maximizing the serving information rate between the relay and the destination, and maximizing the energy stored at the relay, respectively.

11.3 Maximization of the serving information rate

In this section, we aim at maximizing the serving information data between the selected relay and the destination, by taking into consideration the caching capacity at the relay, while ensuring the predefined QoS constraint and that the total transmit power at the source does not exceed the limit. The corresponding optimization problem (P1) is stated as follows:

$$(P1) : \underset{i \in \mathcal{K}, \alpha_i, \beta_i, P_{R_i}}{\text{maximize}} \quad (1-\beta_i)\hat{R}_{R_i,D} \tag{11.7}$$

$$\text{subject to: } (C1) : (1-\alpha_i)\beta_i(\hat{R}_{S,R_i} + (\delta_i \cdot r)) \geq (1-\beta_i)\hat{R}_{R_i,D}, \tag{11.8}$$

$$(C2) : (1-\beta_i)P_{R_i} \leq E_{R_i} + E_{ext}, \tag{11.9}$$

$$(C3) : 0 < P_S \leq P^{\star}, \tag{11.10}$$

$$(C4) : 0 \leq \alpha_i \leq 1, \tag{11.11}$$

$$(C5) : 0 \leq \beta_i \leq 1, \tag{11.12}$$

where $\mathcal{K} = \{1, \ldots, K\}$, E_{R_i} is given in Eq. (11.6), E_{ext} is the external energy required at the relay for further transmission of the signal, $P^{\star}$ is the maximum power limit at the transmitter, and r is the QoS constraint. The objective in (11.7) is to maximize the transfered data to the destination, since the relay-destination link is active only in $1-\beta_i$ (active times normalized by T). Constraint (11.8) is to assure non-empty buffer at the relay. Constraint (11.9) is to ensure that the used energy at the relay cannot exceed the input.

This is a mixed-integer programming problem implying that relay selection along with joint computations of α_i, β_i, and P_{R_i} is a difficult task. Therefore, we recast $(P1)$ into a pair of coupled optimization problems, namely, outer optimization for choosing the best relay and inner optimization for joint computations of α_i, β_i, and P_{R_i}. In the following sections, we address the optimal solutions to the inner and outer optimizations, respectively.

11.3.1 Optimization of TS Factors and the Relay Transmit Power

In this subsection, we address the inner optimization problem of $(P1)$ involving joint computations of α_i, β_i, and P_{R_i}, assuming that the ith relay is active. The result subproblem is formulated as follows:

$$(P2): \underset{\alpha_i,\beta_i,P_{R_i}}{\text{maximize}} \quad (1-\beta_i)\hat{R}_{R_i,D} \tag{11.13}$$

$$\text{subject to:} \quad (11.8)-(11.12)$$

This is a non-linear programming problem involving joint computations of α_i, β_i, and P_{R_i}, which is difficult in terms of finding the exact solution. Since the constraints are partially convex on each variable while fixing the others, we propose to solve this problem using the KKT conditions.

The Lagrangian corresponding to $(P2)$ can be denoted as follows:

$$\begin{aligned}\mathcal{L}(\alpha_i,\beta_i,P_{R_i};\lambda_1,\lambda_2,\lambda_3,\lambda_4) &= F(\alpha_i,\beta_i,P_{R_i}) - \lambda_1 \quad G(\alpha_i,\beta_i,P_{R_i}) \\ &-\lambda_2 \quad H(\alpha_i,\beta_i,P_{R_i}) - \lambda_3 \quad I(\alpha_i,\beta_i,P_{R_i}) - \lambda_4 \quad J(\alpha_i,\beta_i,P_{R_i}),\end{aligned} \tag{11.14}$$

where

$$F(\alpha_i,\beta_i,P_{R_i}) = (1-\beta_i)B\log_2(1+\gamma_{R_i,D}), \tag{11.15}$$

$$G(\alpha_i,\beta_i,P_{R_i}) = (1-\beta_i)B\log_2(1+\gamma_{R_i,D}) - (1-\alpha_i)\beta_i[B\log_2(1+\gamma_{S,R_i}) + (\delta_i\cdot r)] \le 0, \tag{11.16}$$

$$H(\alpha_i,\beta_i,P_{R_i}) = (1-\beta_i)P_{R_i} - \zeta\alpha_i\beta_i(P_S d_{S,R_i}^{-\vartheta}|h_{S,R_i}|^2 + \sigma_{n_{R_i}}^2) - E_{ext} \le 0, \tag{11.17}$$

$$I(\alpha_i,\beta_i,P_{R_i}) = \alpha_i - 1 \le 0, \tag{11.18}$$

$$J(\alpha_i,\beta_i,P_{R_i}) = \beta_i - 1 \le 0. \tag{11.19}$$

For (local) optimality, it must hold that $\nabla\mathcal{L}(\alpha_i,\beta_i,P_{R_i};\lambda_1,\lambda_2,\lambda_3,\lambda_4) = 0$. Thus, we can represent the equations for satisfying the optimality conditions as

$$\begin{aligned}\frac{\partial\mathcal{L}(\alpha_i,\beta_i,P_{R_i};\lambda_1,\lambda_2,\lambda_3,\lambda_4)}{\partial\alpha_i} &\Rightarrow -\lambda_1[\beta_i(B\log_2(1+\gamma_{S,R_i}) + (\delta_i\cdot r))] \\ &-\lambda_2[-\zeta\beta_i(P_S d_{S,R_i}^{-\vartheta}|h_{S,R_i}|^2 + \sigma_{n_{R_i}}^2)] - \lambda_3 = 0,\end{aligned} \tag{11.20}$$

$$\begin{aligned}\frac{\partial\mathcal{L}(\alpha_i,\beta_i,P_{R_i};\lambda_1,\lambda_2,\lambda_3,\lambda_4)}{\partial\beta_i} &\Rightarrow -B\log_2(1+\gamma_{R_i,D}) - \lambda_1[-B\log_2(1+\gamma_{R_i,D})] \\ &-(1-\alpha_i)[B\log_2(1+\gamma_{S,R_i}) + (\delta_i\cdot r)] - \lambda_2[-P_{R_i} - \zeta\alpha_i(P_S d_{S,R_i}^{-\vartheta}|h_{S,R_i}|^2 + \sigma_{n_{R_i}}^2)] \\ &-\lambda_4 = 0,\end{aligned} \tag{11.21}$$

$$\frac{\partial \mathcal{L}(\alpha_i, \beta_i, P_{R_i}; \lambda_1, \lambda_2, \lambda_3, \lambda_4)}{\partial P_{R_i}} \Rightarrow \frac{\ln(2) d_{R_i,D}^{-\vartheta} |h_{R_i,D}|^2}{\sigma_{n_D}^2 + P_{R_i} d_{R_i,D}^{-\vartheta} |h_{R_i,D}|^2}$$
$$-\lambda_1 \left(\frac{\ln(2) d_{R_i,D}^{-\vartheta} |h_{R_i,D}|^2}{\sigma_{n_D}^2 + P_{R_i} d_{R_i,D}^{-\vartheta} |h_{R_i,D}|^2} \right) - \lambda_2 = 0. \tag{11.22}$$

The conditions for feasibility are as expressed in Eqs. (11.16), (11.17), (11.18), and (11.19). Complementary slackness expressions can be represented as follows:

$$\lambda_1 \cdot G(\alpha_i, \beta_i, P_{R_i}) = 0, \tag{11.23}$$
$$\lambda_2 \cdot H(\alpha_i, \beta_i, P_{R_i}) = 0, \tag{11.24}$$
$$\lambda_3 \cdot I(\alpha_i, \beta_i, P_{R_i}) = 0, \tag{11.25}$$
$$\lambda_4 \cdot J(\alpha_i, \beta_i, P_{R_i}) = 0. \tag{11.26}$$

The conditions for non-negativity are $\alpha_i, \beta_i, P_{R_i}, \lambda_1, \lambda_2, \lambda_3, \lambda_4 \geq 0$. It is straightforward to verify that if $\lambda_3 \neq 0$, then $I(\alpha_i, \beta_i, P_{R_i}) = 0$, implying that $\alpha_i = 1$. Since this is not a feasible solution, therefore $\lambda_3 = 0$. Similarly, it can be shown that $\lambda_4 = 0$. From further analysis, we find that the solutions corresponding to the binding cases of KKT yield either zero, negative, or unbounded values of the optimization variables, thereby violating the non-negativity constraints. The final solution can be postulated in the following theorem, following the non-binding case.

Theorem 11.1 For $\lambda_1 \neq 0 \Rightarrow G(x, P_{R_i}) = 0$; $\lambda_2 \neq 0 \Rightarrow H(x, P_{R_i}) = 0$, we obtain the following optimal values:

$$P_{R_i} = (\exp(\mathcal{W}(\mathcal{A} \exp(-\log^2(2)) + \log(2)) + \log^2(2)) - 1) \left(\frac{\sigma_{n_D}^2}{d_{R_i,D}^{-\vartheta} \quad |h_{R_i,D}|^2} \right), \tag{11.27}$$

where $\mathcal{A} = (\ln(2) d_{R_i,D}^{-\vartheta} |h_{R_i,D}|^2)(\zeta(P_S d_{S,R_i}^{-\vartheta} |h_{S,R_i}|^2 + \sigma_{n_{R_i}}^2))/\sigma_{n_D}^2$ and $\mathcal{W}(\cdot)$ is the Lambert W function [22]:

$$\beta_i = \frac{\varphi - E_{ext}(B \log_2(1 + \gamma_{_{S,R_i}}) + (\delta_i \cdot r))}{\varphi - \zeta(P_S d_{S,R_i}^{-\vartheta} |h_{S,R_i}|^2 + \sigma_{n_{R_i}}^2)(B \log_2(1 + \gamma_{_{S,R_i}}) + (\delta_i \cdot r))}, \tag{11.28}$$

where $\varphi = P_{R_i}(B \log_2(1 + \gamma_{_{S,R_i}}) + (\delta_i \cdot r)) - \zeta(P_S d_{S,R_i}^{-\vartheta} |h_{S,R_i}|^2 + \sigma_{n_{R_i}}^2) B \log_2(1 + \gamma_{_{R_i,D}})$; and

$$\alpha_i = \frac{(1 - \beta_i) P_{R_i} - E_{ext}}{\zeta \beta_i (P_S d_{S,R_i}^{-\vartheta} |h_{S,R_i}|^2 + \sigma_{n_{R_i}}^2)}. \tag{11.29}$$

From an economic view-point, the Lagrange multipliers λ_1 and λ_2 can be expounded as the prices for data and energy in cost/bit and cost/Joule, respectively. Leveraging the results from our mathematical analysis, we find that $\lambda_1 = f_1(P_S)/f_2(P_S, \delta_i, r) \cdot \lambda_2$, where $f_1(P_S) = \zeta(P_S d_{S,R_i}^{-\vartheta} |h_{S,R_i}|^2 + \sigma_{n_{R_i}}^2)$ and $f_2(P_S, \delta_i, r) = B \log_2(1 + \gamma_{_{S,R_i}}) + (\delta_i \cdot r)$. Correspondingly, it is clear that if more data rate is demanded by the user provided the caching capacity is fixed, then we are enforced to compensate for the request by using the energy metric per cost unit in order to satisfy the respective data and energy constraints in ($P2$). This action would, however, add more to the energy price. Similarly, it is apparent that increasing the transmit power will readily add to the cost of data transfer in addition to an increased energy price. In the context of caching, it would be needless to mention that *the higher the cache capacity is, the lower will be the prices for data and energy transmissions.*

11.3.2 Relay Selection

In this subsection, we consider optimal selection of a relay to address the solution of outer optimization of ($P1$). Based on that the above developments, we find that the best relay provides maximum throughput corresponding to (11.7). The best relay index is selected as $j^\star = \arg\max_{j\in\{1,\dots,K\}}(1-\beta_j^\star)\hat{R}^\star_{R_j,D}$, where $\beta_j^\star$ and $\hat{R}^\star_{R_j,D}$ are the solutions to problem (11.13). It is worth mentioning that this relay selection is based on exhausted search and provides the best performance with high cost of complexity. Finding a compromise relay selection is of interest in future research.

11.4 Maximization of the Energy Stored at the Relay

In this section, we aim at maximizing the energy stored at the relay. The stored energy is calculated by subtracting the input energy, e.g. E_{ext} plus the harvested energy, by the output energy, e.g. used for forwarding information to the destination. Our motivation behind this section is that the stored energy at the relay can be used to perform an extra processing task, e.g. sensing, or to recharge a battery for future use. In particular, an optimization problem is formulated to jointly select the best relay and maximize the stored energy, while satisfying a given QoS.

For convenience, we introduce a new convention for the fraction of times in the TS scheme, as shown in Figure 11.4. The link between the transmitter and relay with cache is considered to be active for a fraction of $(\theta_i+\phi_i)T$ seconds, while the link between the relay and destination is active for the remaining $(1-(\theta_i+\phi_i))T$, where $0\le\theta_i+\phi_i\le 1$. As mentioned earlier, since the relay adopts a TS type of scheme for SWIPT, we assume that the energy harvesting at the relay takes place for a fraction of $\theta_i T$ seconds and the information decoding at the relay takes place for a fraction of $\phi_i T$ seconds. Similarly, we assume normalized time to use energy and power interchangeably without loss in generality. We must remember that the harvested energy at the relay i is given as

$$E_{R_i} = \zeta\theta_i(P_S d_{S,R_i}^{-\vartheta}|h_{S,R_i}|^2 + \sigma^2_{n_{R_i}}), \tag{11.30}$$

where ζ is the energy conversion efficiency of the receiver.

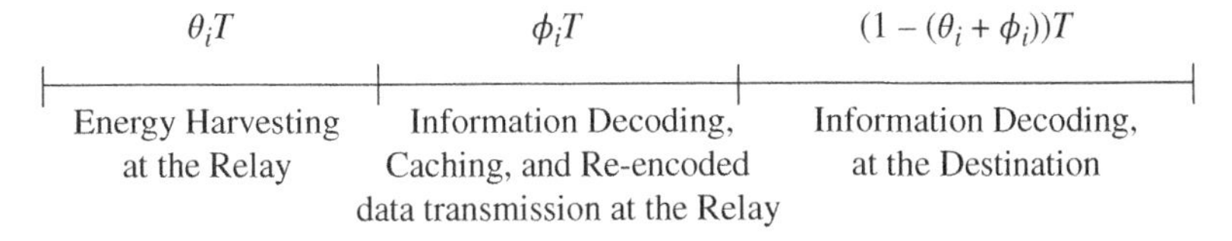

Figure 11.4 Convention assumed for distribution of time to investigate the stored energy maximization problem.

We now consider the problem of relay selection for maximization of the energy stored at the relay, while ensuring that the requested rate between relay-destination is above a given threshold and that the total transmit powers at the transmitter and relay do not exceed a given limit. The corresponding optimization problem (P3) can be expressed as

$$(P3): \underset{i\in\mathcal{K},\theta_i,\phi_i,P_{R_i}}{\text{maximize}} \quad [\zeta\theta_i(P_S d_{S,R_i}^{-\vartheta}|h_{S,R_i}|^2+\sigma_{n_{R_i}}^2)+E_{ext}-(1-(\theta_i+\phi_i))P_{R_i}]^+ \tag{11.31}$$

$$\text{subject to:} \quad (C1): \phi_i(\hat{R}_{S,R_i}+(\delta_i\cdot r))\geq(1-(\theta_i+\phi_i))\hat{R}_{R_i,D}, \tag{11.32}$$

$$(C2): (1-(\theta_i+\phi_i))P_{R_i}\leq E_{R_i}+E_{ext}, \tag{11.33}$$

$$(C3): (1-(\theta_i+\phi_i))\hat{R}_{R_i,D}\geq r, \tag{11.34}$$

$$(C4): 0<P_S\leq P^{\star}, \tag{11.35}$$

$$(C5): 0\leq\theta_i+\phi_i\leq 1, \tag{11.36}$$

where the objective in (11.31) is non-zero and the constraint (11.34) is to satisfy the QoS requirement.

The problem ($P3$) is difficult to solve, since it is a mixed-integer programming problem involving relay selection along with joint computations of θ_i, ϕ_i, and P_{R_i}. Therefore, we recast ($P3$) into a pair of coupled optimization problems for performing the outer optimization to choose the best relay and the inner optimization for joint computations of θ_i, ϕ_i, and P_{R_i}. In the following subsections, we address the optimal solutions to the inner and outer optimizations, respectively.

11.4.1 Optimization of TS Factors and the Relay Transmit Power

In this subsection, we consider the inner optimization problem of (P3). We determine the technique for joint computations of θ_i, ϕ_i, and P_{R_i} for maximizing the energy stored at the relay while ensuring that the requested rate between the relay-destination is above a given threshold and that the total transmit powers at the transmitter and relay do not exceed a given limit. Correspondingly, the subproblem ($P4$) can be formulated as

$$(P4): \underset{\theta_i,\phi_i,P_{R_i}}{\text{maximize}} \quad [\zeta\theta_i(P_S d_{S,R_i}^{-\vartheta}|h_{S,R_i}|^2+\sigma_{n_{R_i}}^2)+E_{ext}-(1-(\theta_i+\phi_i))P_{R_i}]^+ \tag{11.37}$$

$$\text{subject to:} \quad (C1): \phi_i(\hat{R}_{S,R_i}+(\delta_i\cdot r))\geq(1-(\theta_i+\phi_i))\hat{R}_{R_i,D}, \tag{11.38}$$

$$(C2): (1-(\theta_i+\phi_i))P_{R_i}\leq E_{R_i}+E_{ext}, \tag{11.39}$$

$$(C3): (1-(\theta_i+\phi_i))\hat{R}_{R_i,D}\geq r, \tag{11.40}$$

$$(C4): 0<P_S\leq P^{\star}, \tag{11.41}$$

$$(C5): 0\leq\theta_i+\phi_i\leq 1. \tag{11.42}$$

This is a non-linear programming problem involving joint computations of θ_i, ϕ_i, and P_{R_i}, which introduces intractability. Therefore, we propose to solve this problem using the KKT conditions.

The Lagrangian for $(P4)$ can be expressed as follows:

$$\begin{aligned}\mathcal{L}(\theta_i, \phi_i, P_{R_i}, \mu_1, \mu_2, \mu_3, \mu_4) &= F(\theta_i, \phi_i, P_{R_i}) - \mu_1 \quad G(\theta_i, \phi_i, P_{R_i}) \\ &-\mu_2 \quad H(\theta_i, \phi_i, P_{R_i}) - \mu_3 \quad I(\theta_i, \phi_i, P_{R_i}) - \mu_4 \quad J(\theta_i, \phi_i, P_{R_i}),\end{aligned} \tag{11.43}$$

where

$$F(\theta_i, \phi_i, P_{R_i}) = [\zeta\theta_i(P_S d_{S,R_i}^{-\vartheta} |h_{S,R_i}|^2 + \sigma_{n_{R_i}}^2) + E_{ext} - (1 - (\theta_i + \phi_i))P_{R_i}]^+, \tag{11.44}$$

$$G(\theta_i, \phi_i, P_{R_i}) = (1 - (\theta_i + \phi_i))B\log_2(1 + \gamma_{R_i,D}) - \phi_i[B\log_2(1 + \gamma_{S,R_i}) + (\delta_i \cdot r)] \leq 0, \tag{11.45}$$

$$H(\theta_i, \phi_i, P_{R_i}) = (1 - (\theta_i + \phi_i))P_{R_i} - \zeta\theta_i(P_S d_{S,R_i}^{-\vartheta} |h_{S,R_i}|^2 + \sigma_{n_{R_i}}^2) - E_{ext} \leq 0, \tag{11.46}$$

$$I(\theta_i, \phi_i, P_{R_i}) = r - (1 - (\theta_i + \phi_i))B\log_2(1 + \gamma_{R_i,D}) \leq 0, \tag{11.47}$$

$$J(\theta_i, \phi_i, P_{R_i}) = (\theta_i + \phi_i) - 1 \leq 0, \tag{11.48}$$

with $\mu_1, \mu_2, \mu_3, \mu_4$ being the Lagrange multipliers for the corresponding constraints (C1), (C2), (C3), and (C5).

For optimality, $\nabla\mathcal{L}(\theta_i, \phi_i, P_{R_i}; \mu_1, \mu_2, \mu_3, \mu_4) = 0$. Thus, we can represent the equations for satisfying the optimality conditions as

$$\begin{aligned}\frac{\partial\mathcal{L}(\theta_i, \phi_i, P_{R_i}; \mu_1, \mu_2, \mu_3, \mu_4)}{\partial\theta_i} &\Rightarrow [\zeta(P_S d_{S,R_i}^{-\vartheta} |h_{S,R_i}|^2 + \sigma_{n_{R_i}}^2) + P_{R_i}] \\ &-\mu_1[-B\log_2(1 + \gamma_{R_i,D})] - \mu_2[-P_{R_i} - \zeta(P_S d_{S,R_i}^{-\vartheta} |h_{S,R_i}|^2 + \sigma_{n_{R_i}}^2)] \\ &-\mu_3[B\log_2(1 + \gamma_{R_i,D})] - \mu_4 = 0,\end{aligned} \tag{11.49}$$

$$\begin{aligned}\frac{\partial\mathcal{L}(\theta_i, \phi_i, P_{R_i}; \mu_1, \mu_2, \mu_3, \mu_4)}{\partial\phi_i} &\Rightarrow P_{R_i} - \mu_1[-B\log_2(1 + \gamma_{R_i,D}) \\ &-(B\log_2(1 + \gamma_{S,R_i}) + (\delta_i \cdot r))] - \mu_2[-P_{R_i}] - \mu_3[B\log_2(1 + \gamma_{R_i,D})] - \mu_4 = 0,\end{aligned} \tag{11.50}$$

$$\begin{aligned}\frac{\partial\mathcal{L}(\theta_i, \phi_i, P_{R_i}; \mu_1, \mu_2, \mu_3, \mu_4)}{\partial P_{R_i}} &\Rightarrow -(1 - (\theta_i + \phi_i)) \\ &-\mu_1\left[(1 - (\theta_i + \phi_i))\left(\frac{\ln(2)d_{R_i,D}^{-\vartheta}|h_{R_i,D}|^2}{\sigma_{n_D}^2 + P_{R_i}d_{R_i,D}^{-\vartheta}|h_{R_i,D}|^2}\right)\right] \\ &-\mu_2(1 - (\theta_i + \phi_i)) - \mu_3\left[-(1 - (\theta_i + \phi_i))\left(\frac{\ln(2)d_{R_i,D}^{-\vartheta}|h_{R_i,D}|^2}{\sigma_{n_D}^2 + P_{R_i}d_{R_i,D}^{-\vartheta}|h_{R_i,D}|^2}\right)\right] = 0.\end{aligned} \tag{11.51}$$

The conditions for feasibility are as expressed in Eqs. (11.45), (11.46), (11.47), and (11.48). Complementary slackness expressions can be represented as follows:

$$\mu_1 \quad \cdot G(\theta_i, \phi_i, P_{R_i}) = 0, \tag{11.52}$$

$$\mu_2 \cdot H(\theta_i, \phi_i, P_{R_i}) = 0, \tag{11.53}$$

$$\mu_3 \cdot I(\theta_i, \phi_i, P_{R_i}) = 0, \tag{11.54}$$

$$\mu_4 \cdot J(\theta_i, \phi_i, P_{R_i}) = 0. \tag{11.55}$$

The conditions for non-negativity are: $\theta_i, \phi_i, P_{R_i}, \mu_1, \mu_2, \mu_3, \mu_4 \geq 0$. It is clear that if $\mu_4 \neq 0$, then $J(\theta_i, \phi_i, P_{R_i}) = 0$, implying that $\theta_i + \phi_i = 1$. Since this is not a feasible solution, therefore $\mu_4 = 0$. From further analysis, we find that the solutions corresponding to the binding cases of KKT yield either zero, negative, or unbounded values of the optimization variables, thereby violating the non-negativity constraints. The two possible solutions are as mentioned in the following theorems, respectively.

Theorem 11.2 If $\mu_1 \neq 0 \Rightarrow G(\theta_i, \phi_i, P_{R_i}) = 0$; $\mu_2 = 0 \Rightarrow H(\theta_i, \phi_i, P_{R_i}) \neq 0$; $\mu_3 \neq 0 \Rightarrow I(\theta_i, \phi_i, P_{R_i}) = 0$, then we obtain the following optimal values:

$$P_{R_i}^{\dagger} = (v - 1)\left(\frac{\sigma_{n_D}^2}{d_{R_i,D}^{-\vartheta}|h_{R_i,D}|^2}\right), \tag{11.56}$$

$$\phi_i^{\dagger} = \frac{r}{B\log_2(1 + \gamma_{S,R_i}) + (\delta_i \cdot r)}, \tag{11.57}$$

$$\theta_i^{\dagger} = 1 - r\left(\frac{1}{B\log_2(1 + \gamma_{S,R_i}) + (\delta_i \cdot r)} + \frac{1}{B\log_2\left(1 + \frac{P_{R_i}^{\dagger} d_{R_i,D}^{-\vartheta}|h_{R_i,D}|^2}{\sigma_{n_D}^2}\right)}\right), \tag{11.58}$$

where

$$v = \exp(\mathcal{W}(-\mathcal{A}\exp(-\log^2(2)) + \log(2)) + \log^2(2)) \tag{11.59}$$

with $\mathcal{A} = \ln(2) - (\zeta/\sigma_{n_D}^2)(\ln(2) d_{R_i,D}^{-\vartheta}|h_{R_i,D}|^2)(P_S d_{S,R_i}^{-\vartheta}|h_{S,R_i}|^2 + \sigma_{n_{R_i}}^2)$.

Theorem 11.3 If $\mu_1 \neq 0 \Rightarrow G(\theta_i, \phi_i, P_{R_i}) = 0$; $\mu_2 \neq 0 \Rightarrow H(\theta_i, \phi_i, P_{R_i}) = 0$; $\mu_3 \neq 0 \Rightarrow I(\theta_i, \phi_i, P_{R_i}) = 0$, then the following values are optimal:

$$P_{R_i}^{*} = (\eta_L - 1)\left(\frac{\sigma_{n_D}^2}{d_{R_i,D}^{-\vartheta}|h_{R_i,D}|^2}\right), \tag{11.60}$$

$$\phi_i^{*} = \frac{r}{B\log_2(1 + \gamma_{S,R_i}) + (\delta_i \cdot r)}, \tag{11.61}$$

$$\theta_i^{*} = \frac{rP_{R_i}^{*} - E_{ext}B\log_2\left(1 + \frac{P_{R_i}^{*} d_{R_i,D}^{-\vartheta}|h_{R_i,D}|^2}{\sigma_{n_D}^2}\right)}{\zeta(P_S d_{S,R_i}^{-\vartheta}|h_{S,R_i}|^2 + \sigma_{n_{R_i}}^2)}, \tag{11.62}$$

where η_L = largest root of$[\mathcal{A} + B\log_2(\eta)(\mathcal{B} + \mathcal{C}\eta + \mathcal{D}B\log_2(\eta)) = 0]$, with $\mathcal{A} = a \cdot b \cdot r$, $\mathcal{B} = -a \cdot b - b \cdot r \cdot (\sigma_{n_D}^2/d_{R_i,D}^{-\vartheta}|h_{R_i,D}|^2) + a \cdot r$, $\mathcal{C} = b \cdot r \cdot (\sigma_{n_D}^2/d_{R_i,D}^{-\vartheta}|h_{R_i,D}|^2)$, and $\mathcal{D} = -b \cdot E_{ext}$, where $a = \zeta(P_S d_{S,R_i}^{-\vartheta}|h_{S,R_i}|^2 + \sigma_{n_{R_i}}^2)$ and $b = B\log_2(1 + \gamma_{S,R_i}) + (\delta_i \cdot r)$.

To summarize the solutions obtained above, we propose the following algorithm to maximize the stored energy in the relay supporting SWIPT–Caching system (MSE-WC Algorithm).

Algorithm 11.1 MSE-WC Algorithm

Input: The parameters h_{S,R_i}, $h_{R_i,D}$, δ_i, r, and E_{ext}

Output: The maximized value of energy stored at the relay: $\{E_S\}$

1. Initialize: $\zeta \in (0, 1]$, $P_T \in (0, \varepsilon P_{Max}]$, $0.5 < \varepsilon < 1$, $\sigma^2_{n_{R_i}} = 1$, and $\sigma^2_{n_D} = 1$
2. Compute $P^{\dagger}_{R_i}$, $\phi^{\dagger}_i$, and $\theta^{\dagger}_i$ using Eqs. (11.56), (11.57), and (11.58), respectively
3. Define: $E^{\dagger}_S = \zeta\theta^{\dagger}_i(P_S d^{-\vartheta}_{S,R_i}|h_{S,R_i}|^2 + \sigma^2_{n_{R_i}}) + E_{ext} - (1 - (\theta^{\dagger}_i + \phi^{\dagger}_i))P^{\dagger}_{R_i}$
4. Compute $P^*_{R_i}$, ϕ^*_i, and θ^*_i using Eqs. (11.60), (11.61), and (11.62) respectively
5. Define: $E^*_S = \zeta\theta^*_i(P_S d^{-\vartheta}_{S,R_i}|h_{S,R_i}|^2 + \sigma^2_{n_{R_i}}) + E_{ext} - (1 - (\theta^*_i + \phi^*_i))P^*_{R_i}$
6. $E_S = \max(E^{\dagger}_S, E^*_S)$
7. **return** E_S

The algorithm proposed above returns the maximized value of the objective function as its output. First, we initialize all the necessary values as indicated in 1. Then, we compute the optimal values of $P^{\dagger}_{R_i}$, $\phi^{\dagger}_i$, and $\theta^{\dagger}_i$ in 2, and define the energy stored at the relay in 3, where 2 and 3 correspond to the solutions obtained for Case VI during the analysis. Similarly, we find the optimal values of $P^*_{R_i}$, ϕ^*_i, and θ^*_i in 4 and define the energy stored at the relay in 5 accordingly, where 4 and 5 correspond to the solutions obtained for Case VIII during the analysis. Next, we find the maximum of the two computed local optimal solutions for the energy stored at the relay, which in turn maximizes the objective function. It should also be noted that the solutions proposed in (P2) for maximizing the energy stored at the relay are not necessarily a global optimum, as the problem is non-linear in nature. However, the KKT conditions guarantee the local optimal solutions.

In order to analyze the proposed approach from an economic perspective, we denote the equivalent relationship between the Lagrange multipliers μ_1 and μ_3 as μ_R and rename μ_2 as μ_E, which corresponds to the the prices for data and energy, respectively, in cost units. Using the results from our mathematical analysis, we find that $\mu_E = f_1(P_S, \delta_i, r)/f_2(P_S) \cdot \mu_R$, where $f_1(P_S, \delta_i, r)$, and $f_2(P_S)$ are functions computed as per the illustrated technique. Correspondingly, if more energy is required by the relay with fixed caching capacity, then we are forced to compensate for the request by using the energy metric per cost unit at the source in order to satisfy the respective data and energy constraints in ($P2$). This action would, however, add more to the data price as well. Similarly, it is apparent that increasing the transmit power will readily add to the cost of energy transfer in addition to an increased data price. Furthermore, in the context of caching, it is worth mentioning that *extra cache capacity implies subordinate prices for data and energy transmissions per cost unit.*

11.4.2 Relay Selection

From the methods proposed above, optimal TS ratios and the relay transmit power can be computed easily. Herein, we propose to find the best relay that provides maximized harvested power corresponding to (11.31). In this context, the index of the optimally selected relay can be expressed as $j^\star = \arg\max_{j\in\{1,\ldots,K\}} E^\star_{S_j}$, where $E^\star_{S_j}$ is the optimal energy stored at the jth relay as the solution of problem (11.37).

11.5 Numerical Results

In this section, we evaluate the performance of the proposed system for the solutions presented in this chapter. We consider a total bandwidth of $B = 1$ MHz, $\zeta = 0.80$, $\sigma^2_{n_{R_i}} = 0$ dBW, and $\sigma^2_{n_D} = 0$ dBW [23]. Throughout the simulations, we assume that the channel coefficients are i.i.d. and follow Rayleigh distribution. The path loss is considered to be 3 dB [24]. Additionally, all the relays have the same caching coefficient, i.e., $\delta_i = \delta, \forall i$. All the results are evaluated over 500 Monte Carlo random channel realizations. The proposed architecture is compared with a reference scheme using a fix time-splitting. The reference scheme spends the first half period for information broadcasting and energy harvesting, and uses the second half period for relaying.

Figure 11.5 plots the performance of the cache-aided SWIPT as a function of the total number of relays. The result is calculated based on the best relay selected as in Section 11.3 and Section 11.4. In both cases, $\delta = 0.5$, $E_{ext} = 30$ Joules. It is observed from Figure 11.5a that the proposed architecture significantly outperforms the reference and the gain is larger as the number of available relays increases. In particular, the proposed architecture achieves a performance gain of 15% for $P_S = 25$ dBW and 20% for $P_S = 30$ dBW over the reference scheme. This result confirms the effectiveness of the proposed optimization framework. It is also shown that having more relays results in a better serving rate thanks to the inherent diversity gain brought by the relays. The harvested energy comparison between the proposed and the reference is plotted in Figure 11.5b. A similar conclusion is observed where the proposed architecture surpasses the reference in all cases. In addition, having more available relays improves the harvested energy.

Figure 11.6 illustrates the results corresponding to the solutions proposed for the rate maximization problem, assuming that an optimal relay is chosen as per the solutions corresponding to the outer optimization of (11.7). It is observed from the figure that the source transmit power has a significant influence on the achievable rate. In particular, by increasing the source transmit power by 5 dBW, the serving rate is increased by 30%. This result can be explained by the fact that for a given caching coefficient, the source does not have to send the whole requested content to the relay. In this case, having a larger source power results in more harvested energy, which in turn increases the relay's transmit power. Especially, this observation is also obtained when there is not an external energy source, e.g. $E_{ext} = 0$, which shows the effectiveness of the proposed cache-aided SWIPT architecture.

Figure 11.7 plots the maximum throughput as a function of the caching gain coefficient with $E_{ext} = 30$ Joules and $r = 3$ Mbps. It is observed that the caching gain has a similar impact on the achievable throughput for different values of P_S. In general, a larger cache size (or equivalent larger caching coefficient) results in a higher serving rate. This result together with result in Figure 11.6 suggest an interactive role of the caching capacity and the transmit power. In particular, a smaller source power system can still achieve the same throughput by increasing the cache size.

Figure 11.8 presents the stored energy at the chosen relay, according to the solution of outer optimization of (11.31), as a function of the source's transmit power and different external energy values. It is shown from the results that the source transmit power has a large impact on the stored energy at the relay. In particular, increasing the source's transmit power by 2 dBW will double the stored energy at the relay. It is also observed

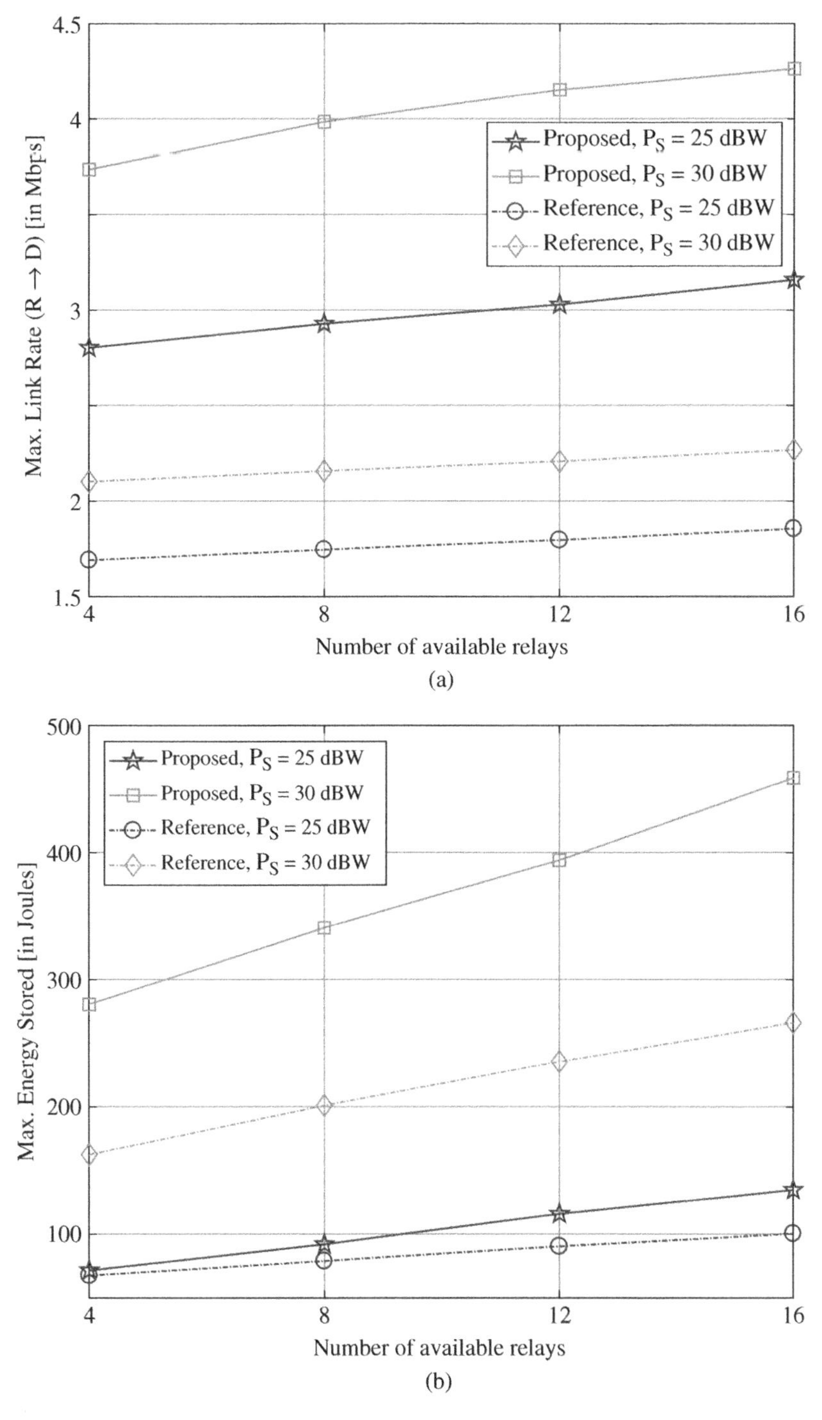

Figure 11.5 Performance comparison between the proposed cache-aided SWIPT and the reference scheme for different numbers of available relays with $\delta = 0.5$ and $E_{ext} = 30$ Joules. (a) Link rate performance, $r = 1$ Mbps. (b) Stored energy performance, $r = 1$ Mbps.

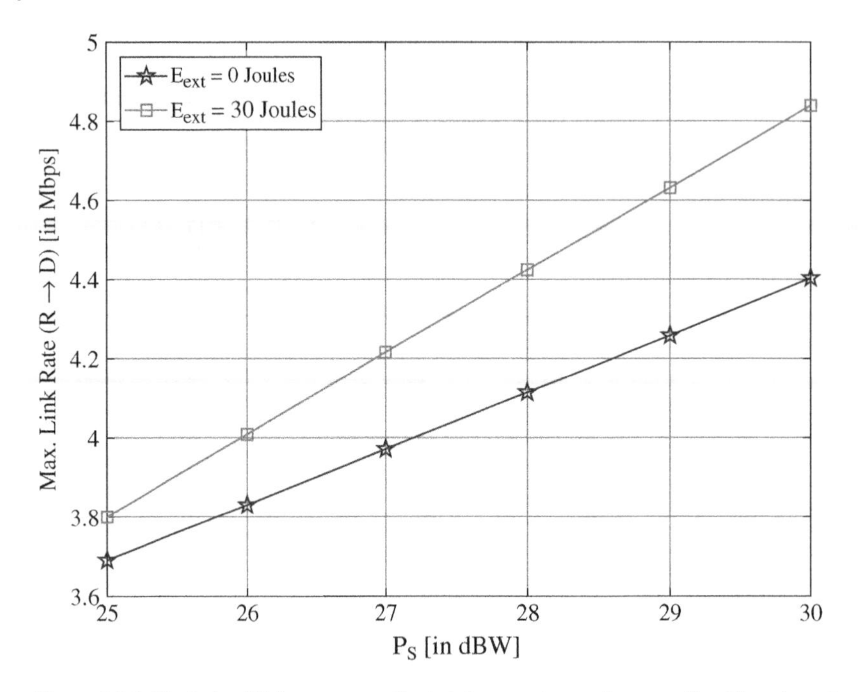

Figure 11.6 Maximized link rate versus the total source transmit power P_S with various E_{ext}. The total number of available relays $K = 8$, caching coefficient $\delta = 0.5$, and $r = 1$ Mbps.

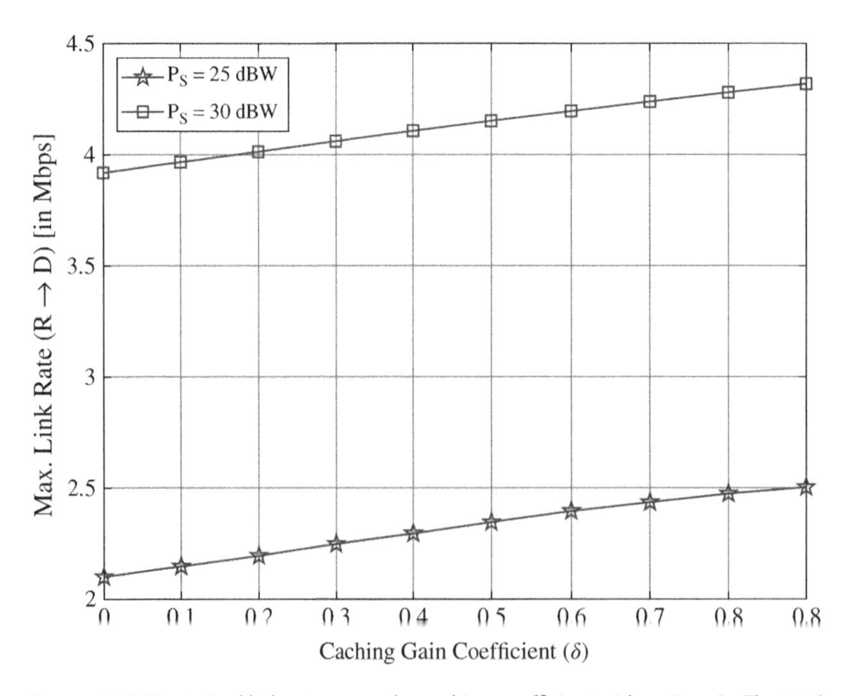

Figure 11.7 Maximized link rate versus the caching coefficient with various P_S. The total number of available relays $K = 8$, $E_{ext} = 30$ Joules, and $r = 3$ Mbps.

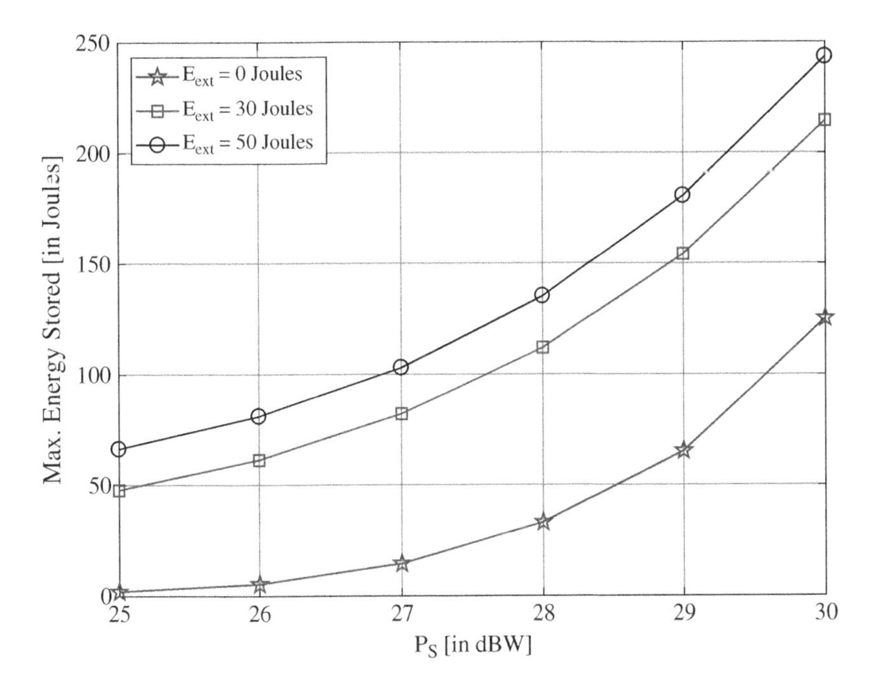

Figure 11.8 Stored energy performance as a function of the source transmit power for various values of E_{ext}. $K = 8$ relays, $\delta = 0.5$, and $r = 3$ Mbps.

that increasing the external energy can significantly improve the stored energy at high P_S values. However, when P_S is small, increasing E_{ext} does not bring considerable improvement. This is because at low P_S values, most of the time is used for information transfer from the source to the relay.

Figure 11.9 depicts the plot of the energy stored at the selected relay as a function of the cache capacity δ, with $E_{ext} = 30$ Joules and $r = 3$ Mbps. The case with $\delta = 0$ implies that there is no caching at the relay. It is shown that caching helps to increase the saved energy at the relay for all P_S values and the increased stored energy is almost similar for different P_S. This is because of the linear model of the caching system.

Figure 11.10 presents an evaluation of the energy stored at the chosen relay against the increasing values of r. It is seen that the energy stored at the relay decreases with increasing values of the requested rate (r), for $\delta = 0.5$ and $E_{ext} = 30$ Joules. On the other hand, it is clear that with increasing values of P_S, the energy stored at the relay increases non-linearly. The former variation is due to the fact that in order to meet the demand of the requested rate at the destination, more energy would be required for resource allocation at the relay that utilizes the harvested energy.

11.6 Conclusion

In this chapter, we proposed and investigated relay selection strategy in a novel time switching (TS) based SWIPT with caching architecture involving half duplex relaying systems where the relay employs the DF protocol. We addressed the problem of

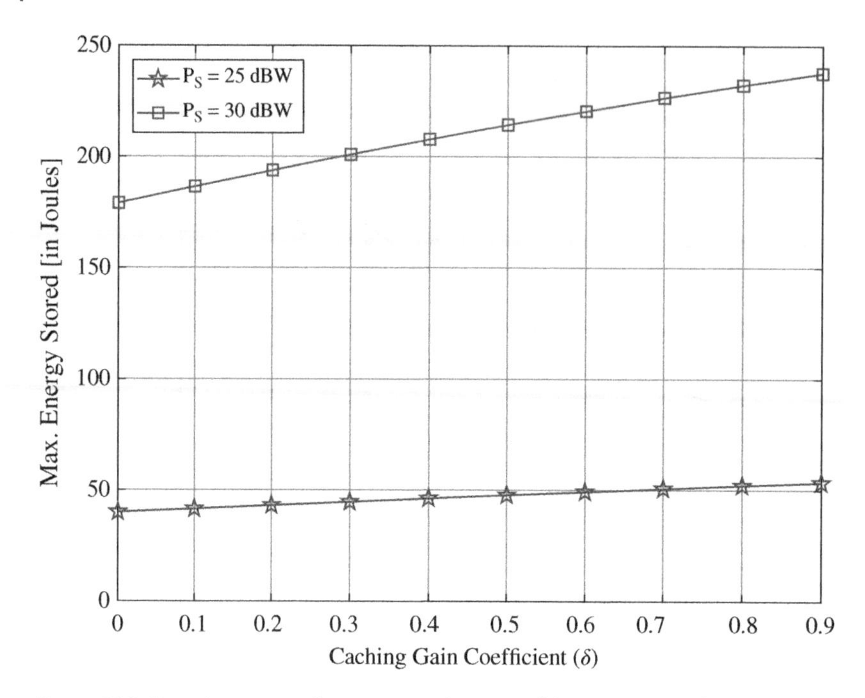

Figure 11.9 Stored energy performance as a function of the caching coefficient for various values of P_S. $K = 8$ relays, $E_{ext} = 30$ Joules, and $r = 3$ Mbps.

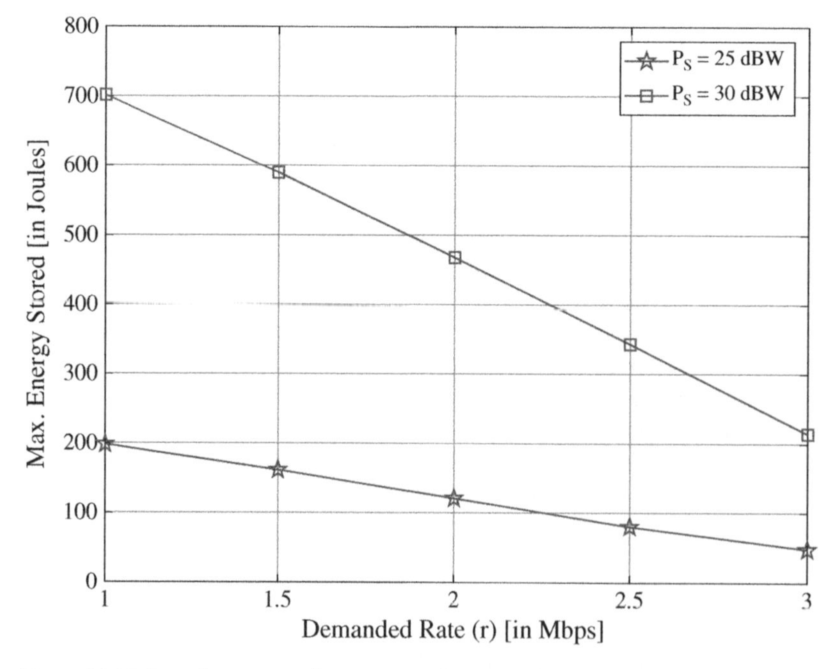

Figure 11.10 Stored energy performance as a function of QoS requirements with different source transmit power. $K = 8$ relays, $\delta = 0.5$, and $E_{ext} = 30$ Joules.

relay selection to maximize the data throughput between the relay and destination under constraint on minimum energy stored at the relay and also relay selection for maximizing the energy stored at the relay under constraints on the minimum rate and harvested energy, guaranteeing a good performance in both cases with regards to the QoS constraints. In addition, both problems were formulated according to two separate yet distinct conventions over the time period. We presented the closed-form solutions for the proposed relay system to enable SWIPT with caching. With the help of simulations, we illustrated the results corresponding to the solutions obtained for the aforementioned problems with parameter variations. This work can be further extended in many fascinating directions, such as multi-user and multi-carrier scenarios and relaying with a full duplexing mode.

Acknowledgment

This work was supported, in parts, by the Luxembourg National Research Fund (FNR) under the project FNR CORE ProCAST, FNR-FNRS bitateral InWIP-NET and, in parts, by the European Research Council (ERC) under project AGNOSTIC.

Bibliography

1 S. Gautam, E. Lagunas, S.K. Sharma, S. Chatzinotas, and B Ottersten, "Relay selection strategies for SWIPT-enabled cooperative wireless systems," in *28th Annual IEEE International Symposium on Personal, Indoor and Mobile Radio Communications (PIMRC)*, Oct. 2017.
2 S. Zhou, J. Gong, Z. Zhou, W. Chen, and Z. Niu, "GreenDelivery: proactive content caching and push with energy-harvesting-based small cells," *IEEE Commun. Mag.*, vol. 53, no. 4, pp. 142–149, Apr. 2015.
3 ERICSSON White Paper, "More than 50 Billion Connected Devices," 2011.
4 S. Gautam and P. Ubaidulla. "Simultaneous transmission of information and energy in OFDM systems," in *Proc. 18th Wireless Personal Multimedia Communications (WPMC)*, Dec. 2015.
5 Q. Gu, G. Wang, R. Fan, Z. Zhong, K. Yang, and H. Jiang, "Rate-energy tradeoff in simultaneous wireless information and power transfer over fading channels with uncertain distribution", *IEEE Trans. Veh. Technol.*, vol. 67, no. 4, pp. 3663–3668, 2018.
6 J. Park, B. Clerckx, C. Song, and Y. Wu. "An analysis of the optimum node density for simultaneous wireless information and power transfer in ad hoc networks," *IEEE Trans. Veh. Technol.*, vol. 67, no. 3, pp. 2713–2726, 2018.
7 B. Clerckx, "Waveform and transceiver design for simultaneous wireless information and power transfer," *CoRR*, abs/1607.05602, 2016.
8 I. Bang, S.M. Kim, and D.K. Sung, "Adaptive multiuser scheduling for simultaneous wireless information and power transfer in a multicell environment," *IEEE Trans. Wireless Commun.*, vol. 16, no. 11, pp. 7460–7474, Nov. 2017.

9 S. Mahama, D.K.P. Asiedu, and K.J. Lee, "Simultaneous wireless information and power transfer for cooperative relay networks with battery," *IEEE Access*, vol. 5, pp. 13171–13178, 2017.
10 X. Zhou, R. Zhang, and C.K. Ho, "Wireless information and power transfer in multiuser OFDM systems," *IEEE Trans. Wireless Commun.*, vol. 13, no. 4, pp. 2282–2294, Apr. 2014.
11 X. Zhou, R. Zhang, and C.K. Ho, "Wireless information and power transfer: architecture design and rate-energy tradeoff," *IEEE Trans. Commun.*, vol. 61, no. 11, pp. 4754–4767, Nov. 2013.
12 J. Huang, C.C. Xing, and C. Wang. "Simultaneous wireless information and power transfer: technologies, applications, and research challenges," *IEEE Commun. Mag.*, vol. 55, no. 11, pp. 26–32, Nov. 2017.
13 S. Borst, V. Gupta, and A. Walid. "Distributed caching algorithms for content distribution networks," in *Proc. IEEE Int. Conf. Comput. Commun.*, pp. 1–9, Mar. 2010.
14 M.A. Maddah-Ali and U. Niesen. "Fundamental limits of caching," *IEEE Trans. Inf. Theory*, vol. 60, no. 5, pp. 2856–2867, May 2014.
15 T.X. Vu, S. Chatzinotas, and B. Ottersten. "Edge-caching wireless networks: performance analysis and optimization," *IEEE Trans. Wireless Commun.*, vol. 17, no. 4, pp. 2827–2839, 2018.
16 W. Peng, M. Li, Y. Li, W. Gao, and T. Jiang. "Ultra-dense heterogeneous relay networks: a non-uniform traffic hotspot case," *IEEE Network*, vol. 31, no. 4, pp. 22–27, Jul. 2017.
17 C. Yang, J. Xiao, J. Li, X. Shao, A. Anpalagan, Q. Ni, and M. Guizani, "Disco: interference-aware distributed cooperation with incentive mechanism for 5g heterogeneous ultra-dense networks," *IEEE Comm. Mag.*, vol. 56, no. 7, pp. 198–204, 2018.
18 A. Kumar and W. Saad. "On the tradeoff between energy harvesting and caching in wireless networks," in *2015 IEEE International Conference on Communication Workshop (ICCW)*, pp. 1976–1981, Jun. 2015.
19 D. Niyato, D.I. Kim, P. Wang, and L. Song. "A novel caching mechanism for Internet of Things (IoT) sensing service with energy harvesting," in *Proc. IEEE Int. Conf. Commun.*, pp. 1–6, May 2016.
20 S. Zhang, N. Zhang, X. Fang, P. Yang, and X.S. Shen, "Cost-effective vehicular network planning with cache-enabled green roadside units," in *Proc. IEEE Int. Conf. Commun.*, pp. 1–6, May 2017.
21 S. Lohani, R.A. Loodaricheh, E. Hossain, and V.K. Bhargava, "On multiuser resource allocation in relay-based wireless-powered uplink cellular networks," *IEEE Trans. Wireless Commun.*, vol. 15, no. 3, pp. 1851–1865, Mar. 2016.
22 R.M. Corless, G.H. Gonnet, D.E.G. Hare, D.J. Jeffrey, and D.E. Knuth, "On the LambertW function," *Advances in Computational Mathematics*, vol. 5, no. 1, pp. 329–359, 1996.
23 V. Havary-Nassab, S. Shahbazpanahi, A. Grami, and Z.Q. Luo, "Distributed beamforming for relay networks based on second-order statistics of the channel state information," *IEEE Trans. Signal Process.*, vol. 56, no. 9, pp. 4306–4316, Sep. 2008.
24 D. Green, Z. Yun, and M.F. Iskander, "Path loss characteristics in urban environments using ray-tracing methods," *IEEE Antennas Wireless Propag. Lett.*, vol. 16, pp. 3063–3066, 2017.

12

Cooperative Video Streaming in Ultra-dense Networks with D2D Caching

Nguyen-Son Vo[1] *and Trung Q. Duong*[2]

[1] *Institute of Fundamental and Applied Sciences, Duy Tan University, Vietnam*
[2] *School of Electronics, Electrical Engineering and Computer Science, Queen's University Belfast, UK*

12.1 Introduction

The world has been in a new era of industry, namely the fourth industrial revolution, where by 2020 there will be about 50 billion connected devices communicating with each other in the context of the Internet of Things [1]. In this context, 5G networks will be facing the challenge of a massive number of mobile users (MUs) requesting an explosive amount of traffic. In other words, this may cause 5G networks to be degraded due to the traffic congestion problem in the backhaul links of macro base stations (MBSs). Though ultra-dense networks (UDNs) hold considerable promise as a new architecture innovation to meet the requirements of 5G networks, e.g. a thousand-fold system capacity and less than one millisecond end-to-end latency [2], the development of UDNs requires more disruptive techniques and optimization designs that can provide the surge in MUs with high data rate services, e.g. video streaming services (VSSs), at high quality of experience (QoE) and high resource saving.

To cope with the challenge, many techniques and optimization designs for UDNs have been focusing on how to efficiently utilize the spatial, spectrum, bandwidth, and energy resources as well as how to bring the advanced services closer to the MUs. For example, massive MIMO and mmWave communication technologies have been studied to improve the spectrum efficiency and expand the transmission bandwidth [3–5]; UDNs can be combined with mmWave technology, interference-aware, power control in D2D networks to increase the energy and spectral efficiency [6–8]; and D2D communications [9–11] and especially D2D caching techniques [12–14] for proximity services can be applied to UDNs to maximize the MUs capacity while mitigating the congestion in the backhaul links of the MBSs.

Though the aforementioned solutions can serve a massive number of MUs, high performance in terms of a high data rate at high resource efficiency and high quality of services, they cannot be directly applied to VSSs for a high QoE. The reason is that the QoE of VSSs is sensitive to many aspects of wireless transmission environment, MUs' behavior and resource, video characteristics, and video coding techniques. Basically, the QoE for the MUs, i.e. the requesters, who request VSSs can be represented by three features: (1) punctual arrival for continuous playback, (2) high playback quality, and (3) low

Ultra-dense Networks for 5G and Beyond: Modelling, Analysis, and Applications,
First Edition. Edited by Trung Q. Duong, Xiaoli Chu and Himal A. Suraweera.

quality fluctuation for smooth playback. It has been observed recently that though many video streaming techniques for D2D communications have been studied in the literature [15–17], simultaneously satisfying high QoE, saving D2D communications energy, and preserving target signal-interference plus noise ratio (SINR) of the MUs, i.e. the cellular users (CUs), who share their available downlink resources for D2D communications, have been unsolved challenges.

In this chapter, we exploit the caching storage resources in the MUs (i.e. D2D helpers) and the downlink resources in the CUs to cooperatively support the MBSs in VSSs. This means that the requesters can retrieve VSSs from both the D2D helpers with offloading D2D communications and the MBSs in 5G networks. In this way, thanks to close proximity D2D communications, the requesters punctually receive the video segments for continuous playback at a higher capacity. Furthermore, by taking into account the lossy features of wireless channels, video characteristics, and users' behaviour, such as the rate-distortion (RD) model and intra-popularity of video segments [18], D2D helpers energy, and the impact of cointerference caused by D2D communications on the CUs, we formulate a joint encoding rate allocation and description distribution optimization (RDO) problem [19]. The RDO is solved for higher performance of video streaming services by finding two optimal results. First, the optimal encoding rates are found to partition each video segment into different layers. These layers are then packetized into multiple descriptions for transmission using layered multiple description coding with embedded forward error correction (MDC-FEC) [18, 20, 21]. This is to ensure that video streaming services are protected against the error-prone transmission and are adaptively sent over the diverse transceiver bandwidth and playback capacities of mobile devices. Second, the optimal numbers of descriptions distributed to each D2D helper and to the MBSs are found for cooperative transmission. Consequently, the average reconstructed distortion of received video segments is minimized for high playback quality. The RDO also provides the requesters, the D2D helpers, and the CUs with low quality fluctuations of VSSs, low energy consumption, and high SINR guarantee, respectively.

The chapter is organized as follows. Section 12.2 proposes the system model for cooperative video streaming in 5G networks with dense D2D caching. In Section 12.3, we formulate the RDO problem and introduce its solution. Section 12.4 is dedicated to presenting simulation results for performance evaluation. Finally, we conclude the chapter in Section 12.5.

12.2 5G Network with Dense D2D Caching for Video Streaming

In this section, we first introduce our system and some important notations and assumptions. Afterwards, a cooperative transmission strategy (CTS) for video streaming in a 5G network with dense D2D caching is proposed to show how the system works in detail. Finally, we present the source video encoding model to packetize the video (i.e. video segments) into multiple descriptions for transmission over the system.

12.2.1 System Model and Assumptions

We consider a typical system model for cooperative video streaming in a 5G network with dense D2D communications, as shown in Figure 12.1. The system works on the assumptions given as follows:

- Thanks to dense D2D helpers, many of them, which have the same wireless channel characteristic to a particular requester, are categorized into a group, namely a group of D2D helpers (GoH), by the MBS. In each GoH, a D2D helper that is equipped with the highest available energy and processing capacity, is elected to the D2D head.
- All D2D helpers in a GoH are scheduled with their own time slots for transmission over the same subchannel shared by a CU.
- The wireless channel features remain unchanged during sending a video segment from the GoHs and MBS to the requesters.

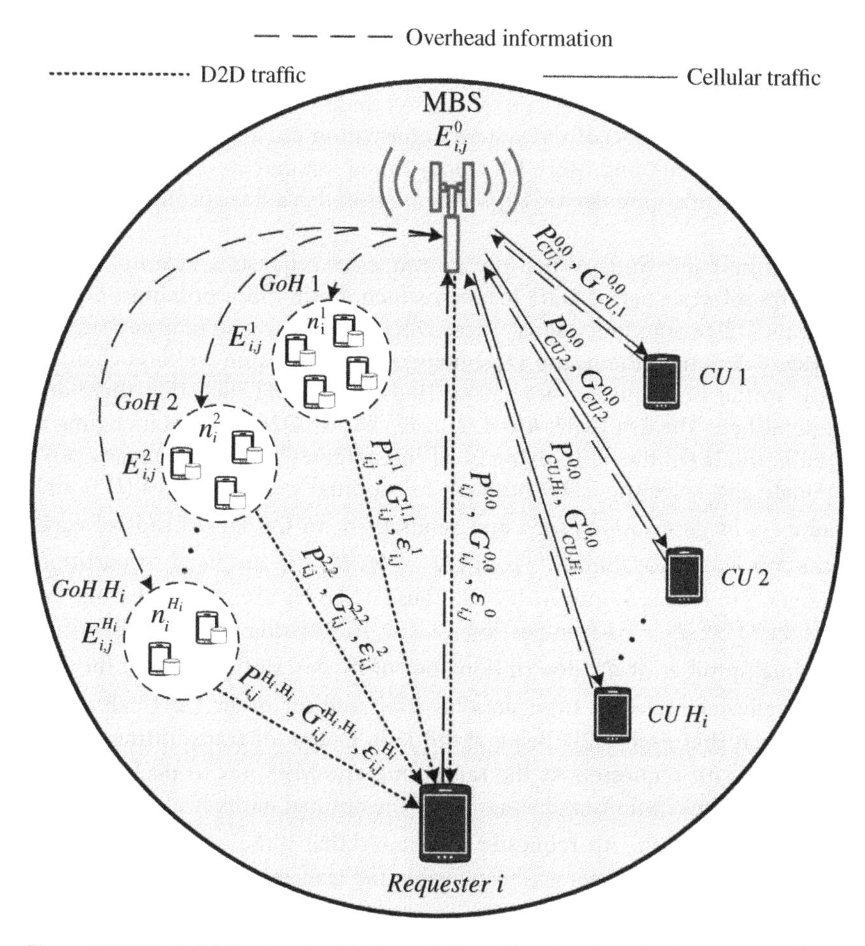

Figure 12.1 Typical 5G network with dense D2D caching. Reprinted with permission from [19].

In particular, the system consists of one MBS and a massive number of MUs categorized into three types: CUs, D2D helpers, and requesters. When the ith requester, $i = 1, 2, ..., I$, requests a video of S segments, the MBS and H_i GoHs (each has n_i^h D2D helpers, $h = 1, 2, ..., H_i$) will collaborate in streaming the video, i.e. segment by segment, to the ith requester. There are also H_i CUs that share their subchannels with H_i GoHs for D2D communications between the D2D helpers and the ith requester. The system parameters are defined in Table 12.1. The detailed operation of the proposed system is presented in the CTS as below.

12.2.2 Cooperative Transmission Strategy

The system works by relying on the CTS with integrated RDO solution installed in the MBS. The CTS includes three stages: initial stage, cooperative transmission stage, and updated stage, given as follows:

- *Initial Stage* The MBS collects all the system parameters in its cell such as the remaining energy of D2D helpers, lossy and gain of wireless channel, bandwidth and additive white Gaussian noise (AWGN) of the system, information of videos (i.e. RD model, size, and cache placement), and user's behaviour to request and view the videos (i.e. inter-popularity and intra-popularity [19]). They are stored in a stack profile for future use [22].
- *Cooperative Transmission Stage* When the ith requester requests a video of S segments, the MBS selects a set of D2D helpers, which are in close proximity to the ith requester for D2D communications. These D2D helpers have already cached the requested video. Amongst them, the D2D helpers with the same wireless channel characteristic to the ith requester are categorized into a GoH such that there are H_i GoHs established. The hth GoH, $h = 1, 2, ..., H_i$, has n_i^h D2D helpers including a D2D elected head. Then, the MBS extracts all the parameters from the stack profile to formulate and solve the RDO problem for optimal encoding rates (R_j^M) and optimal numbers of descriptions ($d_{i,j}^h$) and sends them to the D2D heads of each GoH. At the hth D2D head, the R_j^M, $m = 1, 2, ..., M$, $R_j^0 = 0$, are used to partition the jth segment, $j = 1, 2, ..., S$, into M layers. These M layers are packetized into M descriptions, i.e. $\sum_{h=0}^{H_i} d_{i,j}^h = M$ (see Section 12.2.3). Afterwards, the hth D2D head takes its optimal number of $d_{i,j}^h$ descriptions out of M descriptions. Next, the hth D2D head equally schedules n_i^h time slots for itself and for other D2D helpers in the hth GoH such that each D2D helper/head is in charge of transmitting $d_{i,j}^h/n_i^h$ descriptions to the ith requester. At the same time, the MBS has to packetize the jth segment into M descriptions and assign itself the optimal number of $d_{i,j}^0$ descriptions for transmission to the ith requester. It is noted that if the quotient of $d_{i,j}^h/n_i^h$ is not an integer, the D2D heads are responsible for transmitting the remaining descriptions.
- *Updated Stage* To serve the next requester, the MBS keeps updating the system parameters by checking if there are any significant changes. In addition, the MBS will remove the information about a video from the stack profile if the video is not popular.

Table 12.1 Notation.

Symbols	Descriptions
I	Number of requesters for the considered video
H_i	Number of GoHs already cached the requested video and connected to the ith requester ($i = 1, 2, ..., I$) for D2D communications; H_i is also the number of CUs that share their downlink resources with the GoHs
n_i^h	Number of D2D helpers in the hth GoH connected to the ith requester, $h = 1, 2, ..., H_i$; $h = 0$ is used for the MBS
S	Number of segments of the requested video
V,v	Size of video, size of each segment (bits), i.e. $v = V/S$
M	Number of layers (or descriptions) of a segment
W	System bandwidth
R_j^M	Maximum encoding rate of the jth segment
$D_j(R_j^m)$	Source distortion of the jth segment at rate R_j^m (kbps), $m = 1, 2, ..., M$
$RS(M, m)$	Reed-Solomon erasure code applied across m blocks of the mth layer of the jth segment to yield the codewords of length M blocks with embedded FEC
$d_{i,j}^h$	Number of descriptions of the jth segment sent to the ith requester by the hth GoH ($h > 0$) and the MBS ($h = 0$)
$E_{i,j}^h$	Energy consumption of the hth GoH and the MBS to send $d_{i,j}^h$ descriptions of the jth segment to the ith requester
$P_{i,j}^{h,h}$	Common transmission power of the hth GoH and transmission power of the MBS ($h = 0$) drained from the battery to send $d_{i,j}^h$ descriptions of the jth segment to the ith requester
$P_{CU,h}^{0,0}$	Transmission power of the MBS to communicate with the hth CU that shares the downlink resource with the hth GoH
$\overline{E}$	Average energy consumed per D2D helper to send the requested video
$t_{i,j}^h$	Time to send $d_{i,j}^h$ descriptions of the jth segment by the hth GoH and the MBS ($h = 0$)
$G_{i,j}^{h,h}$	Common channel gain (or channel gain) between the hth GoH (or the MBS) and the ith requester
$G_{CU,h}^{0,0}$	Channel gain between the MBS and the hth CU
r_j	Access rate/intra-popularity of the jth segment
$P_{i,j,M}^m$	Probability of correctly receiving m out of M descriptions of the jth segment at the ith requester
$P_{d_{i,j}^h}^{m_h}$	Probability of correctly receiving m_h out of $d_{i,j}^h$ descriptions of the jth segment sent by the hth GoH and the MBS ($h = 0$),$\{m_h\} : \sum_{h=0}^{H_i} m_h = m$
$\varepsilon_{i,j}^h$	Description error rate of the jth segment over the common channel (the channel) between the hth GoH (the MBS) and the ith requester
$\overline{D}$	Average reconstructed distortion of a received video at the requesters

12.2.3 Source Video Packetization Model

Consider a video of size V bits including S segments; the source video packetization of the jth segment, $j = 1, 2, ..., S$, for transmission is shown in Figure 12.2 and presented in three steps given below [18, 20, 23, 24].

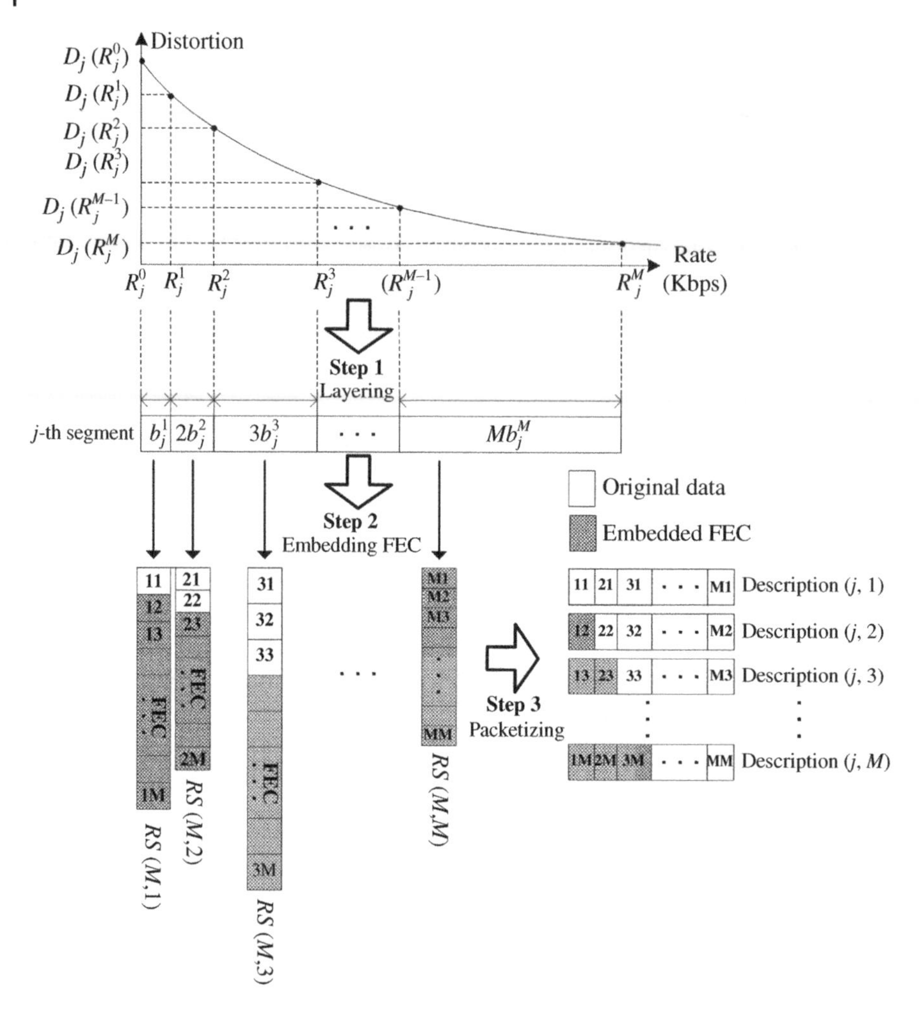

Figure 12.2 Source video packetization model. Reprinted with permission from [19]).

In the layering step, the jth segment is divided into M layers. The mth layer, $m = 1, 2, ..., M$, is limited by the lower encoding rate R_j^{m-1} and the upper encoding rate R_j^m, where $0 = R_j^0 \leq R_j^1 \leq R_j^2 \leq \cdots \leq R_j^M$. The mth layer is partitioned into m blocks, each having the same size, given by

$$b_j^m = \lfloor \frac{v\left(R_j^m - R_j^{m-1}\right)}{mR_j^M} \rfloor, \tag{12.1}$$

where $v = V/S$ is the size of each segment. The operator $\lfloor . \rfloor$ is used to round the size of each block to the nearest lower integer. For convenience, we do not dwell on complementing the bits truncated by this operator.

In the embedding FEC step, the FEC of size $(M - m)$ blocks is embedded in the m blocks of the mth layer by using the Reed-Solomon (RS) code. This results in the mth layer with an embedded FEC, i.e. the codeword $RS(M, m)$ of length M blocks. In this way, the $RS(M, m)$ can correct up to $(M - m)$ erroneous blocks.

In the packetizing step, the mth blocks are selected from each layer with an embedded FEC to packetize the m-th description for transmission. Finally, we have M descriptions of the jth segment, each having the same size of

$$d_j^{FEC} = \sum_{m=1}^{M} b_j^m. \tag{12.2}$$

In this way, the reconstructed distortion of the jth segment depends on the number of correctly received descriptions regardless of the order of the descriptions. If there are m out of M descriptions correctly received, all the first m layers of the jth segment are decoded, and thus the jth segment is played back at the reconstructed distortion $D_j(R_j^m)$.

12.3 Problem Formulation and Solution

In this section, by following the given system model and taking the characteristic of the lossy wireless channel into account, the average reconstructed distortion of received video segments is derived as the objective function of the optimization problem. Then, the energy consumption of D2D helpers and co-channel interference guarantee are expressed as the two important constraints of the RDO problem. Finally, we formulate the RDO problem and solve it by using genetic algorithms (GAs).

12.3.1 System Parameters Formulation

12.3.1.1 Average Reconstructed Distortion

For the sake of formulation, we apply a simple decaying exponential function to the RD model [21, 25], given as

$$D_j(R_j^m) = \gamma_j (R_j^m)^{\beta_j}, \tag{12.3}$$

where γ_j and β_j are the sequence-dependent parameters. It is noted that by finding the proper values of γ_j and β_j, Eq. (12.3) is closely equivalent to the experimental RD model [21, 25].

Following Eq. (12.3) over a lossy wireless channel, let $P_{i,j,M}^m$ be the correct probability to receive m out of M descriptions; the average reconstructed distortion of received video is then calculated by

$$\overline{D} = \frac{1}{I} \sum_{i=1}^{I} \sum_{j=1}^{S} r_j \gamma_j \left[\sum_{m=0}^{M} P_{i,j,M}^m R_j^m \right]^{\beta_j}, \tag{12.4}$$

where I is the number of requesters, S is the number of segments of the considered video, and r_j is the access rate of the jth segment (i.e. the intra-popularity of the jth segment). The statistical study in [26] shows that the access rates of different segments can be defined by Zipf-like distribution as

$$r_j = \frac{j^{-\alpha}}{\sum_{j=1}^{S} j^{-\alpha}}. \tag{12.5}$$

Here α is the skewed coefficient used to represent the behavior of the requesters, i.e. the higher the value of α, the first less number of segments of a video the requesters access.

We can see in Eq. (12.4) that to have m descriptions correctly received, the hth GoH ($h > 0$) and the MBS ($h = 0$) cooperatively send $d^h_{i,j}$ descriptions and $d^0_{i,j}$ descriptions, respectively. Let $P^{m_h}_{d^h_{i,j}}$ be the correct probability to receive m_h out of $d^h_{i,j}$ descriptions such that $\sum_{h=0}^{H_i} m_h = m$. Then $P^m_{i,j,M}$ in Eq. (12.4) is computed by

$$P^m_{i,j,M} = \sum_{\{m_h\}:\sum_{h=0}^{H_i} m_h = m} \prod_{h=0}^{H_i} P^{m_h}_{d^h_{i,j}}, \tag{12.6}$$

where

$$P^{m_h}_{d^h_{i,j}} = \binom{m_h}{d^h_{i,j}} \left(1 - \varepsilon^h_{i,j}\right)^{m_h} \left(\varepsilon^h_{i,j}\right)^{d^h_{i,j} - m_h}, \tag{12.7}$$

where $\varepsilon^h_{i,j}$ is the description error rate over the common channel (or the channel) between the hth GoH (or the MBS) and the ith requester.

12.3.1.2 Energy Consumption Guarantee

Because the energy at the MBS is extremely higher than that at the D2D helpers, we only focus on the energy consumed by the hth GoH to transmit $d^h_{i,j}$ descriptions of the jth segment to the ith requester, which is given by

$$E^h_{i,j} = t^h_{i,j} P^{h,h}_{i,j}, \tag{12.8}$$

where the transmission time is

$$t^h_{i,j} = \frac{d^h_{i,j} d^{FEC}_j}{10^3 R^M_j}, \tag{12.9}$$

and $P^{h,h}_{i,j}$, which is the common transmission power of the hth GoH ($h > 0$) and the transmission power of the MBS ($h = 0$), is obtained by relying on the Gaussian channel between the hth GoH/MBS and the ith requester with Shannon-like capacity [27, 28]:

$$P^{h,h}_{i,j} = \begin{cases} \left(G^{h,h}_{i,j}\right)^{-1} \left(2^{\frac{R^M_j}{W}} - 1\right) \left(P^{0,0}_{CU,h} G^{0,0}_{CU,h} + \sum_{l=0, l \neq h}^{H_i} P^{l,l}_{i,j} G^{l,l}_{i,j} + N^0\right), h \geq 1, \\ \left(G^{0,0}_{i,j}\right)^{-1} \left(2^{\frac{R^M_j}{W}} - 1\right) N^0, h = 0, \end{cases} \tag{12.10}$$

where W, $P^{l,l}_{i,j}$, $P^{0,0}_{i,j}$, $P^{0,0}_{CU,h}$, $G^{l,l}_{i,j}$, $G^{0,0}_{i,j}$, and $G^{0,0}_{CU,h}$ are defined in Table 12.1, N^0 is the AWGN power at the ith requester, and R^m_j is the transmission rate. The channel gain $G^{h,h}_{i,j}$ is represented by the pathloss $g^{h,h}_{i,j}$ and Rayleigh fading $|c^{h,h}_{i,j}|^2$ with $\mathbb{E}[|c^{h,h}_{i,j}|^2] = 1$ as

$$G^{h,h}_{i,j} = g^{h,h}_{i,j} |c^{h,h}_{i,j}|^2. \tag{12.11}$$

Finally, the average energy consumption per D2D helper to transmit S video segments to I requesters is expressed by

$$\overline{E} = \sum_{i=1}^{I}\sum_{j=1}^{S}\frac{r_j}{H_i}\sum_{h=1}^{H_i}\frac{E_{i,j}^h}{n_i^h}, \tag{12.12}$$

where n_i^h is the number of D2D helpers in the hth GoH.

To ensure reasonable energy consumption at each D2D helper, Eq. (12.12) is guaranteed by a given energy threshold E^* as

$$\overline{E} \leq E^*. \tag{12.13}$$

12.3.1.3 Co-channel Interference Guarantee

As mentioned earlier, the transmissions of D2D helpers in the GoHs using the downlink resources of the CUs can degrade the target SINR of the CUs. Therefore, a threshold γ^0 for the instantaneous SINR should be considered to guarantee the target SINR of the CUs. Based on this threshold, the transmission power of the D2D helpers in the hth GoH of Eq. (12.10) is limited by

$$P_{i,j}^{h,h}G_{i,j}^{h,h} \leq \frac{P_{CU,h}^{0,0}G_{CU,h}^{0,0}}{\gamma^0} - N^0. \tag{12.14}$$

12.3.2 RDO Problem

So far, the above formulations enable us to formulate the RDO problem, which is solved to minimize $\overline{D}$ by finding the optimal matrices of encoding rate points $\mathbf{R}_{S\times M} = \{R_j^m\}$ $(j = 1, 2, ..., S; m = 1, 2, ..., M)$ and the numbers of descriptions $\mathbf{d}_{S\times I\times H_i} = \{d_{i,j}^h\}$ $(i = 1, 2, ..., I; h = 0, 2, ..., H_i)$. Mathematically, the RDO problem is written as follows:

$$\min_{\mathbf{R},\mathbf{d}} \overline{D} \tag{12.15}$$

$$\text{s.t.}\begin{cases} 0 = R_j^0 \leq R_j^1 \leq \cdots \leq R_j^M, j = 1, 2, ..., S \\ \sum_{j=1}^{S}\sum_{m=1}^{M} mb_j^m \leq \delta V, 0 < \delta \leq 1 \\ \sum_{m=1}^{M} mb_j^m \leq v, j = 1, 2, ..., S \\ \sum_{h=0}^{H_i} d_{i,j}^h \leq M, i = 1, 2, ..., I, j = 1, 2, ..., S \\ \overline{E} \leq E^* \\ P_{i,j}^{h,h}G_{i,j}^{h,h} \leq \frac{P_{CU,h}^{0,0}G_{CU,h}^{0,0}}{\gamma^0} - N^0, i = 1, 2, ..., I, j = 1, 2, ..., S, h = 1, 2, ..., H_i. \end{cases} \tag{12.16}$$

In (12.16), the first constraint is used to ensure that the optimal $\mathbf{R}$ strictly follows the encoding rate distribution as discussed in Section 12.2.3; δ in the second constraint is added to adjust the encoded video size to satisfy the common display resolution of the requesters. Obviously, the encoded segment size cannot exceed the original segment size v as given in the third constraint. The fourth constraint is to make sure that the total optimal number of descriptions $\mathbf{d}$ assigned to all the GoHs and the MBS cannot be greater than the given number of descriptions of a segment M. In this constraint, the equality holds or not depending on the remaining energy of D2D helpers limited by the fifth constraint. Finally, the last constraint is used to guarantee the target SINR of the CUs.

12.3.3 GAs Solution

Solving Eqs. (12.15) and (12.16) is not feasible by using the Lagrange optimization approach or that of Karush-Kuhn-Tucker because the gradient terms with respect to $\mathbf{d}_{S\times I\times H_i}$ in the objective function cannot be computed [18, 19]. In this chapter, GAs, i.e. a family of adaptive heuristic searching algorithms within reasonable processing time at low computational resource, are used to solve the proposed RDO. The reason is that GAs can avoid being trapped in the local optimum by exploiting the random property of the evolutionary principles of nature selection and genetic variation to search many optimal peaks in parallel for an exact or approximate optimal solution, regardless of unimodal or multimodal searching space [18, 29]. The detailed implementation of GAs to solve the RDO problem is presented in [19].

12.4 Performance Evaluation

12.4.1 D2D Caching

To evaluate the performance of D2D caching, we consider a typical cell of 5G networks including one MBS and two CUs that share their downlink resources with ten D2D pairs (i.e. each has a D2D helper and a D2D requester). The system bandwidth is 5 MHz. The channel gains between the MBS, CUs, D2D helpers, and D2D requesters are modeled by the exponential power fading coefficient and the standard power law pathloss function [30]. The distances (1) from the MBS to the CUs, D2D helpers, and D2D requesters d(MBS; CUs, D2D) (2) from the D2D helpers to the CUs d(D2D;CUs), and (3) from the D2D helpers to D2D requesters for D2D communications d(D2D;D2D) are randomly distributed from 100 m to 500 m, from 50 m to 100 m, and from 1 m to 50 m, respectively. A D2D helper decides to cache a video depending on its percentage of available storage and the popularity of the video. We also take into account the target SINR of the CUs to preserve the interference from D2D communications on the performance of CUs. In addition, the transmit power of MBS is 10 W and the transmit power of the D2D helpers is randomly allocated from 0.01 W to 0.1 W.

Figure 12.3 plots the performance of D2D caching versus the number of D2D pairs. We set the target SINR of the CUs to 0 dB and compare three schemes: (1) maximum D2D caching (Max-D2D), (2) without D2D caching (None-D2D), and (3) minimum D2D caching (Min-D2D). In Max-D2D, the objective is to maximize the system capacity by finding the best set of the D2D helpers that have cached the videos and of the CUs that are willing to share their downlink resources to serve the D2D requesters. In None-D2D, D2D caching is not considered, while in Min-D2D, the worst set of D2D helpers and CUs are selected. The results in Figure 12.3 show that the higher the number of D2D pairs the system establishes, the higher the capacity the system gains. It can also be seen that if we increase the number of CUs, more proper CUs to share their downlink resources are selected to further enhance the system capacity. In all cases, the Max-D2D outperforms the None-D2D and Min-D2D in terms of system capacity improvement.

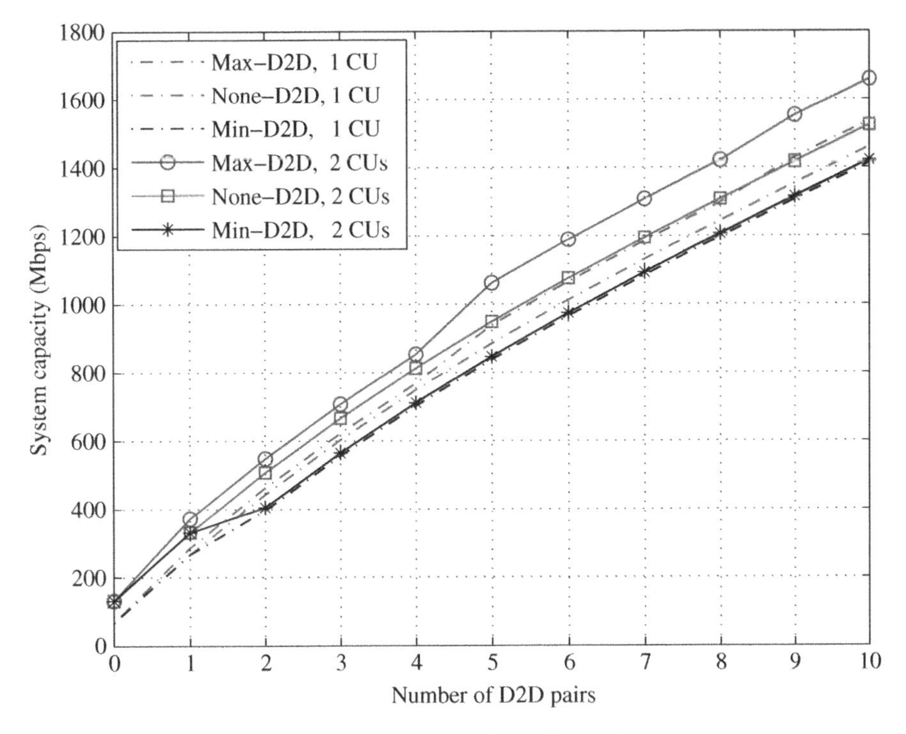

Figure 12.3 D2D caching performance versus number of D2D pairs.

We also investigate the effect of the target SINR of the CUs on the system capacity performance, as shown in Figure 12.4. It can be observed that if we set a higher priority on the CUs, i.e. increasing the target SINR while not exploiting the close proximity D2D caching, the system capacity of Max-D2D decreases. It should be noted that when the target SINR increases, more sets of D2D helpers and CUs that make the system capacity worse are removed so that the system capacity of Min-D2D increases. Obviously, the system capacity of None-D2D is not affected by the target SINR.

12.4.2 RDO

12.4.2.1 Simulation Setup

We evaluate the performance of the proposed RDO by setting the system parameters given in Table 12.2. The considered video is the Foreman sequence. We assume that all the segments of the Foreman sequence have a similar experimental RD model which is obtained by using the HM version 12.0 [31] to generate γ_j and β_j. The performance metrics include playback quality, playback quality fluctuation, and average energy consumption. The playback quality and playback quality fluctuation are represented by the peak signal-to-noise ratio (PSNR) and standard deviation of PSNR, which can be derived from the reconstructed distortion measured in the mean squared error [18]. The proposed RDO is compared to the other two schemes without optimization, namely EQU and RAN. For EQU, the encoding rate and the number of descriptions are equally set as $R_{i,j}^{m,EQU} = \frac{R_j^M \times m}{M}$ and $d_{i,j}^{h,EQU} = \frac{\sum_{h=0}^{H_i} d_{i,j}^h}{H_i+1}$, respectively. For RAN, the $R_{i,j}^{m,RAN}$ and $d_{i,j}^{h,RAN}$ are

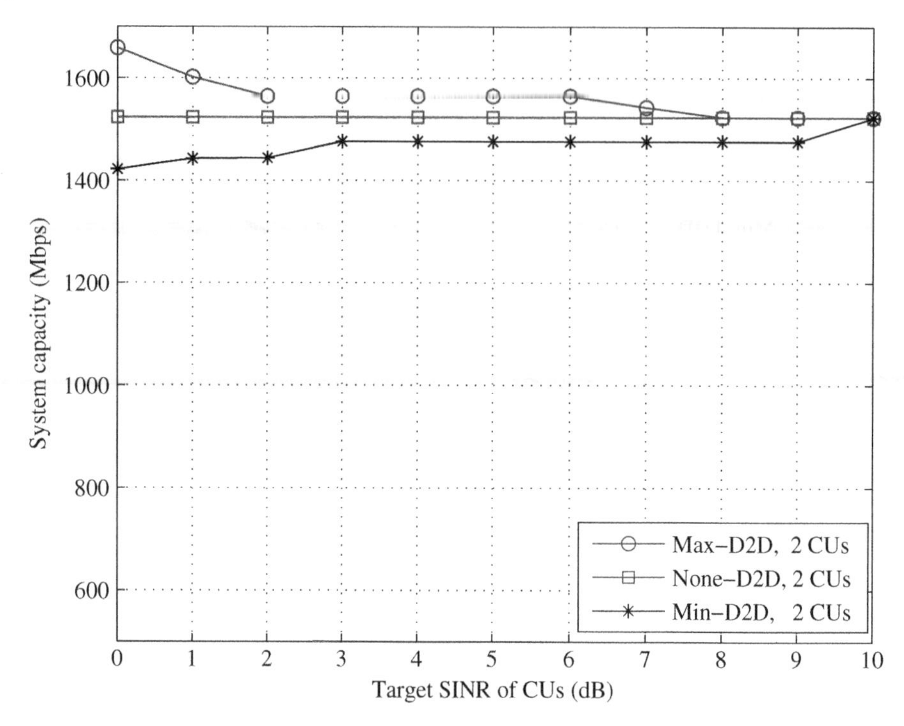

Figure 12.4 D2D caching performance versus target SINR of CUs.

Table 12.2 Parameters Setting.

Symbols	Specifications
I	1 requester
$H_i = H$	1 to 5 GoHs, due to $I = 1$
$n_i^h = n$	10 D2D helpers per each GoH
S	8 segments of CIF (352 × 288) Foreman sequence
v	50 Kbytes
M	64 layers (or descriptions) of a segment
γ_j, β_j	$1.914 \times 10^4, -1.20515$ [21, 25]
$P_{i,j}^{h,h}$	0.5 W
$P_{CU,h}^{0,0}$	15 W
$\varepsilon_{i,j}^h$	Randomly distributed from 0.05 to 0.5 [18, 29, 32]
δ	75% of full size of Foreman sequence
E^*	2.5 to 100 Joules [33, 34]
W	5 MHz
N^0	10^{-13} W [32, 35]
α	0.75 [26]
d(MBS;D2D)	300 m
d(D2D;D2D)	Randomly from 50 m to 100 m

randomly distributed in the range of $(0, R_j^M]$ and $(0, \sum_{h=0}^{H_i} d_{i,j}^h]$, respectively. Here R_j^m and $d_{i,j}^h$ are the optimal results found by solving the RDO problem. It is noted in the following simulation that all the results in Figures 12.5, 12.6, 12.7, and 12.8 are performed without the effect of the interference constraint (12.14) on the RDO problem, but this constraint is included in the results in Figure 12.9.

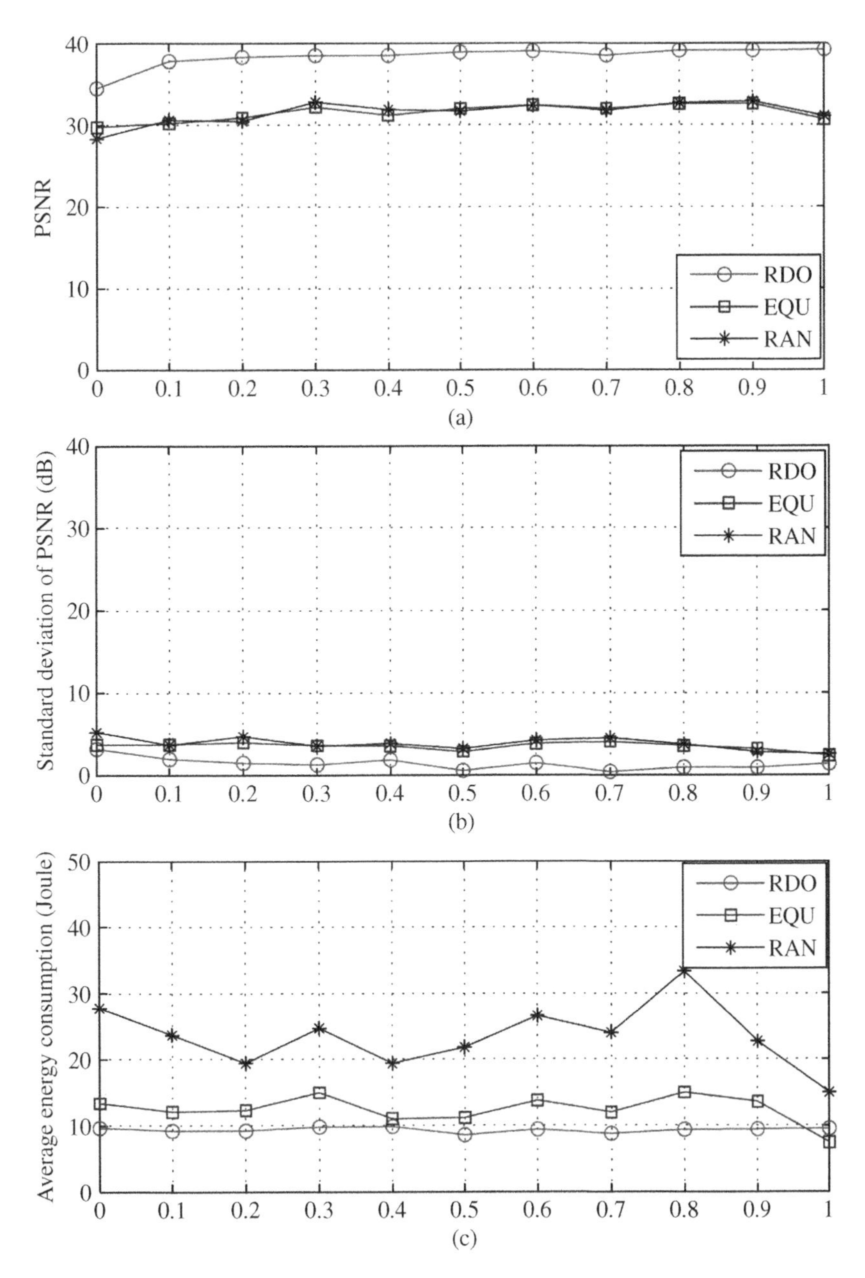

Figure 12.5 Performance metrics versus α. Reprinted with permission from [19].

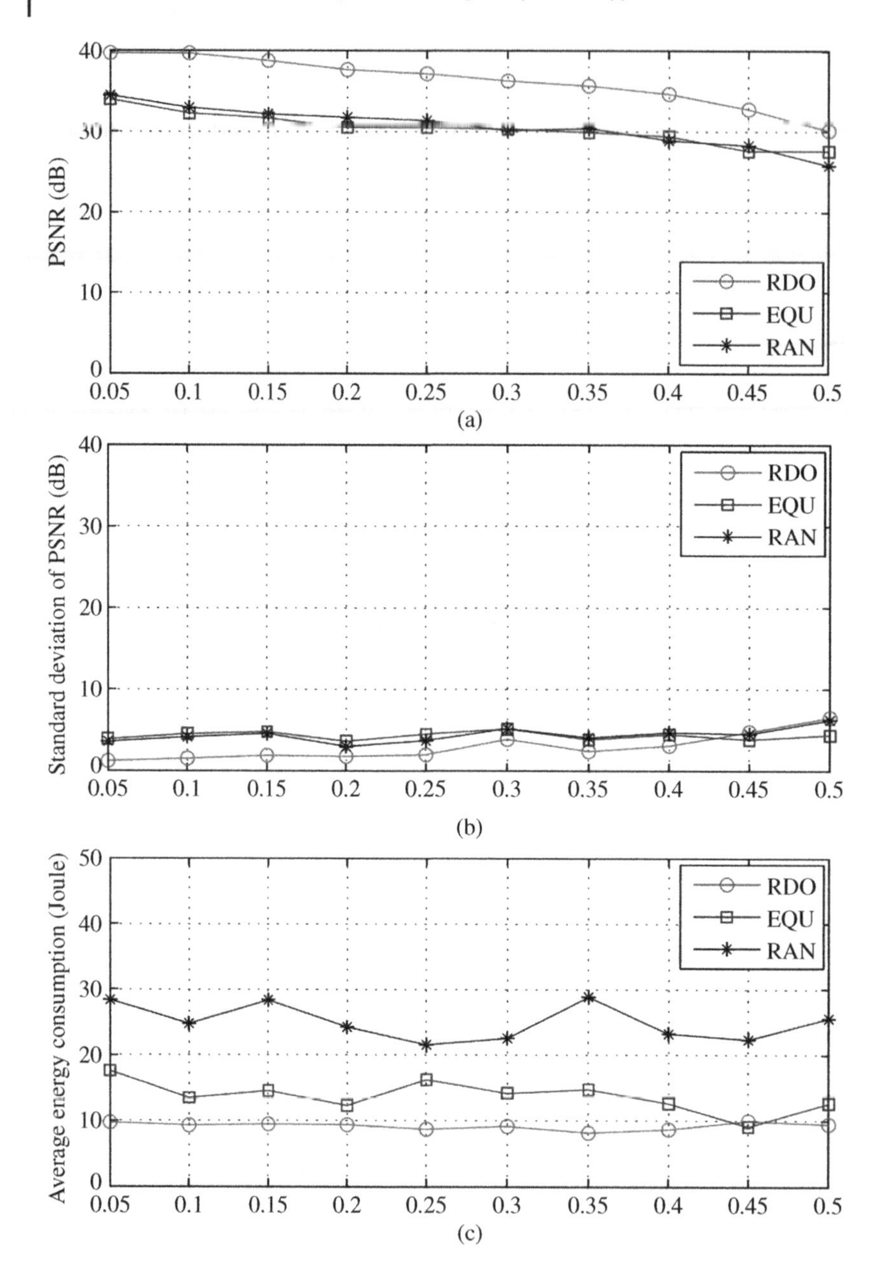

Figure 12.6 Performance metrics versus $\varepsilon_{i,j}^h$. Reprinted with permission from [19].

12.4.2.2 Performance Metrics

Performance Metrics versus α Figure 12.5 plots the performance metrics versus the skewed access rate α in the range from 0 to 1, $\varepsilon_{i,j}^h = 0.1$, $E^* = 10$ Joules, and $H = 3$. We can see in Figure 12.5a and b that the RDO outperforms the EQU and RAN in terms of PSNR and the standard deviation of PSNR. This means that the RDO provides the

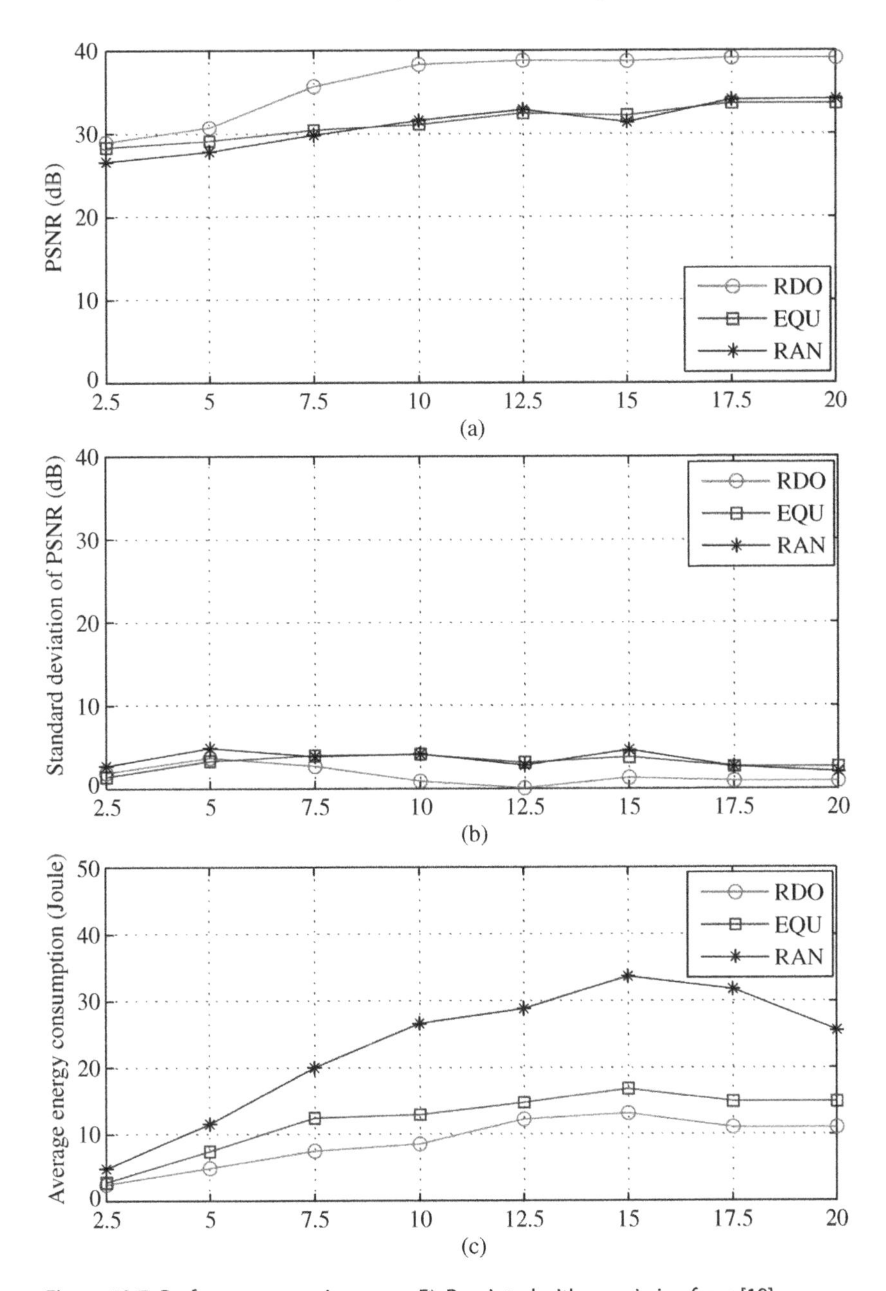

Figure 12.7 Performance metrics versus E^*. Reprinted with permission from [19].

requester with a higher QoE, i.e. higher playback quality and lower playback quality fluctuations. Obviously, the performance of the RDO increases with respect to the increase of the skewed access rate α. In addition, the results in Figure 12.5c show that the RDO stringently embraces the energy constraint such that $\overline{E} \leq E^* = 10$ Joules. This remains the average energy consumption of RDO lower than the EQU and RAN.

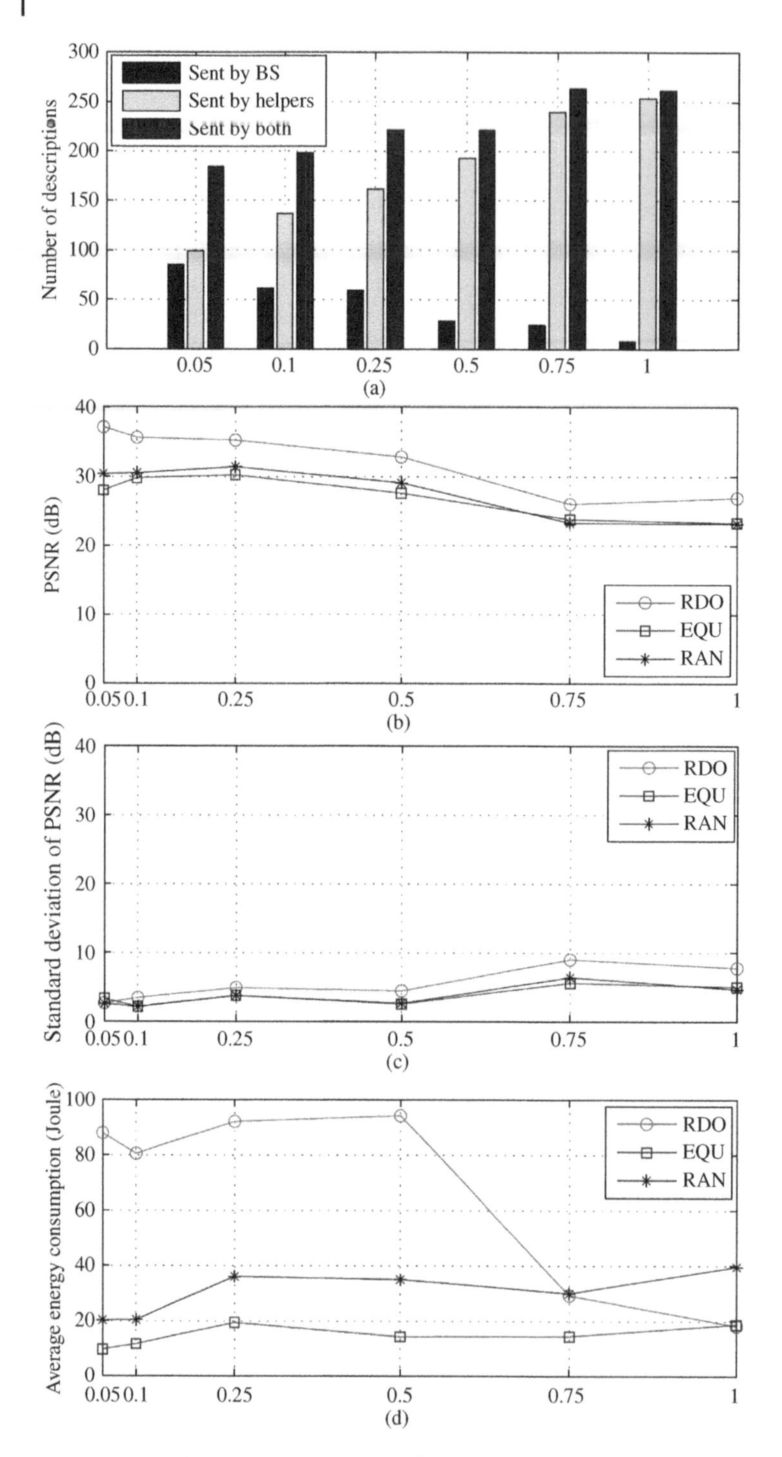

Figure 12.8 Performance metrics versus $\varepsilon_{i,j}^{0}$. Reprinted with permission from [19].

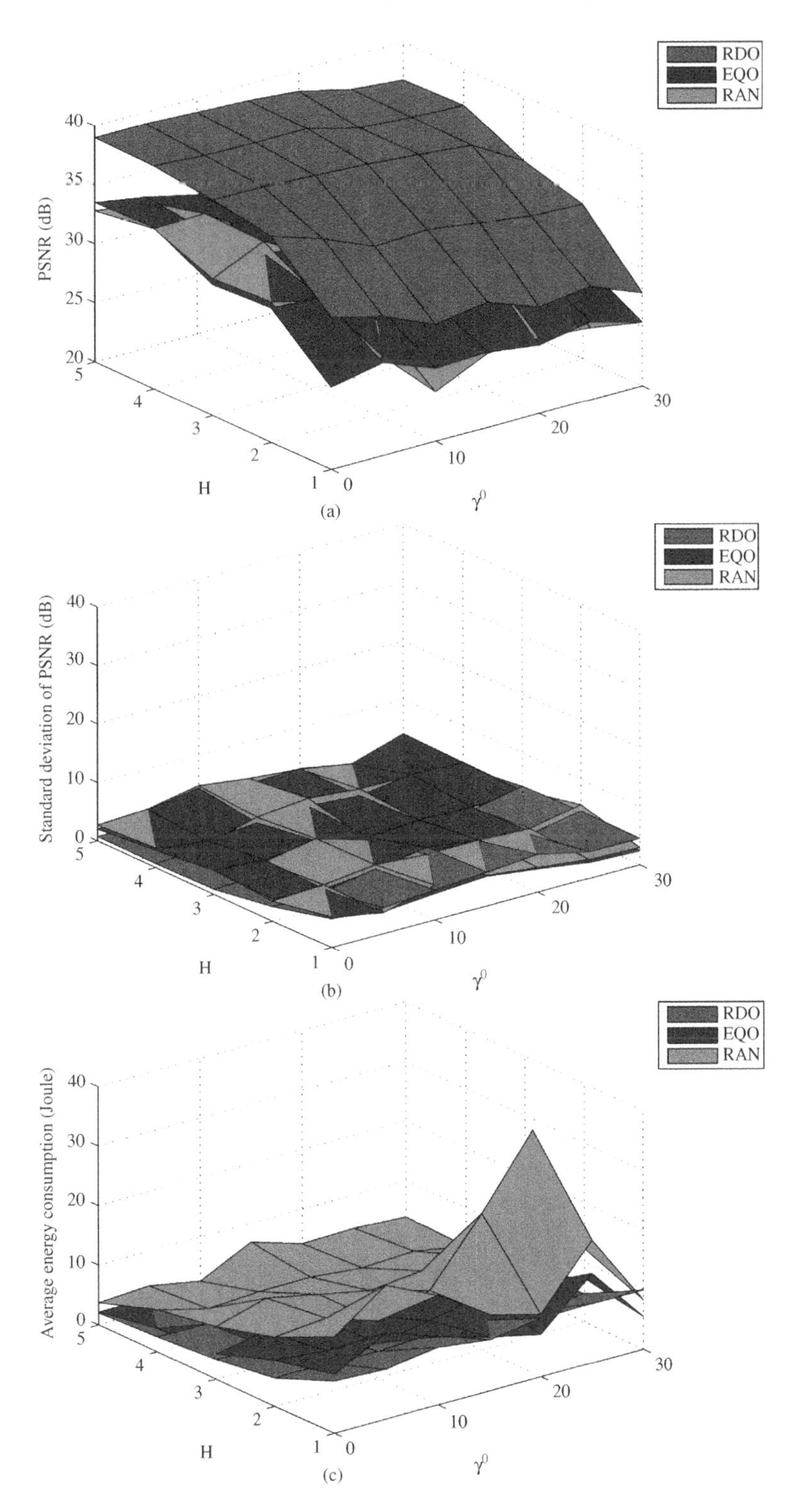

Figure 12.9 Performance metrics versus γ^0 (dB) and H. Reprinted with permission from [19].

Performance Metrics versus $\varepsilon_{i,j}^{h}$ Figure 12.6 plots the performance metrics versus $\varepsilon_{i,j}^{h}$ in the range from 0.05 to 0.5 while keeping $\alpha = 0.75$, $E^{*} = 10$ Joules, and $H = 3$. The RDO gains a higher QoE (Figure 12.6a and b) and lower energy consumption (Figure 12.6c) than the EQU and RAN. It is clear that the QoE of the RDO, EQU, and RAN decreases when the wireless channels become worse with a higher description error rate $\varepsilon_{i,j}^{h}$. Importantly, if $\varepsilon_{i,j}^{h}$ is too high, i.e. $\varepsilon_{i,j}^{h} \geq 0.45$ (Figure 12.6b), the RDO cannot guarantee a lower quality fluctuation than the EQU and RAN because we have not taken into account the constraint on quality fluctuations in the RDO problem.

Performance Metrics versus E^{*} To evaluate the performance metrics under the effect of the energy consumption constraint, we change E^{*} in the range from 2.5 to 20 Joules and set $\alpha = 0.75$, $\varepsilon_{i,j}^{h} = 0.1$, and $H = 3$. As shown in Figure 12.7, the increase E^{*} enables the D2D helpers in the GoHs and the MBS to transmit more descriptions to the requester for a higher QoE. The RDO always surpasses the EQU and RAN in terms of QoE and energy savings. Importantly, our proposed RDO can strictly follow the energy consumption constraint, i.e. the remaining energy of the D2D helpers, to flexibly serve the requester at the corresponding QoE.

Performance Metrics versus $\varepsilon_{i,j}^{0}$ We evaluate the performance metrics versus the description error rate $\varepsilon_{i,j}^{0}$ of the channel between the MBS and the requester to see how the RDO can coordinate the D2D helpers in the GoHs and the MBS to serve the requester in an efficient cooperative transmission manner. To do so, we change $\varepsilon_{i,j}^{0}$ from 0.05 to 1 and set $H = 3$, $\varepsilon_{i,j}^{h} = 0.75$ $(h > 0)$, and $\alpha = 0.75$. In addition, we relax the energy constraint by setting $E^{*} = 100$ Joules to insightfully understand this cooperative transmission.

Figure 12.8a shows that when $\varepsilon_{i,j}^{0}$ decreases, i.e. the quality of channel between the MBS and the requester is worse, the number of descriptions transmitted by the MBS decreases. Because the energy consumed to transmit a description by the MBS is much higher than by the D2D helpers, a small decrease in the number of descriptions transmitted by the MBS is altered by a large increase in the number of descriptions transmitted by the D2D helpers. This results in an increase in the total number of descriptions transmitted by both the MBS and the D2D. However, the increase in the total number of descriptions transmitted cannot ensure a higher QoE achieved as shown in Figure 12.8b and c. In comparison, the RDO gains a higher playback quality but higher playback quality fluctuations compared to the EQU and RAN. The results in Figure 12.8d demonstrate that the RDO consumes higher energy than the EQU and RAN when $\varepsilon_{i,j}^{0} < \varepsilon_{i,j}^{h} = 0.75$ $(h > 0)$. The reason is that the RDO focuses on the energy consumed by the D2D helpers, while the number of descriptions transmitted by the D2D helpers is not high enough if $\varepsilon_{i,j}^{0} < \varepsilon_{i,j}^{h} = 0.75$. However, if the channel between the MBS and the requester is worse than the channels between the D2D helpers and the requester $(\varepsilon_{i,j}^{0} \geq \varepsilon_{i,j}^{h})$, the RDO consumes much less energy (from 20 to 30 Joules) even when the energy consumption constraint is high (100 Joules). This is because all the descriptions are mostly transmitted by the D2D helpers.

Performance Metrics versus γ^{0} ***and*** H We now investigate the impact of the target SINR of the CUs and the number of GoHs on the performance metrics of the system. To do so, the interference constraint (12.14) is added to the RDO problem in (12.15) and (12.16). We change γ^0 from 0 dB do 30 dB, H from 1 to 5 while keeping $\alpha = 0.75$, $E^* = 10$ Joules, and $\varepsilon_{i,j}^{h} = 0.1$.

The results in Figure 12.9a show that if γ^0 increases, the transmission power of D2D helpers decreases to limit the interference effect on the performance of the CUs. This causes the playback quality of the received video to decrease. Meanwhile, as shown in Figure 12.9b and c, the increase of γ^0 does not significantly affect the performance of quality fluctuation and energy consumption. Importantly, in dense D2D caching, increasing H provides the system with more opportunities to select the best GoHs with the best channels to the requester for a higher QoE (Figure 12.9a and b). Obviously, if there are many GoHs (by increasing H), the QoE performance becomes saturated and the average energy consumption decreases thanks to more D2D helpers sharing out their energy resources. In comparison, the RDO obtains a higher QoE and higher energy savings than the EQU and RAN.

12.4.2.3 Discussions

We can observe from the aforementioned results that our RDO provides the requester with the higher QoE and energy savings compared to the EQU and RAN. The reason is that the RDO can minimize $\overline{D}$ in Eq. (12.4) for high quality playback by finding the optimal encoding rate and the number of descriptions in accordance with the condition of the wireless channel, the characteristic of video, and the users' behavior, while the EQU and RAN cannot. Furthermore, the optimal results can also satisfy the constraints of average energy consumption of all D2D helpers and the target SINR of CUs for energy saving and co-channel interference limitation, respectively. In addition, it is noted that the playback quality fluctuation should be considered in the RDO problem if the wireless channels become extremely worse to ensure a high QoE.

12.5 Conclusion

In this chapter, we have proposed an optimal cooperative video streaming in 5G networks with dense D2D caching. In particular, we have designed a joint rate allocation and description distribution optimization (RDO). The RDO coordinates the D2D helpers and the MBS in 5G networks to serve the requesters high QoE represented by (1) continuous playback thanks to close proximity of dense D2D caching, (2) high playback quality, and (3) smooth playback. Furthermore, the RDO gains a high energy saving while having the ability to preserve the target SINR of the CUs that share their downlink resources for D2D communications.

Bibliography

1 B. Hammi, R. Khatoun, S. Zeadally, A. Fayad, and L. Khoukhi, "IoT technologies for smart cities," *IET Networks*, vol. 7, pp. 1–13, 2017.

2 A. Gupta and R.K. Jha, "A survey of 5G network: architecture and emerging technologies," *IEEE Access*, vol. 3, pp. 1206–1232, 2015.

3 P. Kela, M. Costa, J. Turkka, K. Leppänen, and R. Jäntti, "Flexible backhauling with massive MIMO for ultra-dense networks," *IEEE Access*, vol. 4, pp. 9625–9634, 2016.
4 V. Petrov, D. Solomitckii, A. Samuylov, M.A. Lema, M. Gapeyenko, D. Moltchanov, S. Andreev, V. Naumov, K. Samouylov, M. Dohler, and Y. Koucheryavy, "Dynamic multi-connectivity performance in ultra-dense urban mmwave deployments," *IEEE Journal on Selected Areas in Communications*, vol. 35, pp. 2038–2055, 2017.
5 Z. Gao et al., "Mmwave massive-MIMO-based wireless backhaul for the 5G ultra-dense network," *IEEE Wireless Communications*, vol. 25, pp. 13–21, 2015.
6 H. Zhang, S. Huang, C.X. Jiang, K. Long, V.C.M. Leung, and H.V. Poor, "Energy efficient user association and power allocation in millimeter-wave-based ultra dense networks with energy harvesting base stations," *IEEE Journal on Selected Areas in Communications*, vol. 35, pp. 1936–1947, 2017.
7 C.G. Yang, J.D. Li, Q. Ni, A. Anpalagan, and M. Guizani, "Interference-aware energy efficiency maximization in 5G ultra-dense networks," *IEEE Transactions on Communications*, vol. 65, pp. 728–739, 2017a.
8 C. G. Yang, J.D. Li, P. Semasinghe, E. Hossain, S.M. Perlaza, and Z. Han, "Distributed interference and energy-aware power control for ultra-dense D2D networks: a mean field game," *IEEE Transactions on Wireless Communications*, vol. 16, pp. 1205–1217, 2017b.
9 Y. Liu, G.Y. Li, and W. Han, "D2D enabled cooperation in massive MIMO systems with cascaded precoding," *IEEE Wireless Communications Letters*, vol. 6, pp. 238–241, 2017.
10 N. Lee, X. Lin, J.G. Andrews, and R.W. Heath, "Power control for D2D underlaid cellular networks: modeling, algorithms, and analysis," *IEEE Journal on Selected Areas in Communications*, vol. 33, pp. 1–13, 2015.
11 C. Xu, L. Song, Z. Han, Q. Zhao, X. Wang, X. Cheng, and B. Jiao, "Efficiency resource allocation for D2D underlay communication systems: a reverse iterative combinatorial auction based approach," *IEEE Journal on Selected Areas in Communications*, vol. 31, pp. 348–358, 2013.
12 Z. Chen, N. Pappas, and M. Kountouris, "Probabilistic caching inwireless D2D networks: cache hit optimal versus throughput optimal," *IEEE Communications Letters*, vol. 21, pp. 584–587, 2017.
13 X. Song, Y. Geng, X. Meng, J. Liu, W. Lei, and Y. Wen, "Cache-enabled device to device networks with contention-based multimedia delivery," *IEEE Access*, vol. 5, pp. 3228–3239, 2017.
14 R. Wang, J. Zhang, S.H. Song, and K.B. Letaief, "Mobility-aware caching in D2D networks," *IEEE Transactions on Wireless Communications*, vol. 16, pp. 5001–5015, 2017.
15 D. Wu, J. Wang, R.Q. Hu, Y. Cai, and L. Zhou, "Energy-efficient resource sharing for mobile device-to-device multimedia communications," *IEEE Transactions on Vehicular Technology*, vol. 63, pp. 2093–2103, 2014.
16 Y. Shen, C. Jiang, T.Q.S. Quek, and Y. Ren, "Device-to-device-assisted communications in cellular networks: an energy efficient approach in downlink video sharing scenario," *IEEE Transactions on Wireless Communications*, vol. 15, pp. 1575–1587, 2016.
17 H. Zhu, Y. Cao, W. Wang, B. Liu, and T. Jiang, "QoE-aware resource allocation for adaptive device-to-device video streaming," *IEEE Network*, vol. 29, pp. 6–12, 2015.

18 X. Du, N.-S. Vo, T.Q. Duong, and L. Shu, "Joint replication density and rate allocation optimization for VoD systems over wireless mesh networks," *IEEE Transactions on Circuits and Systems for Video Technology*, vol. 23, pp. 1260–1273, 2013.
19 N.-S. Vo, T.Q. Duong, H.T. Duong, and A. Kortun, "Optimal video streaming in dense 5G networks with D2D communications," *IEEE Access*, vol. 6, pp. 209–223, 2017.
20 P.A. Chou, H.J. Wang, and V.N. Padmanabhan, "Layered multiple description coding." in *Proc. Packet Video Workshop*, pp. 1–7, 2003.
21 W. Xiang, C. Zhu, C.K. Siew, Y. Xu, and M. Liu, "Forward error correction-based 2-D layered multiple description coding for error resilient H.264 SVC video transmission," *IEEE Transactions on Circuits and Systems for Video Technology*, vol. 19, pp. 1730–1738, 2009.
22 N.-S. Vo, T.Q. Duong, L. Shu Shu, X. Du, H.-J. Zepernick, and W. Cheng, "Cross-layer design for video replication strategy over multihop wireless networks," in *Proc. IEEE Int. Conf. Commun.*, pp. 1–6, 2011.
23 A. Albanese, J. Blomer, J. Edmonds, M. Luby, and M. Sudan, "Priority encoding transmission," *IEEE Transactions on Information Theory*, vol. 42, pp. 1737–1744, 1996.
24 R. Puri and K. Ramchandran, "Multiple description source coding using forward error correction codes," in *Proc. Asilomar Conf. Signals, Syst., Comput.*, pp. 342–346, 1999.
25 D. Jurca, P. Frossard, and A. Jovanovic, "Forward error correction for multipath media streaming," *IEEE Transactions on Circuits and Systems for Video Technology*, vol. 19, pp. 1315–1326, 2009.
26 L. Breslau, P. Cao, L. Fan, G. Phillips, and S. Shenker, "Web cachingand Zipf-like distributions: evidence and implications," in *Proc. IEEE INFOCOM*, pp. 126–134, 1999.
27 S. Shalmashi, G. Miao, and S. B. Slimane, "Interference management for multiple device-to-device communications underlaying cellular networks," in *Proc. IEEE 24th Int. Symp. Pers., Indoor Mobile Radio Commun.*, pp. 223–227, 2013.
28 L.B. Le, "Fair resource allocation for device-to-device communications in wireless cellular networks," in *Proc. IEEE Global Commun. Conf.*, pp. 5451–5456, 2012.
29 T. Fang and L.-P. Chau, "GOP-based channel rate allocation using genetic algorithm for scalable video streaming over error-prone networks," *IEEE Transactions on Image Processing*, vol. 15, pp. 1323–1330, 2006.
30 W.C. Cheung, T.Q.S. Quek, and M. Kountouris, "Throughput optimization, spectrum allocation, and access control in two-tier femtocell networks," *IEEE Journal on Selected Areas in Communications*, vol. 30, pp. 561–574, 2012.
31 HEVC HM reference software version 12.0. https://hevc.hhi.fraunhofer.de.
32 N.-S. Vo, T.Q. Duong, and M. Guizani, "QoE-oriented resource efficiency for 5G two-tier cellular networks: a femtocaching framework," in *Proc. IEEE Global Commun. Conf.*, pp. 1–6, 2016.
33 J. Wu, B. Cheng, M. Wang, and J. Chen, "Energy-aware concurrent multipath transfer for real-time video streaming over heterogeneous wireless networks," *IEEE Transactions on Circuits and Systems for Video Technology*, vol. 28, no. 8, pp. 2007–2023.

34 C. Bezerra, A.D. Carvalho, D. Borges, N. Barbosa, J. Pontes, and E. Tavares, "QoE and energy consumption evaluation of adaptive video streaming on mobile device," in *Proc. IEEE Annu. Consum. Commun. Netw. Conf.*, pp. 1–6, 2017.

35 N. Eshraghi, V. Shah-Mansouri, and B. Maham, "QoE-aware power allocation for device-to-device video transmissions," in *Proc. IEEE 27th Annu. Int. Symp. Pers., Indoor Mobile Radio Commun.*, pp. 1–5, 2016.

Index

Ultra-dense Networks for 5G and Beyond: Modelling, Analysis, and Applications,
First Edition. Edited by Trung Q. Duong, Xiaoli Chu and Himal A. Suraweera.

Printed and bound by CPI Group (UK) Ltd, Croydon, CR0 4YY

16/04/2025

14658471-0003